International Series in Operations Research and Management Science

For other titles in this series, go to:
www.springer.com/series/6161

Jerzy A. Filar • Alain Haurie
Editors

Uncertainty and Environmental Decision Making

A Handbook of Research and Best Practice

Springer

Editors
Jerzy A. Filar
University of South Australia ISST
Mawson Lakes SA 5095
Australia
jerzy.filar@unisa.edu.au

Alain Haurie
ORDECSYS/C-ORDEE
Place de L'Etrier 4
1224 Chene-Bougeries
Switzerland

ISBN 978-1-4419-1128-5 e-ISBN 978-1-4419-1129-2
DOI 10.1007/978-1-4419-1129-2
Springer New York Dordrecht Heidelberg London

Library of Congress Control Number: 2009940575

Printed on acid-free paper

Springer is part of Springer Science+Business Media (www.springer.com)

Preface

The 21st century promises to be an era dominated by international response to certain global environmental challenges such as climate change, depleting biodiversity and biocapacity as well as general atmospheric, water and soil pollution problems. Consequently, Environmental decision making (EDM) is a socially important field of development for Operations Research and Management Science (OR/MS). Uncertainty is an important feature of these decision problems and it intervenes at very different time and space scales. The *Handbook on "Uncertainty and Environmental Decision Making"* provides a guided tour of selected methods and tools that OR/MS offer to deal with these issues. Below, we briefly introduce, peer reviewed, chapters of this handbook and the topics that are treated by the invited authors.

The first chapter is a general introduction to the challenges of environmental decision making, the use of OR/MS techniques and a range of tools that are used to deal with uncertainty in this domain.

Chapter 1. Filar and Haurie present a broad overview of Operations Research and Management Science methods and models used to support Environmental Decision Making under uncertainty. They first outline challenges and pitfalls of OR/MS applications to EDM. Subsequently, they classify different sources of uncertainty and show how stochastic reasoning pervades some fundamental issues raised in EDM and environmental economics. In particular, they discuss issues related to discounting and intergenerational equity. A selection of concepts and techniques is surveyed that enable the modeler to better understand and manage uncertainty in EDM. Finally, it is shown how the methods of stochastic control, stochastic programming, robust optimization and statistical emulation in meta-modeling can be used to shed some light on difficult issues arising in EDM under uncertainty.

The next two chapters present application of stochastic or robust programming methods to techno-economic modeling of energy/environment interactions.

Chapter 2. Labriet, Loulou and Kanudia consider a large scale, partial equilibrium, technology rich global 15-region TIMES Integrated Assessment Model (ETSAP-TIAM). They apply the well-known method of stochastic programming in extensive form to assess climate policies in a very uncertain world. The main uncertainties considered are those of the Climate Sensitivity parameter, and of the rate of economic development. They argue that the stochastic programming approach is well suited to the treatment of major uncertainties, in spite of the limitation inherent to this technique due to increased model size when many outcomes are modeled. The main advantage of the approach is to obtain a single hedging strategy while uncertainty prevails, unlike in the case of classical scenario analysis. Furthermore, the hedging strategy has the very desirable property of attenuating the (in)famous "razor edge" effect of Linear Programming. It thus arrives at a more robust mix of technologies to attain the desired climate target. Although the example treated uses the classical expected cost criterion, the authors also presents, and argue in favor of, altering this criterion to introduce risk considerations, by means of a linearized semi-variance term, or by using the Savage criterion. Risk considerations are arguably even more important in situations where the random events are of a "one-shot" nature and involve large costs or payoffs, as is the case in the modeling of global climate strategies. The article presents methodological details of the modeling approach, and uses realistic instances of the ETSAP-TIAM model to illustrate the technique and to analyze the resulting hedging strategies.

Chapter 3. Babonneau, Vial and Apparigliato show how traditional approaches to optimization under uncertainty, in particular stochastic programming, chance-constrained programming or stochastic dynamic programming, encounter most severe numerical difficulties. This is because models in this area are large and complex, already in their deterministic formulation. In this chapter the authors introduce a relatively new method, known as robust optimization, as an alternative to traditional methods and formulations. Through an illustrative example, they suggest ways of putting robust optimization to work in environmental and energy optimization models.

Traditionally, both short term weather and longer term climatic conditions have been regarded as sources of uncertainty in great many human endeavors. Nowadays, it is widely believed that anthropogenic induced climate change will impact not only long term trends such as global warming and sea level rise but also frequency and severity of the southern oscillation effect as well of extreme events such as hurricanes, droughts and floods. Consequently, the next three chapters focus on certain important consequences of uncertainty inherent in weather patterns, the El-Nino phenomenon and the anticipated climate change.

Chapter 4. Naylor and Mastrandrea provide a framework for using climate information in the design of policy to manage risks for agriculture, rural economic growth, and food security. They highlight several tools of analysis that can be applied in the context of both climate variability and global climate change. The concepts are developed through a case study of the rice sector in Indonesia a country directly affected by climate variability related to El Nino Southern Oscillation events. The risk assessment model is based on the probability of climate events, critical thresholds of damage related to those events, and the role of policy in reducing climate-related impacts on agricultural systems. Because risk assessment involves estimations of both the probability of climate events and the expected consequences of those climate events, Bayesian analysis is applied to show how climate information can be used to update subjective probabilities over short- and long- time scales. Bayesian updating can help reduce the chances that policymakers will make the wrong policy decision given all of the available information. However, the chapter demonstrates that - even with the help of these tools - Type I and Type II errors in policymaking will always be present and hence need to be estimated and incorporated in policy planning.

Chapter 5. Barrieu and Scaillet give a short introduction to weather derivatives. They discuss the purposes for which they were created, describe the markets on which they are exchanged, and how they are used to promote agricultural risk transfer in developing countries via the World Bank program. They also treat some specific issues such as basis risk, pricing and design.

Chapter 6. Ambrosi, Hourcade, Hallegatte, Lecocq, Dumas and Minh Ha Duong examine the consequences of various attitudes towards climate damages through a family of stochastic optimal control models (RESPONSE): cost-effectiveness for a given temperature ceiling; cost-benefit analysis with a pure preference for current climate regime and full cost-benefit analysis. The choice of a given proxy for climate change risks is regarded as more than a technical option. It is essentially motivated by a degree of mistrust of the legitimacy of an assessment of climate damages and the possibility of providing, in due course, reliable and non controversial estimates. The authors' results demonstrate that: (a) for the early decades of abatement, the difference between various decision-making frameworks appears to matter less than the difference between stochastic and non stochastic approach given the cascade of uncertainty from emissions to damages; (b) in a stochastic approach, the possibility of non-catastrophic singularities in the damage function is sufficient to significantly increase earlier optimal abatements; (c) a window of opportunity for action exists up to 2040: abatements further delayed may induce significant regret in case of bad news about climate response or singularities in damages.

One fundamental problem underlying attempts to apply OR/MS methods to environmental decision making stems from the fact that human development processes, typically, operate on much shorter time scales than natural processes of the biosphere. Consequently, the next two chapters exploit tools of decision analysis, utility theory and optimal control theory to model the ensuing difficulties that are caused by the need to make environmental policy decisions that affect disparate periods of time and under conditions where new information becomes available during the periods affected by these policies.

Chapter 7. Bahn, Haurie and Malhamé present an application of stochastic control or stochastic game methods to the modeling of climate policy timing. The authors first propose a stochastic control approach for a cost-effectiveness model where two sources of uncertainty are included. In a second part they use a similar stochastic control approach for a cost-benefit model where only the uncertainty on the access to a clean technology is taken into account. Finally, they show how these models could be extended to a game theoretic framework, assuming non-cooperative behavior of two groups of countries, under a treaty imposing a coupled constraint on atmospheric concentrations of greenhouse gases.

Chapter 8. De Lara and Gilotte consider an agent taking two successive decisions under uncertainty. After the agent's first decision, a signal is revealed providing information about the state of nature and then the second decision is taken accordingly. Suppose that the agent is an expected utility maximizer. The precautionary effect holds when, in the prospect of future information, his optimal initial decision is smaller (more conservative?) than without such a prospect. Indeed, the first decision is usually a scalar representing consumption, so that precaution is identified with less consumption. The authors introduce the second-period value of information as a function of the initial decision and show that, when this function is decreasing, namely, the precautionary effect holds true. More generally the condition enables the comparison of optimal decisions related to different signals, not necessarily comparable. It also ties together and clarifies many conditions for the precautionary effect that are scattered in the environmental economics literature. A practical illustration with Nordhaus's DICE model is included.

Whenever stochastic phenomena impact on a decision making processes, probability distributions, moments, or other key parameters of the most important random variables need to be modeled. Consequently, the final three chapters present methods that combine statistical and decision analyses that can support a variety of environmental management problems.

Chapter 9 Chiera, Filar, Gordon and Zachary present an analysis of two separate single-indicator forecasting methods for the El Nino Southern Oscillation phenomenon, based on the oscillation persistence. The authors use the southern oscillation index (SOI) of pressure, to forecast in short time scales of 4 – 8 months. A Bayesian approach is used in order to explore SOI persistence and compare results to a Taylor Series expansion (control method). The authors find that signal persistence is important when forecasting more than a few months and the techniques presented provide a relatively simple approach to environmental risk forecasting in situations where the underlying phenomenon exhibits substantial amount of persistence.

Chapter 10. Boland develops a model to generate synthetic sequences of half hourly electricity demand. The generated sequences represent possible realizations of electricity load that may occur in a region under consideration. Each of the components included in the model has a physical interpretation. These components are yearly and daily seasonality which were modelled using Fourier series, weekly seasonality modelled with dummy variables, and the relationship with current temperature described by polynomial functions of temperature. Finally the stochastic component was modelled with ARMA processes. The temperature series was modelled in a similar fashion. The stochastic modelling was performed to build probability distributions of the outputs to calculate probabilistic forecasts. As one application several summers of half hourly electricity demand were generated and from them the value of demand that is not expected to be exceeded more than once in ten years was calculated. Additionally, the bivariate temperature and demand model was used in software designed to optimize the orientation of photovoltaic cells to match demand.

Chapter 11. Gabriel, Vilalai, Sahakij, Ramirez, Peot describe recent modeling efforts to identify factors that lead to high biosolids odor levels associated with advanced waste water treatment plants (AWTP). These factors can be broken down into two groups: (i) those that are beyond the control of the AWTP such as ambient temperature, (ii) those that are controllable such as the number of centrifuges in operation, the amount of lime used, etc. The authors summarize their findings relative to different statistical models that predict biosolids odor levels based on either subjective or analytic measurements from the District of Columbia Water and Sewer Author (DCWASA). These models take into account a host of factors to predict biosolids odor levels and are then used to generate a relevant probability distribution for odor levels using Monte Carlo simulation. Such probability distributions will guide AWTP managers relative to where to send the biosolids products taking into account the likelihood of high levels and thus indirectly, possible complaints from those living or working near the reuse sites. The authors also describe recent efforts in also optimizing the operations of the AWTP and distribution network to balance both biosolids odors and costs. The resulting multiobjective optimization models are computationally challenging due to their size and non-convexities. The discussion presented also shows how to handle stochasticity directly within such optimization models.

Adelaide (Australia), Chêne Bougeries (Switzerland), June 2009

Jerzy Filar
Alain Haurie

Acknowledgements

We are indebted to the series editor, Professor Hillier, for encouraging us to undertake the task of producing this handbook and for his patience during the unexpectedly long manuscript preparation period. We also gratefully acknowledge partial support from the Australian Research Council's Discovery Grant # DP0987148 and from the University of South Australia's ITEE Division's Research Grant 2008. We finally thank the publisher of the journal *Environmental Modeling and Assessment* for permitting us to reproduce the paper by Ambrosi et al.

Contents

Chapter 1
OR/MS and Environmental Decision Making under Uncertainty

Jerzy A. Filar and Alain B. Haurie

Abstract In this chapter we present a broad overview of Operations Research and Management Science (OR/MS) methods and models used to support Environmental Decision Making (EDM) under uncertainty. We first survey the challenges and pitfalls frequently accompanying OR/MS applications to problems involving environmental issues. We consider and classify generic sources of uncertainty in quantitative models involving life support systems. We show how stochastic reasoning pervades some fundamental issues affecting decision making pertaining to the natural environment and environmental economics, in particular those related to discounting and intergenerational equity. We then discuss a selection of concepts and techniques that enable us to better understand and manage uncertainty in EDM. Finally, we indicate how the methods of stochastic control, stochastic programming, robust optimization and statistical emulation in meta-modeling can be used to shed light on some difficult issues arising in environmental decision making under uncertainty. This general discussion constitutes a preparation for the forthcoming chapters of this book.

1.1 Introduction

Since its inception during the second world war, the subject of Operations Research has grown rapidly and adapted itself repeatedly in order to answer an ever growing array of important problems. Links with Management Science broadened its reach beyond military logistics and engineering applications to those of business and management. Subsequently, aided by the explosive growth of computing power

J.A. Filar
University of South Australia, Adelaide SA, Australia e-mail: Jerzy.Filar@unisa.edu.au

A.B. Haurie
GERAD-HEC Montréal, Canada and ORDECSYS, Chêne Bougeries, Switzerland e-mail: ahaurie@ordecsys.com

J.A. Filar and A. Haurie (eds.), *Uncertainty and Environmental Decision Making*,
International Series in Operations Research & Management Science 138,
DOI 10.1007/978-1-4419-1129-2_1,

and information technology, its domain of interest expanded even further to include problems of manufacturing systems, communications, operations management, decision support systems, financial systems and many others. Loosely speaking, we shall refer to the multi-disciplinary subject that had so evolved as OR/MS (short for Operations Research & Management Science).

For the purpose of this contribution, we shall think of OR/MS as a body of largely quantitative models and techniques aimed at identifying a best (or, at least, an adequate) course of action out of a, possibly enormous, set of alternative courses of action. In this context, the course of action adopted influences the performance of some complex "system" and may lead to benefits such as profits, efficiency, enhanced productivity as well as to a wide range of possible costs (e.g., operational costs, failures, or waste).

In recent years, concerns about our society's impact on natural environment – fuelled by global problems such as climate change, loss of biodiversity and spread of pollution – have opened many new opportunities for OR/MS practitioners. As industries and regulatory agencies are beginning to show real interest in minimizing adverse impacts of human development processes on the natural environment a whole new set of problems waiting to be analyzed has emerged. While some of these problems are readily amenable to existing OR/MS techniques, many possess novel characteristics that require special attention and development of new techniques to produce meaningful results when analyzed from the OR/MS perspective. While some of these characteristics will be discussed in more detail in subsequent sections, two that form a recurrent theme throughout this volume are highlighted at this preliminary stage.

U: *Uncertainty.* Typically, a coupling of environmental impacts with a business or an engineering decision making problem dramatically increases uncertainty concerning a number of important issues. These range from uncertainty concerning the natural variability of certain biological and physical phenomena, through uncertainty related to the understanding of impacts of the selected courses of action on these phenomena, to the uncertainty generated by a combination of disparate analytical tools to the study of the problem of interest.

TSA: *Time Scales' Asynchrony.* Arguably, at the core of most controversies regarding environmental protection lies the problem of lack of "synchronization" between the time scales of human development processes and the natural processes of the biosphere. Typically, the former are relatively short time scales while the latter are much longer.

The bulk of this volume is devoted to quantitative, fundamentally mathematical, modelling paradigms capable of capturing certain essential interactions between human development activities and the natural environment. Hence, it is worthwhile to briefly discuss whether the problems encountered are, indeed, suitable for mathematical modeling.

A relatively recent survey article (36) presents the minimal requirements that situations studied should possess in order to make mathematical modeling meaningful. The characterization of these requirements is not simple because the boundaries of what can be achieved with mathematical models are constantly pushed back. Nonetheless, three principles were identified capturing what might be called an analyst's common sense which, if violated, raise questions as to whether mathematical modelling is appropriate in these situations. These criteria are:

- The presence of (at least some) variables that can be quantified and (ideally) observed and measured, or of data from which such variables could be extracted
- The presence of some understanding of relations between quantifiable variables or of, at least, a need to discover such relations empirically
- The presence of data, experimental designs, or other procedures to be used to validate the model.

The preceding requirements may appear to be so obvious as to be taken for granted and yet, there are many important situations where one or more of these requirements are very difficult to satisfy. Fortunately, in our case, the first two requirements are generally – albeit separately – satisfied for both the business/industrial activities and the environmental phenomena of interest.

For instance, the operations of airlines and airports constitute a classic example of success of OR/MS techniques from the strictly business/industrial application perspective. Hence, there is ample information available concerning demand for flights, optimal scheduling and impacts of schedules on airlines' profitability. Furthermore, atmospheric and environmental scientists can supply much information concerning greenhouse gas emissions, energy consumption, waste generated, or embodied energy associated with airlines' operations.

Thus the main problem, in our case, appears to be with the third requirement. If, for instance, we seek an "optimal adaptation or a mitigation strategy" by air travel industry to the problems of, say climate change, we are immediately confronted by lack of historical data as well as the difficulty of testing hypotheses concerning environmental benefits, if any, of the proposed courses of action. Consequently, the problem some OR/MS analysts may initially have considered as a relatively routine, bi-objective, trade-off situation now becomes clouded in uncertainty and confounded by the dramatic difference between the time scale of a typical business planning cycle and that of the response time of the "deep layer" of the oceans!

Despite the preceding, cautionary, introduction the thesis advanced in this contribution and explicitly or implicitly supported in most of the other chapters is that challenges to the successful application of OR/MS techniques to environmental decision making are not insurmountable. Indeed, uncertainty needs to be understood and intelligently managed so as to minimize risks. Similarly, time scales' asynchrony between human development processes and the natural processes of the biosphere needs to be acknowledged, quantified, and used as a basis for building a societal consensus concerning the forthcoming tradeoffs that, hopefully, will be made in a well planned manner that is consistent with an enlightened society.

The remainder of this chapter is organized around the discussion of the following main issues.

1. New challenges and pitfalls to OR/MS applications.
2. The sources of uncertainty in environmental modeling.

3. Stochastic reasoning for some fundamental issues.
4. Different ways to deal with uncertainty: stochastic control and stochastic programming, statistical analysis and statistical emulation, robust optimization.

1.2 New challenges and pitfalls to OR/MS applications to EDM

In this section we present Environmental Decision Making (EDM) and highlight some of the difficulties that have to be overcome when applying OR/MS method to this new domain of application. We give here a list of the main difficulties that we have identified:

- Interactions between physical, biological, economic and technological systems.
- Choice of performance criteria (e.g., public or private welfare, or a mixture of these)
- Susceptibility of extreme events
- Time scales (short to very long)
- Memory effects
- Limitation of statistical analysis
- Danger of inaction for viability
- Ethics of uncertainty and intergenerational equity.

1.2.1 From natural resource management to EDM

Environmental Decision Making (EDM) offers new challenges to OR/MS. Indeed, environmental management is a socially important domain and the situation could be compared to the one, in the early 1940's, at the beginning of modern OR era. At that time the challenge was to help solve the huge logistical problems of war operations around the globe, or to help design better antisubmarine or anti-aircraft weapons (see (59)) . As mentioned in the Introduction, the methods of modern OR developed since WW2 have used powerful mathematical tools, in particular those related to convex analysis to produce efficient decision support tools which were then successfully implemented in private or public organizations and corporate management under the generic name of Management Science (MS). Currently the state of the environment and its possible evolution is viewed as a global threat which requires the development of efficient and operational decision support systems adapted for EDM. Controlling local pollution and improving waste management in mega-cities, managing scarce water resources, controlling ocean pollution, and, of course, mitigating anthropogenic climate change are the big challenges of this century. These are also new challenges to decision analysis which are as encompassing and urgent as the ones of WW2. We believe that OR/MS can address them by adapting its current techniques or by developing radically new ones, using again some recent developments in applied mathematics, in particular those related to the modeling of earth systems and those dealing with uncertainty in optimization.

The development of OR/MS applications to deal with decision problems involving the rational use of natural resources is not a recent phenomenon. In 1781, the French mathematician Gaspard Monge defined the problem of "Déblais & Remblais" (or the moving of earth mass with the least possible amount of work), having in mind some applications to military engineering (building defense walls). This problem is often considered as one of the very early attempts to apply mathematical reasoning to logistics and it was used by Kantorovich in 1942 for one of the first applications of linear programming. Moving earth efficiently would be considered today as an environmental management problem. Since the early days of modern OR/MS, the applications to energy management have been numerous and ever growing. This is a domain which is indeed closely related to environmental management, in particular when one is dealing with hydro systems. The efficient exploitation of water systems involves an environmental system, typically a river basin, which is modeled and its management is optimized using the techniques of OR/MS. However it is only recently that environmental constraints or objectives have been explicitly introduced in decision models for private or public organizations. The great debates concerning the management of air quality or waste in urban communities, and more recently the international negotiations concerning climate policy have triggered the development of an ensemble of decision support systems for EDM using the methods of OR/MS. In particular E^3 models which deal with the three dimensions of energy, economics and the environment have been developed in the realm of activity analysis models over the last few decades.

1.2.2 Cost-effectiveness vs cost-benefit analysis

The first challenge posed by EDM is to formulate models permitting an assessment of environmental policies. Environmental impacts of human activities often arise in the form of a deteriorating "free good", like air or water quality, climate, etc. In order to integrate the environmental dimension in the decision making of firms or public organizations we must, in one way or another, define a cost for environmental degradation or a price for environmental quality. In economic terms this is equivalent to defining a surrogate market for the environmental goods. Optimization models can be used to do that. Economists distinguish two main approaches, called *cost-effectiveness* and *cost-benefit* analysis, respectively. To illustrate these let us consider the problem of assessing climate policies.

In a cost-effectiveness approach, for example, one would define a constraint on the total emissions of greenhouse gases (GHG) that would be compatible with maintaining climate change within tolerable limits. For instance, the official goal of the climate policy in the EU (in 2008) was to limit the global average temperature increase to 2 ^oC, considering that a higher temperature increase would cause irreparable damage to ecosystems and to the economy. To compare climate policies one could then try to compute their relative costs with respect to attaining that goal[1]. In this approach the difficulty lies in the correct modelling of the environmental constraint. For example, to know if an emission program will satisfy the goal of 2 ^{o}C, one must use a climate model that is an imperfect description of the state of nature. Therefore, even if the problem seemed to be well posed, it remains plagued with an important risk; in this particular case it would be the risk of not reaching the goal.

Consider now the cost-benefit approach. It is well illustrated, in the climate change context, by the DICE family of models developed by Nordhaus, in particular by the original DICE94 model (see (20)). In that approach an integrated economic model is proposed where both the cost of abatement and the benefits of reducing climate change are combined in the evaluation of the utility function of a representative decision maker (in that case it is the consumer in the economy). This elegant integrated approach also suffers from many pitfalls that are mainly related to uncertainty and imperfect knowledge. The evolution of climate is a random process, the impact of climate change on economic production is also very uncertain. On top of that, since the change in climate is a process that will extend over decades or even centuries, the utility function of the decision maker should include the welfare of both the constituency of the current generation, and also that of the forthcoming generations. There is therefore a question of inter-generational equity and the attitude of current decision makers toward this equity issue is uncertain.

1.2.3 Activity analysis models and uncertainty

The integration of environmental constraints into energy models generally follows the cost-effectiveness paradigm. In the late 70's as a consequence of the first oil crisis and the creation of the IEA[2] several models of the energy system, based on the paradigm of linear activity analysis, have been developed to assess the new technologies and energy forms that could be used to replace oil and oil-based technologies in a perspective of a durable shortage. A few years later the shortage of oil was less an urgent problem, but the global environmental impact of energy systems became an important concern.

The bottom-up E^3 models integrate within a large scale linear programming formulation, a technology rich description of energy supply, a representation of the emissions of the main atmospheric pollutants and GHGs, a representation of the adaptation of the demand for energy services to the energy prices (price elasticity of demands). Among the most successful models in that class one may cite MARKAL, developed by the ETSAP[3] consortium under the aegis of the IEA and MESSAGE, developed at the IIASA[4]. In its most recent avatars, MARKAL has been succeeded

[1] More precisely one would consider the addition in cost compared with a business as usual situation where no environmental concern would have to be taken into consideration.

[2] International Energy Agency which is run by the OECD.

[3] Energy Technology Systems Analysis Program.

[4] International Institute of Applied Systems Analysis, Laxenburg, Austria.

by TIMES[5] and TIAM[6] which allow the use of stochastic programming to deal with uncertainty, as shown in one chapter of this book[7].

1.2.4 Meta-modeling: interactions between physical, bio-chemical, economic and technological systems

In EDM, one typically has to organize a production or a logistic process so that it supplies the requested service while the environmental impact of these activities is maintained in a tolerable region. The modeling of the logistic part is similar to what has been developed since the early days of OR/MS; the difficulty lies in the representation of environmental impacts, the delimitation of a "tolerable region" and the integration of this concept as a meaningful set of constraints in the EDM. The representation of air quality or climate change dynamics, for example, rely on very complex numerical models which represent atmospheric dynamics and chemistry. The inclusion of this information in an EDM requires the development of what has been called a "meta-model" in (50) and (21) whose authors proposed an EDM for air quality management in urban regions. The problem to solve could be formulated as follows:

> Find the optimal development and use of the energy system to satisfy the demand for energy services in a given urban region (transportation, residential and commercial heat and cooling, industrial heat, electricity usage etc.) which would maintain the air quality at an acceptable level, according to meaningful indicators.

The proposed "meta-model" is combining an activity analysis model describing the energy and technology choices made to provide the needed services in an urban region and an air quality simulation model representing some typical ozone pollution episodes that might occur. The latter are typically caused by the emission of precursor pollutants, like NO_x and $VOCs$ and are generated by these technology choices. The meta-model is set in an optimization framework corresponding to a cost-effectiveness approach. In the rest of this section we outline the main features of this approach.

1.2.4.1 Modeling atmospheric dynamics and chemistry

Consider first the air quality model (AQM). Air quality is the result of a complicated process. Pollution episodes are due to many different factors (e.g., atmospheric dynamics, complex chemistry, generally incomplete knowledge about urban and biogenic emissions and others). Since numerical air quality models are capable of in-

[5] The Integrated MARKAL-EFOM System.

[6] The TIMES Integrated Assessment Model.

[7] M.Labriet, M. Loulou and A. Kanudia, Modeling Uncertainty in a Large scale integrated Energy-Climate Model, this book.

corporating many of these factors, they constitute a reasonable approach to first understanding and, ultimately, controlling air pollution. The processes represented in a typical comprehensive air quality model include the chemistry leading to the formation and destruction of pollutants, the dispersion (transport and diffusion) and the deposition. The inputs of such models are the emissions of harmful pollutants and their precursors, meteorological fields (wind speed and direction, turbulent coefficients, temperature and density), and ground characteristics (topography, land use etc.). The outputs are the concentrations of harmful air pollutants, such as ozone and particulate matter, the deposition of acids on land and water, visibility degradation, and, eventually, the potential exposure of humans to various species of interest. This is summarized in Figure 1.1.

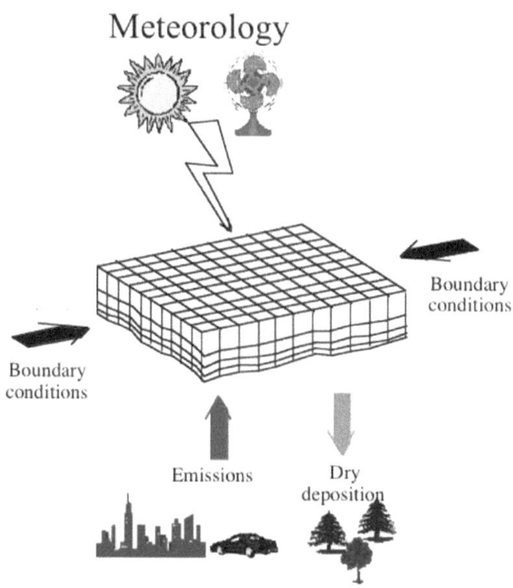

Fig. 1.1 : Input and output of a typical comprehensive air quality model

A typical air quality model is the atmospheric diffusion model found in the California/Carnegie Institute of Technology (CIT) airshed model (82), (80) or the *Transport and Photochemistry Mesoscale Model* (TAPOM) implemented at the Swiss Federal Institute of Technology at Lausanne (EPFL), and at the Joint Research Center of Ispra (JRC-Ispra) (83).

These models solve a mass conservation equation for each pollutant which can be summarized by the distributed parameter system for all species i

$$\frac{\partial c_i}{\partial t} + \nabla \cdot (Uc_i) = R_i(T,J,H,\mathbf{c}) + \sum_{j=1}^{3} \frac{\partial}{\partial s_j}(\sum_{\ell=1}^{3} K_{j\ell}(s)\frac{\partial c_i}{\partial s_\ell}) + E_i, \qquad (1.1)$$

where c_i is the mean concentration of species i, $\mathbf{c}$ denotes the aggregate concentration vector of all species[8], U is the wind velocity vector, $\nabla \cdot (Uc_i)$ denotes the divergence in the space variables $s = (s_1, s_2, s_3)$, $\bar{K}(s) = (K_{j\ell}(s); j, \ell = 1,2,3)$ is the second order turbulent diffusivity tensor, $R_i(\cdot)$ is the rate of reaction of specie i due to chemistry, T the temperature, J a function defining the solar flux, H the humidity, and the vector of concentrations c. E_i is the elevated source rate of emissions of species i, and t is the time.

These dynamic equations are complemented by the boundary conditions that describe the influence of the rest of the world on the domain $S \subset \mathbb{R}^3$ under consideration. These boundary conditions[9] are expressed as

$$Uc_i - \bar{K}\nabla c_i = Uc_i^b, \quad s_1 \text{ or } s_2 = 0, \quad \text{(horizontal inflow)} \tag{1.2}$$

$$-\nabla c_i = 0, \quad s_1 \text{ or } s_2 = s_{max}, \quad \text{(horizontal outflow)} \tag{1.3}$$

$$v_i^g c_i - K_{33}\frac{\partial c_i}{\partial s_3} = E_i^g, \quad s_3 = 0, \quad \text{(ground emissions dispersion)} \tag{1.4}$$

$$-\frac{\partial c_i}{\partial s_3} = 0, \quad s_3 = H, \quad \text{(top of model domain)} \tag{1.5}$$

where v_i^g is the dry deposition velocity for species i and E_i^g is the ground level emission rate.

1.2.4.2 Air Quality Indicators.

Air quality in a region S depends on the evolution of the concentrations of some chemicals – typically refereed to as "species" – in space and time during critical episodes. We shall focus on ozone (O_3) in this application. Ozone concentrations are functions of the emission schedule of all species, as indicated in the system (1.1-1.5). We summarize this dependence in the notation $c(t,s;\bar{E}^g,\bar{E})$, $t \in [0,\Upsilon]$, $s \in S$ for the concentration of ozone at location s at time t of a critical episode. Here we use the notation $\bar{E}^g = (E_i^g : \text{all species } i)$, $\bar{E} = (E_i : \text{all species } i)$.

To gauge air quality one defines some performance criteria, or indices, based on the concentrations of ozone during representative critical episodes[10]. There are several possibilities to define such criteria. A simple way is to measure the maximum concentration of ozone over the region S during a weather episode of duration $\Upsilon > 0$. That is, we define

$$P^{max}(\bar{E}^g,\bar{E}) = \max_{(t,s)\in[0,\Upsilon]\times S} c(t,s;\bar{E}^g,\bar{E}). \tag{1.6}$$

Another option for air quality performance indices would be to select a threshold level for each specie, say $\theta(s), s \in S$, and consider the average over the threshold

[8] Typically TAPOM considers more than 60 species in its chemistry module

[9] The horizontal inflow and outflow conditions assume that the wind is diagonal across the domain volume, originating from $s_1 = s_2 = 0$ and exiting the domain at $s_1 = s_2 = s_{max}$.

[10] Typically an episode has a duration of a few days with characteristic weather conditions.

(AOT) criterion defined by

$$P^{AOT}(\bar{E}^g,\bar{E}) = \frac{1}{\Upsilon \times |S|}\int_0^{\Upsilon}\int_S \max\left(0,(c(t,s;\bar{E}^g,\bar{E}) - \theta(s))\right) ds\, dt \tag{1.7}$$

in which $|S|$ denotes the volume of the region S. This criterion only considers the regions in which the concentration exceeds the threshold level. Still a third choice could be the mean square AOT. That is,

$$P^{SAOT}(\bar{E}^g,\bar{E}) = \frac{1}{\Upsilon \times |S|}\int_0^{\Upsilon}\int_S |\max\left(0,(c(t,s;\bar{E}^g,\bar{E}) - \theta(s))\right)|^2 ds\, dt. \tag{1.8}$$

We observe that all of the above air quality indices have the form

$$P^*(\bar{E}^g,\bar{E}) = \mathscr{P}(c(\cdot,\cdot;\bar{E}^g,\bar{E})) \tag{1.9}$$

in which $\mathscr{P}$ is a real-valued, convex functional defined on the space of functions that contains the solutions of the dynamical system, that is the distribution over time and space of the concentrations of species. We further notice that in the case when the dynamics described by (1.1) is linear (i.e., when the chemical reactions R are linear in $\mathbf{c}$) then the air quality indices $P^*(\cdot,\cdot)$ ($* = \max$, AOT, or SAOT) are convex functionals of the emission rates $(\bar{E}^g,\bar{E})$.

1.2.4.3 Techno-economic model

Now consider the Techno-economic model (TEM) which describes the logistical aspect of this EDM. Activity analysis models have been developed to assess global long-term energy and environmental policies at national and regional levels. The MARKAL model (1), (21) , which will be used in this study, is an archetypal TEM that has been implemented in more than 16 countries, and in the context of urban environment management in Sweden (85) and Switzerland (42). In these models, the energy system is represented through a network of technologies extracting, transforming and using energy carriers to provide the energy services needed by the economy. The technologies used to satisfy these energy services are the main emitters of primary pollutants. More precisely there is a close relationship between the modeling of energy flows in a production economy and the tracking of pollutant emissions, since they are both determined by the choice of technologies made to satisfy the demand for different products and services. In activity analysis models, economic activities are represented as resources transformers. Each activity is thus characterised by its level of use and the technical coefficients that describe the resources input and output characterising the technology involved in the activity. An activity may bear a cost or contribute some profit (or more generally some utility). Using the appropriate optimization formulation, activity analysis models can be imbedded in different types of market structures (competitive, oligopolistic, and monopolistic).

There have been many successful applications of the above paradigm in the energy sector, starting with the representation of oil refineries, and more recently the representation of the interactions between the energy sector and the rest of the economy (69). These models are particularly well suited for the analysis of environmental management policies and for the design of market based instruments (MBI). As said before, the technologies (activities) are transforming resources and as a byproduct they emit pollutants. A global emission constraint can be imposed on the economic system under consideration. This translates into a higher cost for achieving the production goals of the system.

In brief the structure of the TEM is a mathematical programming model, usually a linear program, that can be written as follows

$$\min g = \gamma x \tag{1.10}$$

$$\text{s.t}$$

$$Ax = a \tag{1.11}$$

$$x \geq 0$$

where x is the vector of activity levels, γ is the vector of economic coefficient (interpreted here as costs). Here A is the matrix of technical coefficients and a is the right hand side of the constraints. Basically, a TEM minimizes an objective function which represents the total discounted system cost. This is the discounted sum over a time horizon (typically 35 to 45 years) of the operation, investment and maintenance cost of all technologies involved in the energy production and usage. The minimization of that function is done subject to four main categories of constraints that express the following principles:

1. an installed capacity (inherited or resulting from investment) is needed to operate a technology;
2. existing capacity is transferred over time periods subject to life duration of equipment;
3. the useful demand has to be satisfied;
4. the energy balance has to be respected (consumption + export = production + import) for each energy form at each time period.

1.2.4.4 Coupling the two types of models

Time-scale issue.

To combine the two approaches (AQM and TEM) we have to deal with the previously mentioned times scales' asynchrony (TSA). This is because techno-economic multi-period model which encompasses a planning horizon of several decades with discrete time steps of five years, while the air quality simulation model has a time horizon of a few days with time steps of usually 15 to 60 minutes. These two time scales have to be made compatible in order to permit a coherent dialogue between the two models.

A way to deal with TSA is to associate an average air quality index with a typical emissions inventory resulting from a technology mix selected by the TEM for each 5 year planning period. The average air quality index is obtained from simulation runs obtained from the AQM over a fixed set of weather episodes. These weather episodes represent the typical meteorological conditions favorable to high air pollution levels (64), (29). Emission control strategies are tested against these selected episodes in order to derive an air quality response that is representative of the average meteorological conditions leading to high levels of pollution. Note that the average air quality indices obtained from these simulations now have a time-scale corresponding to a year. Indeed, we take the statistically relevant weather episodes in a year and we simulate the ozone concentrations resulting from the average daily emission inventory resulting from the activities decided for the TEM.

Therefore a constraint imposed on this average quality index will have a time-scale compatible with the techno-economic model. Indeed there is a considerable latitude in the construction of this average yearly air quality index and in the selection of a sample of weather episodes to be used in simulations. This is a source of uncertainty that will be necessary to handle through sensitivity analyses or some form of robust programming (a concept that will be discussed shortly).

Creating an emission inventory as an input for the AQM

The emissions sources are mostly related to a subset of economic activity sectors. In the TEM, these activities are typically traffic, heating and air conditioning systems and power plants. Let $u_i(k,\alpha,x)$ denote the primary pollutant i emissions level due to the activities in sector α in period k given the decision vector x. Let $f_i(t,s,k,x)$ be the total emissions level of pollutant i at time t at location s in period k. We assume that the distribution in time and space is obtained from the activity pollution level $u_i(k,\alpha,x)$ via an exogenously defined *dynamic emissions map* $\nu(t,s,k,\alpha)$ according to the linear transformation

$$f_i(t,s,k,x) = \sum_{\alpha\in A} \nu(t,s,k,\alpha)u_i(k,\alpha,x). \tag{1.12}$$

The above *dynamic emissions map* converts the global annual emissions for the whole domain into hourly emissions on the spatial mesh of the geographical information system which produces the input for the AQM.

An auxiliary TEM with emission limits

Our aim is to solve a problem where a constraint on air quality is added to the TEM. As shown above, the air quality constraint is complicated and involves simulations done with a detailed numerical model. A much simpler model would be obtained if we could replace the constraint on air quality by a constraint on emission levels for primary pollutants.

For each primary pollutant i, each sector α and each period k, one defines an upper bound vector $U = \{U_i(k,a)\}_{i,k,a}$ where $U_i(k,\alpha)$ is a limit on the total emissions of pollutant i due to activities in sector α, in period k. We can now define the function $h(U)$ as the solution of the following mathematical programming problem:

$$h(U) = \min \gamma x \tag{1.13}$$

$$\text{s.t}$$

$$Ax = a$$

$$u_i(k,\alpha,x) \leq U_i(k,\alpha) \quad k = 1,\ldots,K, \quad \alpha \in A. \tag{1.14}$$

$$x \geq 0.$$

where the constraints are indexed over all primary pollutants i. This mathematical programming problem reduces to an easily solved linear programming problem if one assumes constant emissions rates for the different activities (technologies) entering the model. The function $h(U)$ is thus representing the minimal cost to achieve the program of sectoral emissions of primary pollutants, represented by the upper bounds U.

The Meta-Model as an upper level program.

As we are interested in the air quality resulting from the ozone concentration, we now define a second auxiliary problem, involving an environmental constraint $P(k,U)$ that is computed via the AQM, for the emissions level corresponding to U at period k. This leads us to introduce a nonlinear program that can be written as:

$$\min \quad h(U) \tag{1.15}$$

$$\text{s.t}$$

$$P(k,U) \leq P_{\max}, \tag{1.16}$$

where $h(U)$ has been defined in Eq. (1.13) and the function $P(k,U)$ is typically representing an AOT_{60} or peak ozone index calculated for typical episodes at the successive periods k of the planning horizon. The whole difficulty of the method resides now in the identification of the functions $h(U)$ and $P(k,U)$.

We call the program (1.15)-(1.16) a "Meta-Model" because both parts, the objective function in (1.15) and the constraints in (1.16) are obtained from complex sub-models which are run separately in their own time and geographical scale, either as optimization or numerical simulation tools.

1.2.4.5 How to solve the Meta-Model

In (21) it was shown how solving the program (1.15)-(1.16) could be approached with an oracle-based optimization method (see (43), (10)). The Analytic Center Cutting Plane method is implemented to send queries either to the TEM with emissions constraints to obtain "optimality cuts" or to the AQM to obtain "feasibility cuts" which delineate a shrinking localization set in the U space where the optimum must be found.

In (50), for the same problem, a different approach was used. In particular, a statistical emulation of the response of the complex AQM and an air quality index to an emission program were used to build surrogate constraints that can be tackled directly in the program (1.15)-(1.16).[11]

This rather detailed description of a model of air quality management has permitted us to show some of the critical difficulties of combining techno-economic models and earth science models. We have also seen that recent advances in optimization methods, in particular the oracle-based optimization techniques, provide a way to deal with meta-models where some constraints are obtained from running complex numerical simulation models. Indeed, these simulations are approximate description of the chemical or physical processes that are involved and, therefore should include in the approach an explicit consideration of the possible errors, or more precisely, of the confidence intervals of the indicators obtained.

1.2.4.6 Extension to other EDMs and the possible role of robust optimization

The Meta-Modeling approach described above for a problem of air quality management has also been applied to build an integrated assessment model for global climate policies. In particular, (13), (30) develop a meta-model composed of an optimal economic growth model on one side and a general circulation climate model of intermediate complexity on the other side. In these applications, also reported in (11), the coupling of the optimal economic growth model and the climate 3-D circulation model was also achieved through the use of ACCPM, an oracle based convex optimization method.

A very interesting and promising approach, to deal with the uncertainty which is inherent in these complex simulations of environmental systems, is to build statistical emulations of the response of climate system to some variations in key parameters, using, for example Kriging techniques ((63), (68), (79)).

We refer to (24) for a description of the use of statistical emulation in climate models and to (81) for the implementation of robust optimization in meta-models obtained from complex computer simulations. For robust optimization, which is presented in one chapter of this book[12] we refer to (15) or (17)). It is our contention that robust optimization allied to statistical emulation of complex environmental models will play an important role in the development and implementation of meta-models for EDM, in particular in the domain of air quality management and climate policy assessment.

---

[11] The interested reader can find the implementation and the results obtained through these approaches in (50).

[12] Robust Optimization for Energy-Environment Planning, F. Babonneau, J.-P. Vial, R. Apparigliato. This book.

1.2.5 Choice of time horizon & performance criterion: the ethics of uncertainty in EDM

As we have just seen, the earth system, the biological systems and the economic systems are interacting elements that evolve at very different time scales. Geological transformations take place in the very long run, climate change occurs over centuries but the actions of any living being (plant or animal) are taken on a very short time horizon or a very short expected life duration.

When one designs an environmental policy it is important to identify the constituency for which one is acting: is it the current generation in the population or the sequence (possibly infinite) of generations to come? This is particularly important if decisions taken or postponed have irreversible effects. If the current choices in the energy system may trigger a change of climate particularly harmful in 50 or 100 years, this induces a low cost to the current generation of human beings but a high cost to those who are not yet born. Often, this aspect of the problem is captured by a single parameter entering in most dynamic decision models, namely the *discount rate* also called *pure time preference rate*. Some of the advantages and limitations of this approach are discussed later in this chapter.

1.3 Sources of uncertainty in environmental modeling

Whenever scientists attempt to model a phenomenon they need to be cognisant of both the nature and the magnitude of uncertainties that will, invariably, be embedded in the model's outputs/forecasts.

1.3.1 Classification of uncertainty sources in models

Following (37) and (36) we classify the nature of the uncertainties into five distinct types (see Figure 1.2, below):

1. *Errors in observations.* Uncertainties caused by errors, E_o, in observations. These errors may occur prior to the construction of the model but, to the extent that the observations will be used to estimate the model's parameters, they will also affect the model's eventual outputs.
2. *Errors in estimation of parameters.* Uncertainties caused by errors, E_p, in the estimation of parameters. These errors occur because whatever statistics are used to estimate parameters of interest, these statistics incur some errors.
3. *Errors due to solution algorithms.* Uncertainties caused by errors, E_s, due to the solution algorithm. Nowadays, most models are solved numerically, rather than analytically, and the numerical algorithms almost always involve a degree of approximation.

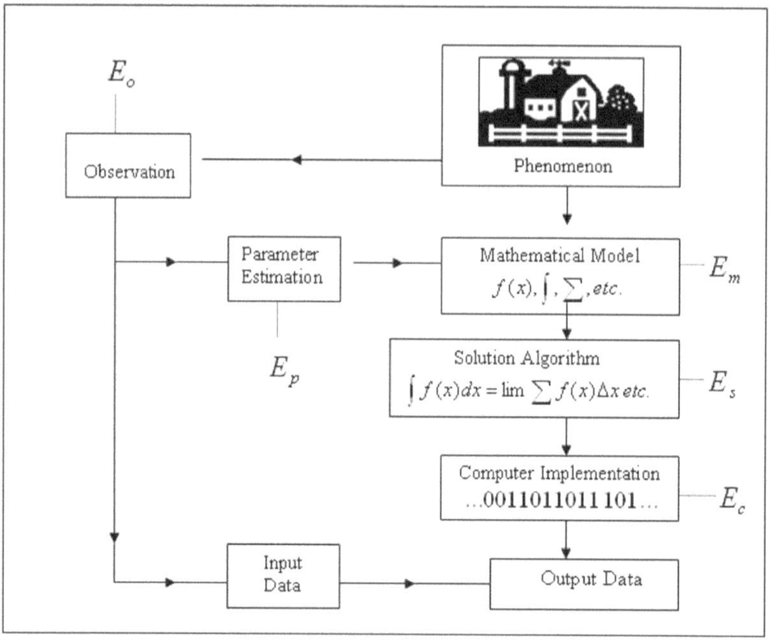

Fig. 1.2 Sources of uncertainty in the outputs of mathematical models.

4. *Errors due to computer implementation.* Uncertainties caused by errors, E_c, due to the computer implementation. These errors occur due to the way that computers perform arithmetic operations and can be strongly influenced by the manner in which various steps of a solution algorithm are programmed.
5. *Errors in the modeling.* Uncertainties caused by errors, E_m, in the modelling. These occur simply because whatever model is constructed, it is only an approximate representation of the underlying natural system that is being modelled.

It can be argued that, of all of the above types of uncertainties, it is the last, or the modelling uncertainty E_m that is most often overlooked or misunderstood. This is because, to a significant extent, each of the first four types of uncertainties is associated with the mainstream activities of a specific professional group whose practitioners have developed methods for estimating bounds on the corresponding types of errors.

Thus, designers of observational equipment (eg. gauges, barometers etc) can tell us about the precision of various instruments, that is place bounds on E_o. Similarly, statisticians can usually tell us about the bounds on E_p, or the estimation errors, numerical analysts can frequently supply bounds on E_s and computer scientists can bound E_c.

However, in the case of the uncertainties caused by the modelling errors E_m, there appears to be no professional group that specializes in their analysis. Perhaps, this is because the understanding of this kind of uncertainty requires an a priori admission from the creators of a model that their creation is, somehow, "flawed".

To be fair to the modelling community, it should be said that, nowadays, most widely used models are based on sound scientific or economic principles and have been calibrated against the available data. Indeed, in the case of many models such a calibration will be accompanied by a statistical measure of the "goodness of fit" of the model's outputs to the observed data. This, in turn, can provide an estimate of the bounds on E_m, if the model is used in the data ranges comparable to the ranges used for calibration purposes.

What makes the modelling error especially challenging in many models is the fact that, often, such models are producing outputs in the ranges which are very different from those of the data that were used for model fitting and/or calibration. In the case of global change models, the outputs that reach 100, 200 or 1000 years into the future clearly step far outside of these ranges. Thus a fundamental question arises:

> Even if the modelling error E_m is small initially, how does the uncertainty due to this error propagate over time, when the model is run over large time horizons?

The inherent difficulty of the preceding question is that analyses that might provide an answer are intimately linked to the mathematical formulation of the model. Thus a model described with the help of partial differential equations may require a different analysis of the uncertainty propagation than one that is described with the help of ordinary differential equations. Furthermore, the form of the functions used to describe various quantities will also affect the manner in which this uncertainty propagates. Since many modelers choose the model description (including functional forms) on the basis of what they perceive is the best current understanding of the phenomenon being modelled, the notion that this description may need to be modified as a result of the uncertainty propagation analysis, may appear as pure heresy.

Fortunately, the above difficulties are usually not insurmountable, once importance of the problem is recognized. In particular, there are many rigorous techniques - primarily statistical or probabilistic in nature - that are potentially adaptable to the problem of estimating both the size of the modelling error and its propagation over time.

1.3.2 Error propagation in an integrated climate model

The propagation of errors in models, discussed above, is now briefly illustrated in the context of an integrated model of climate change. The discussion presented here is based on (86).

One of the best early integrated models for assessing the greenhouse effect was IMAGE 1.0 (Integrated Model to Assess the Greenhouse Effect) (78). This model had a modular structure combining relatively simple models of global carbon cycle, (atmosphere, biosphere, and the ocean) atmospheric chemistry, radiative forcing, and heat exchange in the ocean. These modules were linked with an energy use module, emission and land use modules and temperature change, sea level rise and socio-economic impact modules. In (19) the essential geo-physical processes of IMAGE 1.0 were extracted and represented as a deterministic (155–dimensional) dynamical system of the generic form

$$\dot{X}(t) = F(X(t)) + U(t) \tag{1.17}$$
$$X(t_0) = X_0, \tag{1.18}$$

where the time $t \in [1990, 2100)$ specified the period under investigation, $t_0 = 1990$ was the initial time, $X(t)$ was the vector of 155 state variables, $U(t)$ was the forcing term representing human interference (including CO_2 and other greenhouse gas emissions) with the climate, and $F(X(t))$ described the climate system processes. The solution $X(t)$, as a function of time, was called the trajectory of (1.17).

Essentially, like many integrated models, IMAGE 1.0 functioned by creating scenarios for the forcing term $U(t)$ and numerically simulating the corresponding trajectories $X(t)$ of (1.17). Of course, the resulting trajectories for all state variables (irrespective of whether it would be the concentration of CO_2, or the temperature of the mixed layer of the ocean) could be plotted as smooth functions of the time variable; the latter being the property of the deterministic system of ordinary differential equations (1.17). As such, these curves could exhibit the average trend of the corresponding variables over time – under the selected scenarios - but not their variability.

In the climate change research community there is now a wide range of benchmark scenarios that are frequently used. In the late 1980's when IMAGE 1.0 was developed two of the frequently discussed scenarios were the so-called *business-as-usual (BaU)* and the *forced trends (FT)* scenarios. These were intended to demonstrate the likely impacts of two dramatically different, global, approaches to the problems of climate change. As the name indicates, the former was intended to capture a "hands off" approach where no coordinated reductions in greenhouse gas emissions would be agreed upon, while the latter assumed strong coordinated action to achieve such reductions. Such extreme case scenarios are still widely used even though various specific assumptions comprising them have changed many times.

The innovation in (86) was to demonstrate a methodology whereby stochastic dynamical systems can be exploited to investigate a model's inherent capacity to propagate uncertainty over time. In the case of a model such as IMAGE 1.0 the deterministic system (1.17)-(1.18) is replaced by a corresponding system of stochastic differential equations

$$dX(t) = [F(X(t)) + U(t)]dt + GdW(t) \tag{1.19}$$
$$X(t_0) = C, \tag{1.20}$$

where $X(t), F(X(t)), U(t)$ are as in (1.17)-(1.18), but C is a Gaussian vector containing the initial state uncertainty, and G is a constant, diagonal diffusion matrix corresponding to the vector Wiener process $W(t)$ with independent components. When both the diagonal entries of G and the norm of the variance-covariance matrix of C are small, this could be interpreted as the original deterministic system (1.17)-(1.18) initially being close to a perfect description, except for some small uncertainty in the initial conditions and a small amount of modelling error represented by the "white noise" process $W(t)$.

Therefore, the new representation (1.19)-(1.20) enables us to study the effect of these seemingly benign errors on the uncertainty of future trajectories of the state variable vector. Note that a solution of this new system is now a stochastic process, where at each time t, the state $X(t)$ has a distribution whose moments can be calculated using the tools of stochastic calculus (e.g., (60)).

Of course, the ability to calculate standard deviations $\sigma_i(t)$ of each of the 155 state variables $X_i(t)$, at each time t, enables us to equip the smooth curves from the deterministic model with "uncertainty bands" of, say, plus or minus two standard deviations. Naturally, these can be expected to grow in width as t moves further away from the initial time t_0. The rate at which the widths of these bands grow communicates a lot about the model's capability to propagate errors.

In Figures 1.3 and 1.4 we show these uncertainty bands for the atmospheric CO_2 concentration variable, from IMAGE 1.0, under the two extreme scenarios: BaU and FT. In Figure 1.3 the bands were generated under the assumption that $G = 0$, that is, the small uncertainty about the initial state vector was the only source of stochasticity. In Figure 1.4 the bands were generated under the additional assumption that diagonal entries of G had small positive values, that is, the system now had two sources of stochasticity. The upper and lower uncertainty bands in these figures are represented by the broken curves.

Of course, the mean trend of CO_2 concentration under the BaU scenario always lies above the corresponding mean trend under the FT scenario. These are represented by the solid curves in Figures 1.3 and 1.4 and, naturally, they diverge rapidly. However, the authors of (86) introduced the notion of a "scenario overlap" to refer to the period when the upper band uncertainty band of the FT scenario lies above the lower uncertainty band of the BaU scenario. We see from these figures that, in the case of CO_2 concentration, the scenario overlap persists for a significant period of time after the beginning of the simulation. Arguably, until the scenarios have separated – so that the above overlap no longer applies – it would be hard to tell with a high degree of accuracy whether any future observed changes in CO_2 concentration were due to the effectiveness of the measures contained in the FT scenario or due to the, spurious, model induced variability. Since the period of scenario overlap in Figures 1.3 and 1.4 is of the order of 40 and 45 years, respectively, we can quantify the extent to which model uncertainty and initial state uncertainty degrade the quality of the model's forecasts. However, it should be pointed out that if we decided to narrow the uncertainty bands to be, say, plus or minus one standard deviation, then the periods of scenario overlaps would be considerably shorter.

The usefulness of the above methodology stems from its capacity to identify the variables that account for most of the model's uncertainty. Additional findings reported in (86) were that the uncertainty of most of the system's variables were

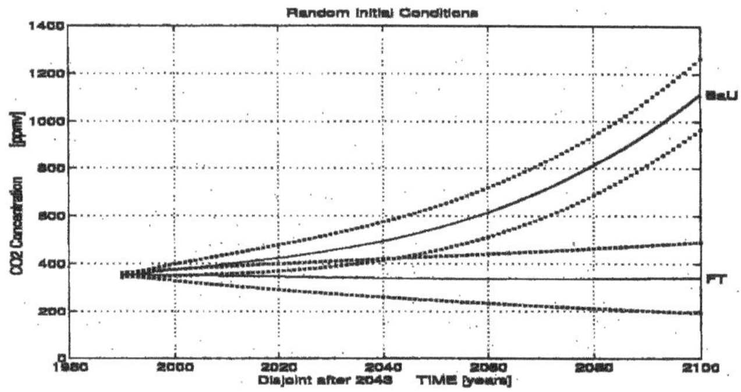

Fig. 1.3 : Overlap of CO_2 concentration predictions for the BaU and FT scenarios. Uncertainty induced only by random initial conditions.

found to be nearly independent of the environmental scenario for the time period under consideration (1990 – 2100). Uncertainties associated with atmospheric concentrations of greenhouse gases other than CO_2 were found to exhibit similar growth patterns. In addition, the uncertainties surrounding these gases strongly influenced the uncertainty of the main variables affecting the climate: temperature change in the mixed layer of the ocean and the atmospheric concentration of carbon dioxide.

1.4 Stochastic reasoning for some fundamental issues

In this section we show how some fundamental issues in EDM, like the proper choice of a discount rate in an environmental policy assessment model involve stochastic or statistical reasoning. The latter open new perspectives on certain classical domains of OR/MS such as infinite horizon dynamic optimization or Markov decision processes (MDPs).

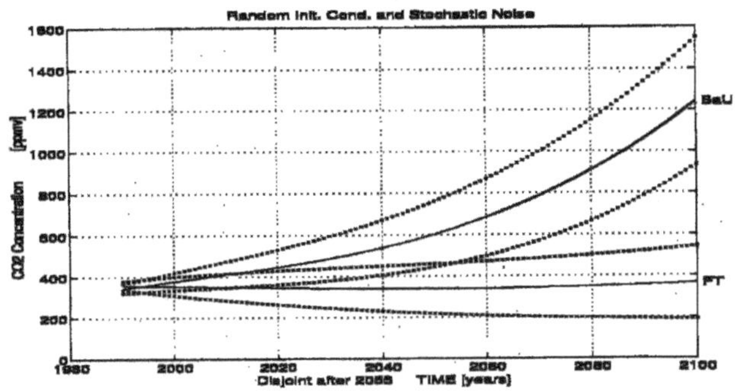

Fig. 1.4 : Overlap of CO_2 concentration predictions for the BaU and FT scenarios. Uncertainty induced by random initial conditions and stochastic system noise.

1.4.1 The uncertain discounting issue

The cost-benefit analysis for global climate change mitigation has triggered a renewed interest in low discounting, or even zero discounting, in economic growth models. In this section[13] we provide a new interpretation of zero discounting, by referring to the theory of turnpikes in infinite horizon optimal control problems. The important contribution of (84) is to explain why one should use the lowest rate to discount distant futures. A related interesting monograph (76) discusses the issue of selecting the proper discount rate for assessing environmental policies that will affect several generations in the future. We may consider this problem from the point of view of infinite horizon optimal control, where an asymptotic behavior known as the *turnpike property* can be observed. When the discount rate is uncertain, it is the turnpike associated with the lowest discount rate that is the eventual attractor of the optimal growth path. The problem becomes more subtle when the decision maker is trading off a discounted criterion and an undiscounted one. Then the trajectory could be attracted first by the turnpike associated with the highest discount rate and the switch to the zero discount turnpike would be postponed forever. This means that an optimal solution does not exist in that case. However the solution will exist if one uses an overtaking optimality criterion with a weighted payoff.

---

[13] The content of this section is largely reproduced from (48) and (49).

1.4.1.1 The optimal control framework

Let $c(t) : t \in [0,\infty)$ be piecewise continuous function representing a stream of consumption and $\mathscr{U}(\cdot)$ a utility function for an economic agent. Let $\rho > 0$ be a discount rate or, more precisely, a *pure rate of time preference* representing impatience of the economic agent. The discounted sum of utilities is defined by

$$\int_0^\infty e^{-\rho t} \mathscr{U}(c(t))\, dt. \tag{1.21}$$

The constant value g_ρ, called the *discounted average per discounted unit of time*, is defined to be the unique scalar that satisfies

$$\int_0^\infty e^{-\rho t} g_\rho\, dt = \int_0^\infty e^{-\rho t} \mathscr{U}(c(t))\, dt. \tag{1.22}$$

Since $\int_0^\infty e^{-\rho t}\, dt = 1/\rho$ one obtains

$$g_\rho = \rho \int_0^\infty e^{-\rho t} \mathscr{U}(c(t))\, dt. \tag{1.23}$$

The problem we address is how to deal with uncertainty about the discount rate. In (84) Weitzman explains why the far distant future should be discounted at the lowest possible rate. The rationale is as follows: Assume that the discount rate is a random variable taking two possible values $\rho_1 > \rho_2$ with respective probability α and $1-\alpha$, $\alpha \in (0,1)$. An expected average discounted utility per unit of discounted time is then given by

$$\alpha g_{\rho_1} + (1-\alpha) g_{\rho_2} = \alpha \rho_1 \int_0^\infty e^{-\rho_1 t} \mathscr{U}(c(t))\, dt + (1-\alpha)\rho_2 \int_0^\infty e^{-\rho_2 t} \mathscr{U}(c(t))\, dt. \tag{1.24}$$

Now the expression (1.24) can be rewritten

$$\int_0^\infty e^{-\rho_2 t} [\alpha \rho_1 e^{-(\rho_1-\rho_2)t} + (1-\alpha)\rho_2] \mathscr{U}(c(t))\, dt. \tag{1.25}$$

As $e^{-(\rho_1-\rho_2)t} \to 0$ when $t \to \infty$, the long term effects are discounted by a factor $(1-\alpha)e^{-\rho_2 t}$, that is, with the lowest possible rate.

A similar conclusion holds if one takes the expected value of the discounted sums instead of the discounted average per unit of discounted time. Instead of (1.25) we obtain the expression

$$\int_0^\infty e^{-\rho_2 t} [\alpha e^{-(\rho_1-\rho_2)t} + (1-\alpha)] \mathscr{U}(c(t))\, dt, \tag{1.26}$$

where, again, the long run is discounted with the lowest rate ρ_2.

This is an interesting observation which comforts the views of those economists who claim that discount rates depend on the period over which the analysis is being conducted. Among the possible *time varying discount rates*, the *hyperbolic* one (see (2), (46), (67)) has been proposed as a convenient formulation.

1.4.1.2 A brief review of turnpike theory

We want to explore the implications of Weitzman's result on the asymptotic behavior of optimal trajectories in infinite horizon optimal control problems that are typical of optimal economic growth modeling. Before proceeding further we have to recall a few results concerning asymptotic control problems.

Turnpikes for discounted infinite horizon control.

We formulate an infinite horizon optimal control problem

$$\max \quad \int_0^\infty e^{-\rho t} L(x(t), u(t))\, dt \tag{1.27}$$

s.t.

$$\dot{x}(t) = f(x(t), u(t)) \tag{1.28}$$

$$u(t) \in U \tag{1.29}$$

$$x(0) = x^o. \tag{1.30}$$

with the usual assumptions of smoothness and concavity of L and f in all their arguments, x being the state variable that is supposed to remain in a bounded set $X \subset \mathbb{R}^n$, and u being the control variable constrained to stay in the compact convex subset of set $U \subset \mathbb{R}^m$. In (20), the asymptotic behavior of the optimal trajectories is studied at length, in particular the existence of a global attractor $\bar{x}$, called a *turnpike*, for all optimal trajectories, whatever the initial state x^o is investigated. Let us recall a sufficient condition for the existence of such an attractor. Form the current value Hamiltonian $\mathscr{H}(x,u,q) = L(x,u) + q' f(x,u)$, where q is called the *current value costate variable*. The Pontryagin Maximum Principle yields the following necessary optimality conditions

$$\mathscr{H}(x^*(t), u^*(t), q^*(t), t) = \max_u \mathscr{H}(x^*(t), u, q^*(t), t) \tag{1.31}$$

$$\dot{q}^*(t) - \rho q^*(t) = -\frac{\partial}{\partial x}\mathscr{H}(x^*(t), u^*(t), q^*(t), t) \tag{1.32}$$

$$\dot{x}^*(t) = \frac{\partial}{\partial q}\mathscr{H}(x^*(t), u^*(t), q^*(t), t) \tag{1.33}$$

$$x^*(0) = x^o. \tag{1.34}$$

These necessary conditions can be given the form of a *quasi-Hamiltonian system*. Let us define

$$\tilde{u}(x,q) = \mathrm{argmax}_{u\in U}\mathscr{H}(x,u,q)$$

and[14]

$$\tilde{\mathscr{H}}(x,q) = \mathscr{H}(x,\tilde{u}(x,q),q). \tag{1.35}$$

Due to the assumptions made we know that $\tilde{\mathscr{H}}(x,q)$ is concave in x and convex in q. The necessary optimality conditions, under these regularity assumptions, imply that the optimal state and costate trajectories are solutions to the *quasi-Hamiltonian system*

$$\dot{x}(t) = \frac{\partial}{\partial q}\tilde{\mathscr{H}}(x(t),q(t)) \tag{1.36}$$

$$\dot{q}(t) - \rho q(t) = -\frac{\partial}{\partial x}\tilde{\mathscr{H}}(x(t),q(t)), \tag{1.37}$$

with initial condition $x(0) = x^o$. The missing terminal conditions, for the costate variables, are obtained from an analysis of the asymptotic stability of the system (1.36)-(1.37). If we assume that the matrix

$$Q(x,p) = \begin{bmatrix} \frac{\partial^2}{\partial x^2}\tilde{\mathscr{H}}(x,q) & -\frac{\rho}{2}I \\ -\frac{\rho}{2}I & -\frac{\partial^2}{\partial q^2}\tilde{\mathscr{H}}(x,q) \end{bmatrix} \tag{1.38}$$

is negative definite for all x,q. Then the pair $(\bar{x},\bar{q})$ that solves the algebraic system

$$0 = \frac{\partial}{\partial q}\tilde{\mathscr{H}}(\bar{x},\bar{q}) \tag{1.39}$$

$$0 = -\frac{\partial}{\partial x}\tilde{\mathscr{H}}(\bar{x},\bar{q}) + \rho\bar{q} \tag{1.40}$$

is an attractor for all the bounded solutions of the system (1.36)-(1.37), that is for all the solutions of the infinite horizon optimal control problems associated with different initial states x^o. In these circumstances, whatever be the initial state x^o, the optimal state $x^*(t)$ and costate $q^*(t)$ are attracted when $t \to \infty$ by $\bar{x}$ and $\bar{q}$ respectively.

In (35) it has been shown that the turnpike $\bar{x}$ would also be characterized as the solution of the *implicit programming problem*[15]

$$\max \quad L(x,u) \tag{1.41}$$

s.t.

$$0 = f(x,u) - \rho(x-\bar{x}) \tag{1.42}$$

$$u \in U. \tag{1.43}$$

[14] For the sake of simplifying the exposition we assume that the function $\tilde{\mathscr{H}}(x,q)$ is twice continuously differentiable in both arguments. Indeed this assumption requires a lot of regularity for this dynamical system. The optimal economic growth models are most of the time sufficiently smooth to allow such a formulation.

[15] The problem is called implicit, because the solution $\bar{x}$ is used in the very definition of the problem.

In summary, under enough concavity/convexity of the Hamiltonian $\tilde{\mathscr{H}}(x,q)$ the discounted infinite horizon optimal control problem admits a unique solution from any initial state $x^o \in X$, and all the optimal trajectories $x^*(\cdot)$ tend to bunch together as $t \to \infty$, by converging toward the turnpike $\bar{x}$. Another interesting aspect of that property is that, even for finite horizon control, the optimal trajectory will spend most of the journey in the vicinity of the turnpike.

How to deal with a zero discount rate.

When the discount rate ρ tends to 0, one may consider different performance criteria. An interesting one is the *overtaking optimality* criterion, where one says that the trajectory $x^*(\cdot)$ generated by the admissible control $u^*(\cdot)$ is overtaking optimal, if for any other trajectory $x(\cdot)$ generated by the control $u(\cdot)$, emanating from the same initial state x^o, the following inequality holds

$$\liminf_{T\to\infty}\left[\int_0^T L(x^*(t),u^*(t))\,dt - \int_0^T L(x(t),u(t))\,dt\right] \geq 0. \tag{1.44}$$

Under the conditions given above, one may expect that an overtaking optimal solution exists, from any x^o in a bounded set X, and that this solution is unique and converges toward the turnpike, defined, in the undiscounted case by the steady-state optimization problem

$$\max \quad L(x,u) \tag{1.45}$$

$$\text{s.t.}$$

$$0 = f(x,u) \tag{1.46}$$

$$u \in U. \tag{1.47}$$

Another criterion often used in the stochastic systems literature is the *limit average reward* criterion where one maximizes the long term reward per unit of time, that is the expression

$$\liminf_{T\to\infty}\frac{1}{T}\int_0^T L(x(t),u(t))\,dt. \tag{1.48}$$

An equivalent formulation to the limit average criterion (1.48) would also be

$$\liminf_{\rho\to 0}\rho\int_0^\infty e^{-\rho t}L(x(t),u(t))\,dt. \tag{1.49}$$

Indeed, with the criterion (1.48) anything that happens during a fixed transient time interval $[0,\theta]$ tends to become negligible when $T\to\infty$. So, even for a system that admits a unique overtaking optimal trajectory from any initial state x^o, with a turnpike $\bar{x}$, there will be a plethora of optimal trajectories according to the criterion (1.48). Any trajectory that reaches eventually the turnpike $\bar{x}$ could be optimal under that criterion. To restrict this abundance of optimality candidates, one may ask that the trajectory by *finitely optimal*. This means that if $x(\theta_1)$ and $x(\theta_2)$ are two points on the optimal trajectories, $\theta_2 > \theta_1$, then the trajectory section $(x^*(t) : \theta_1 \le t \le \theta_2)$ solves the problem

$$\max \quad \int_{\theta_1}^{\theta_2} e^{-\rho t} L(x(t),u(t))\,dt \tag{1.50}$$

s.t.

$$\dot{x}(t) = f(x(t),u(t)) \tag{1.51}$$
$$u(t) \in U \tag{1.52}$$
$$x(\theta_1) = x^*(\theta_1) \tag{1.53}$$
$$x(\theta_2) = x^*(\theta_2). \tag{1.54}$$

It is easy to show that overtaking optimal trajectories are also finitely optimal.

1.4.1.3 A model with a tradeoff between two discount rates

In this section we reconsider the uncertain discount rate problem, in the context of asymptotic (infinite horizon) control and we interpret Weitzman result in terms of *turnpikes*. The consideration of two possible discount rates leads to the following optimization problem

$$\max \quad \Big[\alpha\rho_1 \int_0^{\infty} e^{-\rho_1 t} L(x(t),u(t))\,dt + (1-\alpha)\rho_2 \int_0^{\infty} e^{-\rho_2 t} L(x(t),u(t))\,dt\Big] \tag{1.55}$$

s.t.

$$\dot{x}(t) = f(x(t),u(t)) \tag{1.56}$$
$$u(t) \in U \tag{1.57}$$

where ρ_1 and ρ_2 are two different discount rates, with $\rho_1 > \rho_2$.

Indeed the above criterion (1.55) can be rewritten as

$$\int_0^{\infty} [\alpha\rho_1 e^{-\rho_1 t} + (1-\alpha)\rho_2 e^{-\rho_2 t}] L(x(t),u(t))\,dt.$$

We can write the Pontryagin necessary conditions

$$H(x(t),u(t),\lambda(t),t) = \max_u H(x(t),u,\lambda(t),t) \tag{1.58}$$
$$\dot{\lambda}(t) = -\frac{\partial}{\partial x} H(x(t),u(t),\lambda(t),t) \tag{1.59}$$
$$\dot{x}(t) = \frac{\partial}{\partial \lambda} H(x(t),u(t),\lambda(t),t) \tag{1.60}$$

where

$$H(x(t),u,\lambda(t),t) = [\alpha\rho_1 e^{-\rho_1 t} + (1-\alpha)\rho_2 e^{-\rho_2 t}] L(x(t),u) + \lambda(t)' f(x,u).$$

(1.61)

We consider the current value Hamiltonian, associated with the lower discount rate ρ_2

$$\mathscr{H}(x(t),u,\lambda(t),t) = [\alpha\rho_1\, e^{(\rho_2-\rho_1)t} + (1-\alpha)\rho_2]\, L(x(t),u) + q(t)'\, f(x,u), \tag{1.62}$$

where

$$q(t) = e^{\rho_2 t}\lambda(t). \tag{1.63}$$

The necessary optimality conditions write down as follows

$$\mathscr{H}(x(t),u(t),q(t),t) = \max_u \mathscr{H}(x(t),u,q(t),t) \tag{1.64}$$

$$\dot{q}(t) - \rho_2 q(t) = -\frac{\partial}{\partial x}\mathscr{H}(x(t),u(t),\lambda(t),t) \tag{1.65}$$

$$\dot{x}(t) = \frac{\partial}{\partial q}\mathscr{H}(x(t),u(t),\lambda(t),t). \tag{1.66}$$

As $\rho_2 - \rho_1 < 0$, this system tends, when $t \to \infty$, to take the form of the modified Hamiltonian system

$$\hat{\mathscr{H}}(x(t),u(t),q(t)) = \max_u \hat{\mathscr{H}}(x(t),u,q(t)) \tag{1.67}$$

$$\dot{q}(t) - \rho_2 q(t) = -\frac{\partial}{\partial x}\hat{\mathscr{H}}(x(t),u(t),\lambda(t)) \tag{1.68}$$

$$\dot{x}(t) = \frac{\partial}{\partial q}\hat{\mathscr{H}}(x(t),u(t),\lambda(t)), \tag{1.69}$$

where

$$\hat{\mathscr{H}}(x(t),u(t),q(t)) = (1-\alpha)\rho_2 L(x(t),u) + q(t)'\, f(x,u). \tag{1.70}$$

As usual, in control theory, when the the coefficient in front of the reward rate is positive, it can be given any positive value, without changing the solution. Therefore the turnpike $\bar{x}$ associated with the weighted discount factors is defined by the implicit programming problem

$$\max_{u,x} \quad L(x,u) \tag{1.71}$$

s.t.

$$0 = f(x,u) - \rho_2(x-\bar{x}) \tag{1.72}$$

$$u \in U, \tag{1.73}$$

as for the infinite horizon discounted problem with the lowest discount rate $\rho_2 > 0$.

So we obtain a similar result to Weitzman's; when there is uncertainty about the real discount rate, the optimal trajectory should be driven toward the turnpike corresponding to the lowest discount rate.

1.4.1.4 The case where one rate is equal to zero

A more subtle problem arises when $\rho_2 \to 0$, that is, when the criterion is the result of a tradeoff between a discounted and an undiscounted utility stream. The discounted reward per unit of discounted time is

$$\rho_1 \int_0^{\infty} e^{-\rho_1 t} L(x(t), u(t))\, dt$$

and the long term average reward per unit of time in the undiscounted case is given by

$$\liminf_{T\to\infty} \frac{1}{T} \int_0^T L(x(t), u(t))\, dt.$$

or, equivalently, by

$$\liminf_{\rho_2 \to 0} \rho_2 \int_0^{\infty} e^{-\rho_2 t} L(x(t), u(t))\, dt.$$

Therefore we could envision a decision maker that would try to solve a bicriterion optimization problem, formulated as follows

$$\max \quad \Big[\alpha\rho_1 \int_0^{\infty} e^{-\rho_1 t} L(x(t), u(t))\, dt + (1-\alpha) \liminf_{T\to\infty} \frac{1}{T} \int_0^T L(x(t), u(t))\, dt\Big] \tag{1.74}$$

s.t.

$$x(0) = x^o \tag{1.75}$$

$$\dot{x}(t) = f(x(t), u(t)) \tag{1.76}$$

$$u(t) \in U, \tag{1.77}$$

where $\alpha \in (0, 1)$ is a weight factor

In the other way to deal with a vanishing discount rate we would consider the weighted criterion

$$\max \quad \Big[\alpha\rho_1 \int_0^{\infty} e^{-\rho_1 t} L(x(t), u(t))\, dt + (1-\alpha) \liminf_{\rho_2 \to 0} \rho_2 \int_0^{\infty} e^{-\rho_2 t} L(x(t), u(t))\, dt\Big] \tag{1.78}$$

s.t.

$$\dot{x}(t) = f(x(t), u(t)) \tag{1.79}$$

$$x(0) = x^0 \tag{1.80}$$

$$u(t) \in U. \tag{1.81}$$

We look for a solution to this problem that satisfies also the finite optimality criterion. However we notice that, as $\rho_2 \to 0$, the integral

$$\int_0^{\Theta} [\alpha\rho_1 e^{-\rho_1 t} + (1-\alpha)\rho_2 e^{-\rho_2 t}] L(x(t), u(t))\, dt \tag{1.82}$$

tends to

$$\int_0^{\Theta} \alpha\rho_1 e^{-\rho_1 t} L(x(t),u(t))\, dt \tag{1.83}$$

for any finite horizon Θ. As we have assumed that a turnpike exists for the system with discount rate ρ_1, an optimal trajectory should stay close to $\bar{x}_1$ for most of the journey on any finite horizon Θ. Using this argument for increasing values of Θ we conclude that, for any t large enough the trajectory should be close to $\bar{x}_1$. But this contradicts the fact that the turnpike associated with the lowest discount rate should prevail in the long run[16]. This suggests that it may be impossible to show existence of a finitely optimal solution to the problem formulated as above. Indeed, this is the case for the analogous problem considered in the framework of Markov decision processes outlined in the following subsection.

When one discount rate is equal to 0, the consideration of limit average payoffs per unit of time leads to a difficulty in defining optimal policies. The 0 discount rate component allows the decision maker to always postpone the decision to reach the turnpike which should however be the ultimate attractor of the optimal trajectory.

The overtaking optimality criterion with compromise reward.

The preceding difficulty invites us to consider the overtaking optimal trajectory for a model where a tradeoff between the total discounted and undiscounted payoffs is considered

$$J_T = \int_0^{T} [\alpha e^{-\rho_1 t} + 1 - \alpha] L(x(t),u(t))\, dt \tag{1.84}$$

s.t.

$$\dot{x}(t) = f(x(t),u(t)) \tag{1.85}$$

$$x(0) = x^o \tag{1.86}$$

$$u(t) \in U. \tag{1.87}$$

It can be easily shown that this problem admits a unique overtaking solution, with an asymptotic attractor that coincides with the 0-discount turnpike.

If the decision maker uses overtaking optimality with a compromised reward which is a convex combination of the discounted and undiscounted reward, an optimal trajectory exists that converges toward the 0-discount turnpike.

---

[16] Indeed, when the discount rate is 0, anything that happens on a finite time interval $[0,\Theta]$ has no importance for the long run average. A decision maker can always postpone the time when he/she will start to behave "optimally" on an infinite time interval.

1.4.2 The Weighted Markov Decision Processes

We briefly point out that similar problems to those considered in the preceding sections have also been studied in the context of tradeoffs between the discounted and long-run average criteria in the framework of Markov Decisions Processes (MDP's, for short). The latter are stochastic, sequential, processes where a "decision-maker" has some control over the distributions of future stream of random benefits/costs frequently referred to as "rewards".

For instance, suppose that the decision-makes wishes to influence a stream of expected values of these rewards, denoted by

$$\{E_{sf}(R_t)\}_{t=0}^{\infty},$$

where s is the initial "state", f is the control, R_t is the random reward/benefit at stage t and E denotes the mathematical expectation operator. We assume that specifying, s and f uniquely determines the probability distribution of R_t for every time period t, in the future.

The decision-maker might then wish to choose f so as to maximize either *discounted performance criterion*

$$v_d(s,f) := \sum_{t=0}^{\infty} \beta^t E_{sf}(R_t),$$

where the parameter $\beta \in [0,1)$ is called the *discount factor*, or the *long-run average performance criterion* defined by

$$v_a(s,f) := \liminf_{T\to\infty} \frac{1}{T+1} \sum_{t=0}^{T} E_{sf}(R_t).$$

Note that while the discounted criterion has a natural accounting interpretation as the so-called "present value" of the stream of rewards $\{E_{sf}(R_t)\}_{t=0}^{\infty}$. As the preceding sections indicate, this criterion is so widely accepted by economists that the question is not really whether it ought to be a criterion of choice but only of what value the discount factor β should take[17]. However, it can be argued that most dedicated environmentalists would favor the long-run average criterion as the more likely to guarantee sustainability.

We assume that there are only finitely many states: $1 \ldots \mathscr{S}$ and that in each state there are only finitely many actions a. A choice of such an action in a state s results in a stochastic transition that is Markovian and stationary. That is, there is a probability transition $p(s'|s,a)$ to a state s' that depends only on the current state s and the chosen action a. There is also an immediate reward/benefit of $r(s,a)$ that is accrued in such a "jump".

[17] Of course, this is equivalent to the preceding discussion about the interest rate ρ, as we can simply take $\beta = e^{-\rho}$.

Even in this relatively simple setting some conceptual complications quickly arise. What constitutes a "control" in this dynamic, stochastic setting?

The standard approach is to consider a *history* of the process at time t, namely, $h_t = (s_0, a_0, s_1, a_1, \dots a_{t-1}, s_t)$ and to define a *decision rule* at time t as the map $f_t : h_t \rightarrow f_t(h_t, a) \in [0,1]$. Next, a *control* is defined as a sequence of decision rules, one at each time, denoted by $f := (f_0, f_1, f_2, \dots f_t, \dots)$. Let F_B be the space of *all controls*. If for every t the decision rule f_t depends only on the current state at that stage, then the control f is called *Markov* or *memory-less*. Let F_M be the space of all Markov controls. If $f \in F_M$ and $\forall t, s, a$ the probability of choosing any action a, namely, $f_t(s,a)$ is independent of t, that is, then f is called a *stationary* control. Let $f \in F_{\mathscr{S}}$ (set of stationary controls). Finally, if $\forall s, a$ the probability $f(s,a) \in \{0,1\}$, then f is called a *deterministic* control. Let $F_{\mathscr{D}}$ be the set of deterministic controls, then clearly

$$F_{\mathscr{D}} \subset F_{\mathscr{S}} \subset F_M \subset F_B$$

Next, we consider two "optimal control" problems: (i) find, a simplest control f^0 such that

$$v_d(s) := \max_f v_d(s,f) = v_d(f^0), \tag{1.88}$$

where, $v_d(s)$ will be called the *discounted value* of the corresponding discounted MDP, and (ii) find a simplest control $f^\star$ such that

$$v_a(s) := \max_f v_a(s,f) = v_a(f^\star), \tag{1.89}$$

where $v_a(s)$ will be called the *long-run average value* of the corresponding long-run MDP.

It is well known (e.g., see (77)) that both of these problems have simple solutions in deterministic controls. Namely, there exists $f^0 \in F_{\mathscr{D}}$ optimal $\forall s$ in the discounted problem, as well as $f^\star \in F_{\mathscr{D}}$ optimal $\forall s$ in the long-term average problem. Furthermore, there are "good" algorithms for computing f^0, $v_a(s)$, $f^\star$, $v_a(s)$.

Arguably, the above means that, separately, with regard to time scales favored by either the economist or an environmentalist the ensuing optimal control problem is, in a sense, well solved. However, it is worth considering what happens if we wished to somehow combine these two performance criteria? Surely, the most benign way of attempting to so would be to choose a *weight parameter* $\lambda \in [0,1]$ and to try to find a control f so as to maximise

$$v_\lambda(s,f) := \lambda(1-\beta)v_d(s,f) + (1-\lambda)v_a(s,f), \tag{1.90}$$

thereby creating the so-called *weighted reward criterion* and the corresponding *weighted reward MDP*. Clearly, it follows that

$$\sup_f v_\lambda(s,f) \leq \lambda(1-\beta)v_d(s) + (1-\lambda)v_a(s) \tag{1.91}$$

constitutes the *utopian bound* for this new criterion.

Unfortunately but, perhaps, not surprisingly in (62) it has been shown that:

1. The following inequalities hold

$$\sup_{F_{\mathscr{D}}} v_\lambda(s,f) \leq \sup_{F_{\mathscr{S}}} v_\lambda(s,f) \leq \sup_{F_{\mathscr{M}}} v_\lambda(s,f),$$

   with $<$ possible in both places.
2. In general, an optimal control for $v_\lambda(s,f)$ need not exist. However, for each λ when the discount factor β is sufficiently near 1, then there exists an optimal deterministic control.
3. Nonetheless, it is reassuring that

$$\sup_{F_{\mathscr{M}}} v_\lambda(s,f) = \sup_{F_{\mathscr{B}}} v_\lambda(s,f).$$

4. Given any $\varepsilon > 0$ there exists an $\varepsilon-$optimal control f_ε such that

$$v_\lambda(s,f_\varepsilon) \geq \sup_{F_{\mathscr{M}}} v_\lambda(s,f) - \varepsilon. \tag{1.92}$$

5. In particular there exists a "switching time" $\tau(\varepsilon)$:

$$f_\varepsilon = (\underbrace{f_1, f_2, \ldots f_{\tau(\varepsilon)}}_{\text{be "greedy" for a while}}, \underbrace{f^\star, f^\star, f^\star, \ldots}_{\text{switch to optimal long-run average control}})$$

Of course, $\tau(\varepsilon)$ depends critically on the parameter λ, and the underlying data.

In a sense, properties 1-5 above capture the essence of the conflict between the "industrialist" and the "environmentalist". A conflict that is captured in - but not reconciled by - the dilemma of the "right choice" of the switching time $\tau(\varepsilon)$ in the structure of nearly optimal controls given in item 5 above. After all, that is the time when, after a period of being greedy, the decision-maker switches to the long-run (sustainable development?) type of control.

We conclude this section by pointing out that there is now substantial literature dealing weighted reward criteria (including multiple discount factors) in Markov Decision Processes and Stochastic Games (e.g., see (38) and (33)). For a survey of that interesting topic the reader is referred to (34).

1.4.3 Consequences for GCC policy evaluation

Global Climate Change (GCC) policies involve immediate costly actions in order to prevent damages for distant future generations (150 to 300 years from now). An important question is how to implement Benefit-cost analysis (BCA) to evaluate GCC policies. The BCA makes use of a *damage function and discounting* (DFD) approach. This approach is composed of two distinct steps:

> *(i) the favorable and unfavorable effects of a proposed policy intervention at all future points in time are identified and expressed in dollar terms; (ii) the time streams of future benefits and costs are converted to present values using a single discount rate (or a range of rates when sensitivity analysis is included)...*[18]

The WP3 Report of (58) presents two approaches for selecting the discount rate: the *prescriptive approach* is constructed from ethical principles and results in a discount rate in the range 0.5 to 3.0%, whereas the *descriptive approach* based on the rate of return to capital, produces a rate in excess of 5% in real terms. A compromise for intergenerational discounting is proposed in (28) where, for the initial 30 years one uses the capital opportunity cost with some significant pure time preference rate ($\rho > 0$) whereas, after 30 years one uses a *social rate of time preference* more in line with $\rho = 0$.

Global climate change policies are clearly related to sustainable economic growth. Sustainability is also related with intergenerational equity (see e.g. (46), (54)). Under given level of technology know-how, sustainable or green golden-rule[19] asymptotic levels correspond to turnpikes with low or zero value for ρ. As we have seen above, these prescriptions will result from the use of a weighted discount term involving both the prescriptive and descriptive rates. The approach advocated by Cline (28) could be implemented using the formulation (1.84)-(1.87) with the overtaking optimality criterion. With a weight α relatively big the investment decisions will be initially decided according to the positive pure rate of time preference, but after a period of the order of magnitude $1/\rho$, the social rate of time preference based on $\rho = 0$ will prevail and define the sustainable green golden rule for the distant future. The latter is, of course, consistent with the discussion in the preceding subsection.

1.5 Techniques and methods to deal with uncertainty in EDM

In the previous sections we have seen how EDM can benefit from an OR/MS approach and how this could define new challenging problems for this discipline, in particular when dealing with uncertainty. We will now summarize a selection of relatively standard approaches that can be used to deal with uncertainty in EDM with a special focus on estimating the risk of adverse environmental impacts, including the frequently discussed "extreme events".

---

[18] Quoted from (61).

[19] as discussed in (14), (25), or (9).

1.5.1 Statistical analysis

Indeed, the subject of statistics remains the main battlefront when one has to deal with uncertainty and decision making.

When there are enough data available, perhaps, the most natural approach to invoke is that of statistical modelling. The subject of statistics offers a rich body of modelling techniques (e.g., regressions, time series, principal component analysis, non-parametric methods and others) that are well developed, documented and supported by many highly reliable software packages.

While some would argue that statistical models are only a substitute for the ignorance of the underlying processes, there are a number of benefits from developing statistical models to estimate the probabilities of interest. For instance, even if it is believed that there is sufficient understanding of the dynamics that lead to certain kinds of extreme events and that an alternative - more detailed - methodology could be used to arrive at these probabilities, statistical models can still serve as benchmarks for assessing the quality of the alternative models. In addition, models based on statistical techniques are nearly always equipped with methods for their validation. Finally, domain knowledge is gained by finding out "the story that data tell".

The most natural starting framework of a statistical model would be to introduce a dependent variable $Y(t)$ that may denote, for instance, the number of days in a given period (say, a month) t, in a given city, with the air pollution exceeding a certain healthy level in that period. It is assumed that historical data on such a variable are available or can be collected. Furthermore, it is assumed that there is a pool of data that are available on a (substantial) number of independent, or explanatory, variables denoted by $X_1(t), X_2(t), \cdots, X_K(t)$. For instance, $X_1(t)$ may be the production level of a certain industry during period t, $X_2(t)$ could be the production level of another industry during the same period, and so on. Let the vector of these explanatory variables, at time period t, be denoted by $X(t)$. A most naive "model" that one could postulate would be of the form:

$$Y(t) = f(X(t), X(t-1), \cdots, X(t-k)) + \varepsilon_t, \tag{1.93}$$

where ε_t is the random, "unexplained", error and the lags $t-1, t-2, \cdots, t-k$ indicate that, when it comes to environmental impacts, it could be that a history of certain indicators of neglect, or of "cutting corners" that is a predictor of a dangerous event. It needs to be noted, however, that inclusion of time lags places a model in the well developed branch of statistics known as "time series analysis". We refer the reader to the classic text (18) for a detailed discussion of time series models.

Of course, the above equation is almost certainly going to be much "over fitted" to be seriously considered as a model. However, if the so-called R^2 statistic is sufficiently close to 1, then this is an indication that the pool of explanatory variables is sufficiently rich that there is a good chance of being able to extract a much smaller set of good predictor variables $Z_1(t), Z_2(t), \cdots, Z_m(t)$ from which a parsimonious predictive model of the form

$$Y(t) = g(Z(t), Z(t-1), \cdots, Z(t-k)) + \varepsilon_t, \tag{1.94}$$

could be developed. Note that in this formulation the vector $Z(t)$ will consist of only a small number of predictor variables. Furthermore, it is important to emphasize that while $Z_j(t)'s$ could contain some of the raw original variables $X_i(t)'s$, some of the $Z_j(t)'s$ could be suitably constructed functions of raw variables.

With regard to the latter point we note that the field of multivariate statistical analysis has a number of well established techniques designed for precisely such variable reduction problem. These range from the widely used backward, forward and stepwise variable selection techniques of regression models to principal component analysis and closely related factor analysis. The latter focus on the estimated variance-covariance matrix Σ of the random vector $X(t)$ and its diagonalisation

$$\Lambda := U^T \Sigma U, \tag{1.95}$$

where the entries of Λ consist of eigenvalues of Σ arranged in decreasing order and the orthonormal matrix U consists of the corresponding (normalised) eigenvectors. It is often the case that there is only a small number of "dominant eigenvalues" and that, with the help of their corresponding eigenvectors, it is possible to construct the same small number of new predictor variables (called "principal components", or "factors") that explain most of the variability in the matrix Σ. These principal components are linear functions of the original raw variables, contain a lot of the essential information in the complete data set and are good candidates to be used as new explanatory variables in predictive models such as those mentioned above.

The many successful applications of the above factor variable reduction techniques include psychometric studies where scores on groups of seemingly disparate aptitude questions identify certain generic skills such as a "spatial visualisation ability". Thus a principal component combining these scores would become a single new, spatial ability, explanatory variable and would consitute a compact surrogate for the corresponding group of raw variables. We refer the reader to (31) as just one (of many) comprehensive texts describing these techniques in much more detail.

The subject of statistics is now so well developed that there are many more viable techniques for estimating the probabilities of an adverse environmental impact (that interests us) than could possibly be described in this review paper. The regression, time series and principal component techniques that were mentioned above are simply some of the most natural ones to explore. However, before concluding this sub-section we want to briefly mention a class of regression type techniques that may be particularly well suited to our problem. This is because it is immediately obvious that if we are interested in a probability $P(E)$ of a particular event E, it is possible to associate with it a binary indicator random variable Y that takes on, respectively, values 1 (0) if the event E occurs (does not occur). The so-called "Logistic Regression Methods" were developed especially for the problem of modelling the probability of the event $\{Y = 1|X\}$, where X is the vector of observations of the explanatory variables. The logit transformation associated with this problem is the log of the odds that $Y = 1$ versus $Y = 0$, given the observations, that is

$$h(X) = log\{\frac{P(Y=1|X)}{1-P(Y=1|X)}\}. \tag{1.96}$$

Essentially, logistic regression models are linear models of the form $h(X) = \beta_0 + \beta^T X + \varepsilon$ and, as such, possess many of the familiar properties of standard regression models. However, they are usually fitted with the help of maximum likelihood methods rather than least squares fits. We refer the reader to (56) for a detailed discussion of this subject.

We conclude this subsection by pointing out that the mere possibility of collecting data generated by the phenomenon of interest opens up the possibility of invoking statistical methods to develop mathematical models describing associations or relationships between important variables. Indeed, statistical models such as those alluded to above have acquired a profound role as tools of scientific inquiry. They permit researchers to identify both unanticipated relationships among variables and, in some cases, to confirm the presence of anticipated relationships.

1.5.2 Extreme event analysis

By virtue of their exceptionality some environmental disasters (e.g., droughts, floods) certainly constitute "extreme events" in the common sense meaning of that phrase. Consequently, it is prudent to examine what the now classical "extremal value theory" has to offer in our context of modelling the probability of such a disaster. The origins of this theory - that has evolved out of the twin subjects of Statistics and Stochastic Processes - date back to the seminal work of (41) in first half of the last century. By now, this challenging subject has grown enormously with researchers following a number of fruitful lines of investigations. For a comprehensive modern text we refer the reader to (32).

However, before proceeding, it is worth noting that a substantial portion of the theory of extremal events was motivated by financial considerations such as the "risk of ruin". Thus, to the extent that these techniques and concepts depend on accumulation of losses, they do not correspond very well to the most common type of problem occurring in the context of an environmental disaster. For instance - and without in any way advocating the underlying connotations - the phrase "a miss is as good as a mile" captures some of the above distinction. Thus, a levee that proves adequate to a "100 year flood" is not significantly more likely to be breached in the near future, then just prior to that flood. However, an investor who only just avoided ruin when the market had its last downturn has probably suffered such losses that his or her likelihood of failing to avoid ruin in the next downturn is severely reduced.

Due to the above considerations, and for the sake of simplicity of exposition, we will not discuss those aspects of the theory of extremal events that deal with sums of random variables exceeding certain thresholds[20] and will focus instead on the aspects that deal with the properties of the so-called "extremal statistics". We shall now introduce some of the notation needed to make the discussion a little more precise.

Consider a sequence $X_1, X_2, \cdots, X_n, \cdots$ of independent identically distributed random variables (iid rv's, for short), all of which are distributed as a given random variable X that has a cumulative distribution function F. The object of interest in the classical extremal value theory is the induced sequence of the *sample maxima* random variables

$$M_n := max(X_1, X_2, \cdots, X_n) \;\; for \;\; n \geq 2.$$

Note that the analysis of the sample minima, defined analogously, is not any different because of the identity $min(X_1, X_2, \cdots, X_n) = -max(-X_1, -X_2, \cdots, -X_n)$ and hence we shall restrict our discussion only to the sequence $\{M_n\}$. It is now clear that the distribution function of M_n is simply

$$P(M_n \leq x) = P(X_1 \leq x, X_2 \leq x, \cdots, X_n \leq x) = F(x)^n. \tag{1.97}$$

Since in this theory we are primarily interested in "failures" corresponding to the sample maxima exceeding certain thresholds we shall be particularly interested in the "tail" of the distribution of M_n, namely in $P(M_n > x) = 1 - F(x)^n$. For our underlying distribution $F(x)$ the tail is defined simply by $\bar{F}(x) := 1 - F(x)$.

Of course, for small n, equation (1.97) provides a means of calculating the tail probabilities for the distribution of M_n, but for large n a direct computation could be very cumbersome. Thus, a major thrust was made to derive asymptotic results that are in the spirit of the celebrated "Central Limit Theorem" of statistics. This naturally led to the question of whether it is possible to find constants $c_n > 0$ and d_n such that for some non-degenerate probability distribution H

$$\frac{M_n - d_n}{c_n} \rightarrow H, \tag{1.98}$$

in distribution, as n tends to infinity. Clearly, if (1.98) holds, then the equation

$$P(\frac{M_n - d_n}{c_n} \leq x) = P(M_n \leq u_n) = \int_{-\infty}^{u_n} dH(x), \tag{1.99}$$

where $u_n = c_n x + d_n$, provides a basis for calculating an approximation of the tail probabilities of interest. The problem contained in equations (1.98) and (1.99) is actually more challenging than the analogous problem for random sums. The validity of these equations seems to require delicate conditions on the tail $\bar{F}$ of the distribution F to ensure the existence of a non trivial limit of $P(M_n \leq u_n)$ as n tends to infinity. In particular, the tail $\bar{F}(u_n)$ needs to decay at an appropriate rate as u_n tends to infinity with n.

---

[20] It should be noted that such random sums could still be of great interest in our context if, for instance, we were trying to analyze the accumulated pollution in a river, or the total amount of CO_2 reflecting the sun's radiation back to earth.

One of the fundamental results is the so-called *Poisson approximation* that states that for any given nonnegative number τ and a sequence $\{u_n\}$ of real numbers the following equivalence holds

$$n\bar{F}(u_n) \rightarrow \tau \quad iff \quad P(M_n \leq u_n) \rightarrow e^{-\tau}, \tag{1.100}$$

as n tends to infinity.

One of the elegant special cases where the above limit can be directly computed is the case when $F(x) = 1 - e^{-x}$, that is the exponential random variable with parameter 1. In this case a direct calculation shows that

$$P(M_n - ln\, n \leq x) = (P(X \leq x + ln\, n))^n = (1 - n^{-1}e^{-x})^n \rightarrow exp\{-e^{-x}\}, \tag{1.101}$$

as n tends to infinity. Of course, $\Lambda(x) := exp\{-e^{-x}\}$ is the well known *Gumbel distribution.* Indeed, the remarkable conclusion of the famous Fisher-Tippet Theorem is that if H is a non-degenerate distribution such that (1.98) holds, then H must belong to one of only three families of distributions: Frechet, Gumbel, or Weibull. Thus these three well known distributions provide a basis for many of the approximations of the probabilities of interest.

1.5.3 Reliability theory

Reliability theory is now a well established branch of both Applied Probability and Systems Engineering. Simply expressed *reliability* is the probability that a system will perform satisfactorily for a given period of time. Hence it immediately becomes obvious that, conceptually, reliability is closely related to the extreme value theory discussed above. However, historically, the two topics developed along different paths with reliability theory becoming much more engineering based and acquiring a systems' orientation while extreme value theory has remained a branch of mathematical statistics and probability.

Let T be the continuous random variable representing the *time to failure* of the system of interest, and let $F(t)$ and $f(t)$ denote its cumulative probability distribution and density functions, respectively. A key concept in reliability theory is that of a *reliability function*, $R(t)$ defined by

$$R(t) := P(T > t) = 1 - F(t) = \int_t^{\infty} f(t)dt, \quad t \geq 0. \tag{1.102}$$

Thus we see that $R(t)$ is simply the probability that the system will perform satisfactorily for, at least, a time t and constitutes the tail of F. The reliability function is sometimes also called the *survival function* a name that, perhaps, suggests its appropriateness for our environmental safety application. Two closely associated notions are the *mean time between failures* (MTBF) and a *mean time to failure* (MTTF). In the context of an industrial/production system operating under environmental emission constraints, "failures" could refer to instances of emissions exceeding certain thresholds.

Nowadays, reliability theory plays an essential part in the design of most complex systems systems. Some of its appeal stems from the elegant formulae for the reliability of a system with components that are configured either "in series" or "in parallel". For instance, suppose that the system consists of two components A and B with individual reliabilities (over a fixed time) of R_A and R_B and such that their corresponding time to failure random variables T_A and T_B are independent. Then the overall reliability R of the system (over the same fixed time) is given by

$$R = R_A \times R_B, \tag{1.103}$$

if the components are configured in series, and

$$R = R_A + R_B - (R_A \times R_B) \quad t \geq 0, \tag{1.104}$$

if they are configured in parallel. Since, as probabilities, reliabilities are numbers between zero and one, the series configuration (1.103) generally reduces the reliability of the system while the parallel configuration (1.104), increases it. Of course, these arguments can be extended to much more complex combinations of components connected either in series or in parallel. Thus the simple formulae (1.103)-(1.104) permit the system designer to manipulate the system's overall reliability.

However, to the extent that in most engineering applications the assumption of independence of components is invoked, the relevance of reliability theory in the context of human behavior leading to an environmental disaster could be limited. For instance, overfishing of a certain species (e.g., a black sea sturgeon) may lead to legal restrictions on its harvest that lead to such a price inflation of caviar which, in turn, stimulates illegal poaching that may well drive the species into extinction.

Indeed, even in the case of purely mechanical systems it is easy to miss the fact that the independence assumption may not hold in a given situation. Perhaps, the most famous instance of the latter was the failure of the two "O-ring" seals on the space shuttle Columbia that resulted in the destruction of the shuttle and the death of all the crew. What occurred in that case was that the failure of the first "O-ring" occurred in the cold followed by heat and stress conditions that ensured that the second "O-ring" would also certainly fail.

1.5.4 Simulations: Synthetic Data and Validation

Simulation and Monte-Carlo analysis will always be available as a way to identify the risk in complex decision problems.

Nowadays, many important situations are modelled with the help of "synthetic" data that are produced by simulations that are intended to imitate the underlying phenomenon in a realistic way. This is necessary in situations where it is very expensive or technically or ethically impossible to collect actual data. Combat simulations and "gaming" in the military are, perhaps, prime examples of this practice. In a more relevant context of coupled atmosphere-ocean models of climate detailed, three dimensional, scientific models are routinely simulated under a range of world greenhouse gas emission scenarios. Even with the computational power of supercomputers these models are sufficiently complex and their simulations so expensive that they can be carried out only by elite research teams. However, the "synthetic data" these simulations generate can be widely used for many studies related to environmental decision making.

Whenever, the use of sophisticated simulation tools is contemplated, every effort must be made to understand the built in assumptions in the techniques that these tools employ (e.g., Poisson arrival times of certain events) and to examine the applicability and implications of these assumptions in the context of the current study and research questions. In some cases the assumptions will disqualify the simulation outputs from being relevant to the phenomenon studied. However, in increasingly many situations modelers must accept that these simulations capture expert knowledge and hence that their outputs represent both:

1. Most realistic, albeit synthetic, data on certain variables that can be obtained, and
2. Most realistic framework for validating new models that are being constructed.

The latter point is, perhaps, not surprising but is still worth emphasising because it allows researchers to imitate the design of classical scientific experiments to validate their models.

For instance, with the help of a classical statistical technique known as "experimental design" it is possible to develop a model capturing the effects of a number of new treatments of tomato seeds on the eventual size and yield of tomato plants. In such a case, it would usually be assumed that the researchers have the luxury of designing an experiment that involves planting a prescribed number of tomatoes in a prescribed number of plots, each having a prescribed selection of treatments performed on the seeds. The untreated seeds would, generally, be assigned to the individual groups based on a suitable random procedure.

There's no reason why a a simulation platform could not be used to implement an analogous experimental design but in a context such as the introduction of new greenhouse gas emission protocols. Just as with tomato plants, researchers could randomly assign a manufacturer (or a city, or even a country) to either a control group (following old protocol) or to the experimental group (following the new protocol) and then run a realistic simulation of random conditions that such an "experimental unit" might encounter. Throughout the simulation scores on various performance variables (economic and environmental indexes) could be tallied. Importantly, this could be repeated many times and at a low cost (in terms of both time and money). The resultant control and experimental samples could be analyzed in much the same way as the corresponding samples of tomato plants. Hypotheses about effectiveness, or otherwise, of the new protocol could be tested and accepted or rejected using standard criteria.

The only essential difference between this and the classical scientific testing and validation in natural sciences such as biology or medicine would be that this would be done in "virtual reality" of the simulated - rather than the real - phenomenon. Without in any way wishing to trivialize this important distinction we wish to claim that in many situations - including, perhaps, the modelling of proposed environmental regulations - this kind of validation is probably the best that can be expected without huge resources being allocated to the project. In addition, the difference between this and the classical validation methods may not be as great as we might, at first, imagine. After all, even the preceding agricultural experiment on tomato plants is, in a sense, virtual because there is likely to be a significant difference between the conditions that these plants experience at a research facility and the conditions they would experience on real commercial farms.

Simulation and Agent Based Modelling.

In the above we have focussed on simulations as both a potential source of synthetic data and a virtual model validation environment. However, nowadays, many researchers think of simulations as a paradigm for modelling complex phenomena. Spearheaded by researchers at the Santa Fe Institute, a new family of related models known as *complex adaptive systems* (or, "agent based modelling") gained prominence since the 1980s. Typically, these models attempt to first reproduce and then simulate into the future the behavior of "complex" phenomena in the social, life, physical or decision sciences. An essential feature of these models is the introduction of individual "agents" in the situation under investigation. The characteristics and activities of these agents are the building blocks of these models. These characteristics are dynamic in the sense that they change over time, as the agents adapt to their environment, learn from their experiences, or experience natural selection in an evolutionary process. The main objective of these models is to capture the macro-level "emergent behavior" that results from the local decisions by individual agents that reflect the assumptions of the local adaptation rules. The agents may be organized into groups that may also influence how the system evolves over time we refer the reader to (55).

Perhaps, one of the most attractive aspects of complex adaptive systems is that they often require only minimal domain knowledge and very few assumptions. They aim to reproduce complex aggregate behavior patterns with the help of rather simple modelling of actions of individual agents. The emergent complexity is seen as a result of replication of these simple actions by many agents and of repetition over many time periods. Of course, a weakness of complex adaptive systems stems from the inherent difficulty of validating these models. It is intuitively clear that, in principle, more than one set of adaptation rules can generate very similar emergent behavior. If so, why should we prefer one over the other?

The proper relationship between the modern simulation based models and mathematical models of complex systems is, perhaps, best captured by the analogy with relationship between exploratory (usually empirical) and theoretical studies. The former aim to identify interesting patterns, the latter seek to explain phenomena by developing theories. Good science needs both types of studies to reinforce one another. Similarly, simulation based models and the more classical mathematical models often need to reinforce one another for proper analysis of complex systems. However, simulation without understanding is not enough. In all likelihood, mathematical modelling of, at least, certain key aspects of the systems will be needed to aid understanding.

1.5.5 Stochastic optimization models

As we have shown in some of the examples developed in this chapter, optimization is a useful paradigm to represent the economic decisions (equilibrium) and to evaluate the cost (explicit or implicit) of constraints due to environmental protection. Stochastic optimization models will, undoubtedly, play an important role in the development of EDMs dealing with uncertainty.

The different approaches in stochastic optimization could be classified as follows:

Dynamic programming and stochastic control: If the uncertainty depends on the decisions we take, perhaps, the most natural rigorous approach is that of Stochastic Dynamic Programming. The latter is one of the methods of stochastic control. A chapter of this book will detail the application of this method to a model of climate policy assessment under uncertain knowledge.

Stochastic programming: The mathematical programming community has developed an approach to deal with the case where the uncertainty is an uncontrolled process. For instance, this is the case if the uncertainty is related to weather conditions. A chapter of this book will explain how this approach can be used in detailed E^3 techno-economic models of energy, economy and environment.

Robust programming: Both dynamic and stochastic programming techniques suffer from the *curse of dimensionality*, although in different degrees. Robust optimization and its implementation in mathematical programming, that we call "robust programming", offers a way to mitigate th curse of dimensionality. A chapter of this book is a tutorial for this approach which will, without doubt, play an important role in the development of EDMs under uncertainty.

1.6 Concluding remarks

We have surveyed a selection of concepts, approaches and modelling techniques relevant to environmental decision making under uncertainty. As indicated in the introduction we believe that challenges to the successful application of OR/MS techniques in this field are not insurmountable.

In fact the survey shows that the established subjects of Statistics, Probability, Stochastic Processes, Optimization and Control, Stochastic Differential Equations, Reliability, Markov Decision Processes, Robust Optimization and Simulation already possess many modeling techniques that can be exploited for the purpose of effective quantification and assessment of many uncertainties arising in the challenging area of EDM. Yet, the capability to bring these tools to bear does not necessarily mean that the challenge associated with the previously identified time scales' asynchrony (TSA) between human development processes and the natural processes of the biosphere will be effectively tackled in the foreseeable future.

The latter challenge is intimately related to societal, cultural and philosophical attitudes towards the notion of *sustainable development*. According to Hawken (51), *sustainability is an economic state where the demands placed upon the environment by people and commerce can be met without reducing the capacity of the environment to provide for future generations.* While, nowadays, many pay lip service to the underlying principle embedded in the preceding statement it is still widely regarded as either utopian or driven by unnecessary pessimism. For instance, the controversial 1972 "The Limits to Growth" publication (70) prepared for the so-called Club of Rome, predicted a bleak future for humanity driven by the dynamics of population growth and environmental exploitation leading to exhaustion of irreplaceable natural resources. A future that many argue never eventuated because it ignored another significant school of thought - sometimes referred to as "free-market environmentalism" - which postulates that economic development, globalization and growth of human capital will eliminate poverty and ultimately lead to sustainable development.

Consequently, a very real – but a philosophical rather than a scientific or a technological – challenge facing OR/MS practitioners wishing to influence environmental decision making is to avoid the "greenwash trap". For instance, building models promoting the use of gasohol as an efficient way to reduce greenhouse gas emissions may fail to recognize the inherent unsustainability of using scarce fertile land for the purpose of driving energy inefficient vehicles. Furthermore, even those OR/MS models that promote the use of energy efficient cars in developing countries, such as, the recently developed Tata Industries' Nano may fall into the trap of the so-called Jevons' Paradox. Despite Nano's apparent fuel efficiency of 20km per liter, if some 200 million units are sold and driven in India and other developing countries in the coming years, both the use of the scarce oil resources and the total amount of greenhouse gas emissions will surely grow dramatically. Similarly, the most efficient flexible logistics and manufacturing systems that OR/MS practitioners may help design may not benefit the environmental cause if they result in shipping ever greater amounts of raw products to be processed or consumed far away from their countries of origin. It is surely, ecologically unsustainable to ship a Californian orange for sale in Australia, a country where oranges are abundant. A logistics-production system that renders this profitable clearly undervalues the environmental cost of the energy used in shipping that orange across the ocean.

There may not be a simple answer to a wide range of issues embedded in the above discussion. Essentially, a societal and even a global consensus is needed to accept the simple notion that the planet's biosphere is finite and in need of protection. Scientific evidence and quantitative models can and should play a role in supporting the many national and international debates that, inevitably, will be needed to develop such a consensus. In this spirit we conclude the present contribution with the following "time scales' conjecture" formally introduced in (40) that, in conceptual terms, postulates that *reasonable notions of sustainability must include a suitable synchronization of time scales of both the processes of human development and those of the natural environment.*

References

[1] ABILOCK H., BERGSTROM, C., BRADY, J., DOERNBERG, A., EK A., FISHBONE L., HILL D., HIRANO M., KARVANAGH, R., KOYAMA, S., LARSSON, K., LEMAN, G., MOY, M., SAILOR, V., SATO, O., SHORE, F., SIRA, T., TEICHMAN AND WENE, C.-O., MARKAL: A multiperiod linear programming model for energy system analysis, in Kavanagh, R. (Editor), Proc. Int. Conf. On Energy System Analysis, 9-11 October 1979, Dublin, Ireland, p482, Reidel, Dordrecht, 1980.

[2] AINSLIE, G. AND HASLAM, N., Hyperbolic discounting, in Loewenstein & Elster (eds) *Choice Over Time*, Russell Sage Foundation, New York, (1992).

[3] ALTMAN A., FEINBERG E. AND SCHWARTZ A.,*Weighted discounted stochastic games with perfect information*, in J.A. FILAR, V. GAITSGORY AND K. MIZUKAMI eds., *Advances in Dynamic Games and Applications*, Annals of the International Socity of Dynamic Games, 2000.

[4] ARNDT, H. W. *The rise and fall of economic growth: a study in contemporary thought*, Longman Cheshire, Sydney, 1978.

[5] ARROW K. AND M. KURZ, *Public investment, the rate of return, and optimal investment policies*, John Hopkins Press, Baltimore, 1970.

[6] ARROW, K, BOLIN, B., COSTANZA, R., DASGUPTA, P., FOLKE, C., HOLLING, C., JANSSON, B-O., LEVIN, S., MALER, K-G., PERRINGS, C., AND PIMENTAL, D., *Economic growth carrying capacity and the environment., Science*, 1995: 268, 520-21.

[7] ARROW, K.J., W.R. CLINE, K.G. MAELER, M. MUNASINGHE, R. SQUITIERI AND J.E. STIGLITZ I*ntertemporal Equity, Discounting, and Economic Efficiency*, in J.P. Bruce, H. Lee and E.F. Haites (eds.), *Climate Change 1995: Economic and Social Dimensions* - Contribution of Working Group III to the Second Assessment Report of the Intergovernmental Panel on Climate Change, pp. 125-144, Cambridge University Press, Cambridge, 1996.

[8] ASHEIM G.B., W. BUCHHOLZ AND B.TUNGODDEN, Justifying sustainability, *Journal of Environmental Economics and Management*, Vol. 41, 2001, pp. 252-268.

[9] AYONG LE KAMA, A., Sustainable growth, renewable resources and pollution, *Journal of Economic Dynamics and Control*, 2001: 25, pp. 1911-1918.

[10] BABONNEAU, F. AND BELTRAN, C. AND HAURIE, A. AND TADONKI, C. AND VIAL, J.-P., Proximal-ACCPM: a versatile oracle based optimization method, In Optimisation, Econometric and Financial Analysis, E. J. Kontoghiorghes editor, vol. 9 of *Advances in Computational Management Science*, 2006.

[11] BAHN, O. AND DROUET, L., EDWARDS, N.R., HAURIE, A., KNUTTI, R., KYPREOS, S., STOCKER, T.F., ANDVIAL, J.-P., (2006), The coupling of optimal economic growth and climate dynamics, *Climatic Change*, Volume 79, Numbers 1-2 /, pp. 103-119.

[12] BAUMOL, W., AND OATES, W., *Economics, environmental policy and the quality of life*, Prentice Hall, Englewood Cliffs, 1979.

[13] Beltran C., L. Drouet, N. R. Edwards, A. Haurie, J.-P. Vial and D. S. Zachary, An Oracle Method to Couple Climate and Economic Dynamics, Chap. 3 in A. Haurie and L. Viguier (eds.), *The Coupling of Climate and Economic Dynamics*, Springer, 2005.

[14] BELTRATTI A., CHICHILNISKY G. AND HEAL G., The green golden rule, *Economic Letters* Vol. 49, 1995, pp. 175-179.

[15] BEN-TAL, A., A. NEMIROVSKI. Robust optimization methodology and applications. *Mathematical Programming* 92(3) 453-480, 2003.

[21] BERGER, C., DUBOIS, R., HAURIE, A.,LESSARD, E., LOULOU R. AND WAAUB, J.-P., Canadian Markal: An advanced linear Programming system for Energy and Environmental modelling, INFOR., 1993. CLAPPIER, A., A correction Method for use in Multidimensionnal Time Splitting Advection Algorithms: Application to Two- and Three- Dimensional Transport, Mon. Wea. Rev, 126, 232-242, 1998.

[17] D. BERTSIMAS, O. NOHADANI, AND K. M. TEO. Robust nonconvex optimization for simulation-based prob- lems. To appear in *Operations Research*.

[18] BOX, G.E.P. AND JENKINS G.M., *Time Series Analysis: Forecasting and Control*. Holden-Day, San Francisco, 1970.

[19] BRADDOCK, R.D. FILAR, J.A., ZAPERT R, DEN ELZEN M.G. ROTMANS J, Mathematical forumulation of the IMAGE greenhouse effect model, *Appl Mathematical Modelling* 18: 234-254, 1994.

[20] CARLSON D., A. HAURIE AND A. LEIZAROWITZ, *Infinite Horizon Optimal Control: Deterministic and stochastic Systems*, Springer Verlag, 1994.

[21] CARLSON D., HAURIE A., J.-P. VIAL AND ZACHARY D.S., Large scale convex optimization methods for air quality policy , *Automatica*, pp. 385395, January 2004.

[22] CHALLENOR, P.. The probability of rapid climate change. *Significance*, 1, (4), 155 - 158. 2004.

[23] CHALLENOR, P.. The probability of rapid climate change: II. *Significance*, 4, (2), 60-62. 2007.

[24] CHALLENOR, P.G., R.K.S. HANKIN AND R. MARSH Towards the Probability of rapid Climate Change. pp 55-63 in Avoiding Dangerous Climate Change Ed Schellnhuber, H.J., W. Cramer, N. Nakicenovic, T. Wigley amd G Yohe. Cambridge University Press, 2006.

[25] CHICHILNISKY G., An axiomatic approach to sustainable development, *Soc. Choice Welf.* Vol. 13, 1996, pp. 231-257.

[26] CHICHILNISKY G., What is sustainable development, *Land Economy* Vol. 73, 1997, pp. 467-491.

[27] CLAPPIER, A., KÜBLER, J., SATHYA, V., MARTILLI, A., PEREGO, S., KRÜGER, B.C., SIMEONOV, V., JEANNERET, F., CALPINI, B. AND VAN DEN BERGH, H., Modélisation du smog photochimique en territoire genevois et Analyse de scénarios de réduction des émissions de polluants, Technical Report for the COST-615 CITAIR program, Swiss NSF, LPAS-DGR/EPFL, CH-1015 Lausanne, 1997.

[28] CLINE, W.R.. Discounting for the Very Long Term, in Portney & Weyant (eds): *Discounting and Intergenerational Equity*, Resources for the Future, Washington DC, 1999.

[29] COHN, R.D., EDER, B.K. AND LEDUC, S.K., An Aggregation and Episode Selection Scheme designed to Support MODELS-3 CMAQ, EPA/600/R-99/030, 1998.

[30] DROUET L., N.R. EDWARDS AND A.HAURIE, Coupling Climate and Economic Models: A Convex Optimization Approach, *Environmental Modeling and Assessment,* Vol.11, pp.101-114, 2006.

[31] DILLON, W.R. AND GOLDSTEIN M., *Multivariate Analysis*, Wiley, New York, 1984.

[32] EMBRECHTS P., KLÜPPELBERG C. AND MIKOSCH T., *Modelling Extremal Events, Applications of Mathematics: Stochastic Modelling and Applied Probability*, Springer, Berlin, 1997.

[33] FEINBERG E.A. AND SCHWARTZ A., Constrained Dynamic Programming with Two Discount Factors: Applications and an Algorithm, *IEEE Transactions on Automatic Control*, 44, 1999, pp. 628-631.

[34] FEINBERG E.A. AND SCHWARTZ A., Mixed Criteria, *Handbook of Markov Decision Processes: Methods and Applications*, Kluwer, pp 209-230, 2002.

[35] FEINSTEIN C.D AND D.G. LUENBERGER, Analysis of the asymptotic behavior of optimal control trajectories, *SIAM Journal on Control and Optimization*, Vol. 19, 1981, pp. 561-585.

[36] FILAR, J.A., "Mathematical Models", In: *Knowledge For Sustainable Systems. An Insight Into the Encyclopedia ofLife Support Systems*, Vol 1., pp. 339-354. UNESCO Publishing-Eolss Publishers, Paris, France, Oxford, UK. 2002.

[37] FILAR, J.A. AND HAURIE A. (1998), Uncertainty in Environmental Models: Dynamic Systems Perspective, in H. Greppin, R. Degli Agosti and C. Penel, eds. *The Co-action between Living Systems and the Planet*, University of Geneva, 1998.

[38] FILAR, J.A. AND VRIEZE, O.J., Weighted Reward Criteria in Competitive Markov Decision Processes, *Zeitschrift für O.R.*, Vol.36, 1992, pp.343-358.

[39] FILAR, J. A. AND ZAPERT R., Uncertainty Analysis of a Greenhouse Effect Model, in: C. Carraro and A. Haurie, eds., *Operations Research and Environmental Management*, Kluwer Academic Press, Dordrecht, 1996.

[40] FILAR, J.A., KRAWCZYK, J.B. AND AGRAWAL,M., On production and abatement time scales in sustainable development. Can we loosen the sustainability screw?, CORE Discussion Paper, forthcoming, 2009.

[41] FISHER R.A. AND TIPPETT T.H.C., Limiting forms of the frequency distribution of the largest or smallest member of a sample. *Proc. Cambridge Philos. Soc.*, 24, pp180-190, 1928.

[42] FRAGNIÈRE, E. AND HAURIE, A., A stochastic programming model for energy/environnement choices under uncertainty, in Gheorghe, A.V. (Editor) Integrated regional health and environmental risk assessement and safety management, Published in *J. Environment and Pollution,* 4-6, 587-603, 1996.

[43] GOFFIN J.-L., HAURIE, A. AND VIAL, J.-P., Decomposition and nondifferentiable optimization with the projective algorithm, Management Science, 38, 2, 284-302, 1992.

[44] GONDZIO J., DU MERLE, O., SARKISSIAN R. AND VIAL, J.-P., ACCPM - A Library for Convex Optimization Based on an Analytic Center Cutting Plane Method, European Journal of Operational Research, 94, 206-211, 1996.

[45] HARTWICK, J. M., Intergenerational equity and the investing of rents from exhaustible resources, *American Economic Review*, 1977: 66, 972-4.

[46] HARVEY, C.M. *The reasonableness of non-constant discounting, Journal of Public Economics*, 1994:53, pp. 31-51.

[47] HAURIE A., *Environmental Coordination in Dynamic Oligopolistic Markets, Group Decision and Negotiation*, Vol. 4, 1995, pp. 39-57.

[48] HAURIE A., Turnpikes in Multidiscount Rate Environments and GCC Policy Evaluation, in G. Zaccour Ed., *Optimal Control and Differential Games*, Kluwer 2002

[49] HAURIE A., *Time, Risk and Conflicts in Economics and Management Science: a Story about Turnpikes*, in G. Zaccour Ed., *Decision and Control in Management Science*, Kluwer 2002.

[50] HAURIE A., J. KÜBLER, A. CLAPPIER AND H. VANDEN BERGH , A Metamodeling Approach for Integrated Assessment of Air Quality Policies, *Environmental Modeling and Assessment,* Vol. 9, pp. 1-12, 2003.

[51] HAWKEN, P., *The Ecology of Commerce: A Declaration of Sustainability.* HarperBusiness, New York, 1994.

[52] HEAL, G. M., *Interpreting sustainability*, mimeo Columbia Business School, New York, May 1996.

[53] HEAL, G. M., *Optimality or sustainability ?*, mimeo Columbia Business School, New York, June 2001.

[54] HEAL, G. M., *Valuing the Future: Economic Theory and Sustainability*, Columbia University Press, New York, 1998.

[55] Holland, J.H (1998), Emergence: From Chaos to Order. Addison-Wesley, Reading, Mass.

[56] HOSMER, D.W. AND LEMESHOW, S., *Applied Logistic Regression*, Wiley, New York, 1989.

[57] HOLLAND, J.H., *Emergence: From Chaos to Order*, Addison-Wesley, Reading, Mass., 1998.

[58] IPCC, *Climate Change 1995: Economic and Social Dimensions of Climate Change*, NY. Cambridge University Pess, 1996.

[59] KIRBY, M. W., *Operational Research in War and Peace*, Imperial College Press, London, 2003

[60] KLOEDEN, P.E., PLATEN,E. AND SCHURZ,H., *Numerical Solutions of SDE Through Computer Experiments*, Springer, New York, 1993.

[61] KOPP, R.J. AND PORTNEY, P.R. Mock Referenda for Intergenerational Decisionmaking in PORTNEY P.R. AND WEYANT J., eds., *Discounting and Intergenerational Effects*, Resources for the Future, Washington, DC, 1999, pp. 87-98.

[62] KRASS, D., FILAR, J.A. AND SINHA, S. A Weighted Markov Decision Process. *Operations Research*, Vol.40, 1992, pp.1180-1187.

[63] KRIGE, D.G, A statistical approach to some mine valuations and allied problems at the Witwatersrand, Master's thesis of the University of Witwatersrand, 1951.

[64] KÜBLER, J., A.G. RUSSEL, HAKAMI, A. CLAPPIER AND VAN DEN BERGH, H., Episode Selection for Ozone Modelling and Control Strategies Analysis on the Swiss Plateau. Submitted March 2001

[65] LOEWENSTEIN, G.; PRELEC, D., Anomalies in Intertemporal Choice: Evidence and an Interpretation, in Loewenstein & Elster (eds) *Choice Over Time*, Russell Sage Foundation, New York, 1992.

[66] LI C.-Z. AND K.-G. LÖFGREN, Justifying sustainability, *Journal of Environmental Economics and Management*, Vol. 41, 2001, pp. 135-165.

[67] LI C.-Z. AND K.-G. LÖFGREN, Economic growth, environmental quality and hyperbolic discounting, *Technical report*, 1Dept. of Economics, Umea University, Sweden, 2001.

[68] MATHERON, G., "Principles of geostatistics", Economic Geology, 58, pp 1246–1266, 1963

[69] MANNE, A.S. AND R. RICHELS, Buying Greenhouse Insurance, MIT Press, Cambridge Mass. 1992.

[70] MEADOWS, D.H, D.L. MEADOWS, J. RANDERS AND W.H. BEYRENS, *The Limits to Growth*, A report for the Club of Rome, Earth Island Ltd, 1972

[20] , NORDHAUS W.D. D. *Managing the Global Commons: The Economics of Climate Change.* Cambridge, MIT Press, 1994.

[21] NORDHAUS W.D., A market based discount rate, in PORTNEY P.R. AND WEYANT J., eds., *Discounting and Intergenerational Effects*, Resources for the Future, Washington, DC, 1999, pp. 145-162.

[73] NORGAARD, R. AND HOWARTH, R. , *Sustainability and discounting the future*, in R. Costanza (Ed) *Ecological economics: the science and management of sustainability*, Columbia University Press, New York, 1991.

[74] PELEG B. AND YAARI M.E., On the existence of consistent course of action when tastes are changing, *Review of Economic Studies*, 1973: 40, pp. 391-401.

[75] PHELPS E.S. AND POLLAK R., On second best national saving and game-equilibrium growth, *Review of Economic Studies*, 1968: 35, pp. 185-199.

[76] PORTNEY P.R. AND WEYANT J., eds., *Discounting and Intergenerational Effects*, Resources for the Future, Washington, DC, 1999.

[77] PUTERMAN M.L., (1994). Markov Decision Processes: Discrete Stochastic Dynamic Programming, Wiley, New York.

[78] ROTMANS J. *IMAGE: An Integrated Model to Assess the Greenhouse Effect*, Kluwer Acad Press, Dordrecht, The Netherlands, 1990.

[79] SACKS, J. AND WELCH, W. J. AND MITCHELL, T. J. AND WYNN, H. P., Design and Analysis of Computer Experiments. 4. Statistical Science. pp. 409435, 1989.

[80] SEINFELD J.H. , *Atmospheric chemistry and physics: from air pollution to climate change*. New York: Wiley, 1998.

[81] STINSTRA E. AND D. DEN HERTOG. Robust optimization using computer experiments. *European Journal of Operational Research*, 191(3):816837, 2008.

[82] STAR Program Final Report : Quantification of Uncertainty in Air Quality Models Used for Analysis of Ozone Control Strategies. National Center for Environmental Research, 1999. http://es.epa.gov/ncer/final/grants/95/air/milford.html

[83] TAPOM (Transport and Air Pollution Model)PAS, EPFL, 2004. http://dgrwww.epfl.ch/PAS/MOD/models.fr.html

[84] WEITZMAN, M.L., "Why the Far-Distant Future Should Be Discounted at Its Lowest Possible Rate", *Journal of Environmental Economics and Management*, 36, pp. 201-208, 1998.

[85] WENE, C.-O. AND RYDEN, B., A comprehensive energy model in the municipal energy planning process, European J. Operational Research, 33, 212-222, 1988.

[86] ZAPERT, R., GAERTNER, P. S. AND FILAR, J. A., Uncertainty Propagation within an Integrated Model of Climate Change, *Energy Economics*, Vol. 20, No. 5-6, 571-598, 1998.

Chapter 2
Modeling Uncertainty in a Large scale integrated Energy-Climate Model

Maryse Labriet, Richard Loulou and Amit Kanudia

Abstract The well-known method of stochastic programming in extensive form is used on the large scale, partial equilibrium, technology rich global 15-region TIMES Integrated Assessment Model (ETSAP-TIAM), to assess climate policies in a very uncertain world. The main uncertainties considered are those of the Climate Sensitivity parameter, and of the rate of economic development. In this research, we argue that the stochastic programming approach is well adapted to the treatment of major uncertainties, in spite of the limitation inherent to this technique due to increased model size when many outcomes are modeled. The main advantage of the approach is to obtain a single hedging strategy while uncertainty prevails, contrary to classical scenario analysis. Furthermore, the hedging strategy has the very desirable property of attenuating the (in)famous 'razor edge' effect of Linear Programming, and thus to propose a more robust mix of technologies to attain the desired climate target. Although the examples treated use the classical expected cost criterion, the paper also presents, and argues in favor of, altering this criterion to introduce risk considerations, by means of a linearized semi-variance term, or by using the Savage criterion. Risk considerations are arguably even more important in situations where the random events are of a 'one-shot' nature and involve large costs or payoffs, as is the case in the modeling of global climate strategies. The article presents methodological details of the modeling approach, and uses realistic instances of the ETSAP-TIAM model to illustrate the technique and to analyze the resulting hedging strategies. The instances modeled and analyzed assume several alternative global temperature targets ranging from less than 2°C to 3°C. The 2.5°C

Maryse Labriet
KANLO and CIEMAT, Madrid, Spain, e-mail: maryse.labriet@gmail.com

Richard Loulou
KANLO, GERAD, and McGill University, Montreal, Canada, e-mail: richardloulou@gmail.com

Amit Kanudia
KanORS, Montreal Canada, and Delhi, India, e-mail: amit@kanors.com

J.A. Filar and A. Haurie (eds.), *Uncertainty and Environmental Decision Making*, International Series in Operations Research & Management Science 138, DOI 10.1007/978-1-4419-1129-2_2,

target is analyzed in some more details.

The paper makes a distinction between random events that induce anticipatory actions, and those that do not. The first type of event deserves full treatment via stochastic programming, while the second may be treated via ordinary sensitivity analysis. The distinction between the two types of event is not always straightforward, and often requires experimentation via trial-and-error. Some examples of such sensitivity analyses are provided as part of the TIAM application.

Key words: Energy modeling; Uncertainty; Stochastic programming; Hedging strategies; Climate policies; Technology.

2.1 Introduction

The Stochastic Programming (S.P.) paradigm (2; 19) is now a fully mature one, and has been used in many applications. It is recognized as a rigorous way to account for risky events while optimizing a particular system. One of the well recognized computational drawbacks of S.P. (at least in its extensive form) is that it quickly leads to large scale instance of the original problem, whenever the number of random events grows and/or the number of outcomes of each event becomes too large. In this article, we argue for a reasoned use of Stochastic Programming in large scale integrated Energy-Climate models such as the ETSAP-TIAM model (10; 12), in which only the main uncertainties are modeled, while others are treated via sensitivity analysis. More precisely, the paper makes a distinction between random events that induce anticipatory actions, and those that do not. The first type of event deserves full treatment via stochastic programming, while the second may be treated via ordinary sensitivity analysis. The distinction between the two types of event is not always straightforward, and often requires experimentation via trial-and-error. Some examples of such sensitivity analyses are provided as part of the TIAM application.

One of the main advantages of the S.P. approach is to obtain an explicit single hedging strategy while uncertainty prevails, contrary to classical scenario analysis. Furthermore, the hedging strategy has the very desirable property of attenuating the (in)famous 'razor edge' effect of Linear Programming, and thus to propose a more robust mix of technologies to attain the desired climate target. In a nutshell, a good hedging strategy takes into account the possible outcomes, and strikes an optimal compromise between the negative effects of the many ways of "guessing wrong" (11).

Although the examples treated uses the classical expected cost criterion, the paper also presents – and argues in favor of, altering this criterion to introduce risk considerations, by means of a linearized semi-variance term, or by using the Savage criterion. Risk considerations are arguably even more important in situations where the random events are of a 'one-shot' nature and involve large costs or payoffs, as is the case in the modeling of global climate strategies.

This article presents methodological details of the modeling approach, and uses realistic instances of the ETSAP-TIAM model to illustrate the technique and to analyze the resulting hedging strategies. The instances modeled and analyzed propose several alternative global temperature targets ranging from less than 2°C to 3°C. The 2.5°C target is analyzed in some more details.

In brief, the main objectives of this work are:

a) to demonstrate the power of stochastic programming in calculating optimal hedging strategies using a large scale, realistic energy-climate integrated model, in the presence major uncertainties on climate policies (climate sensitivity and future economic growth),
b) to analyze hedging strategies, i.e. a set of early robust actions capable of maintaining the global temperature within specified bounds, in spite of the uncertainty. Robust actions are those actions chosen in the hedging strategy but not in the Base case. In fact, hedging is deemed relevant if decisions made prior the resolution of uncertainty are different from those in the base case (otherwise, "*wait and see*" is a good policy). Hedging is even more useful when it is not identical to *any* of the perfect forecast strategies, since such a situation clearly shows that the optimal technology and energy decisions are not easily predictable without an explicit treatment of uncertainty.

Among the results obtained, the fact that no perfect forecast is able to reproduce the hedging strategy confirms the relevance of using stochastic programming in order to analyze preferred climate policies in an uncertain world.

c) To formulate alternate criteria for use with the S.P. approach, such as the adjunction of a semi-variance term or the use of the Savage criterion.

Section 2.2 contains a discussion of climate uncertainties. Section 2.3 describes the TIAM model and the methodology used to represent the uncertainties and to compute hedging strategies with stochastic programming. Sections 2.4 and 2.5 present results, including several sensitivity analyses, Section 2.6 presents alternate criteria, and Section 2.7 concludes the article.

2.2 Uncertainty in Energy-Climate Studies

The impacts of greenhouse gas (GHG) emissions on climate may be sketched as a chain of causal relationships, where GHG emissions provoke an increase in the concentration of GHG's in the atmosphere and in oceans; the increased concentrations provoke an increase of the atmospheric radiative forcing (RF) by the various gases, which in turn has an impact on the global temperature of the atmosphere and oceans. Nordhaus and Boyer (15) proposed simple and well-documented linear recursive equations for calculating CO_2 concentrations and global temperature changes. The climate module of ETSAP-TIAM is based on these equations for the carbon cycle and on two one-box models (simple exponential decay) for the CH_4 and N_2O cycles. The other substances that induce radiative atmospheric forcing (other Kyoto gases, Montreal Protocol gases, aerosols, etc.) are treated via an exogenous forcing trajectory.

In this article, two parameters of the climate equations are considered as highly uncertain: the climate sensitivity (C_s), defined as the equilibrium response of global surface temperature to a doubling of the equivalent CO_2 concentration; and the inverse of the thermal capacity of the atmospheric layer and the upper oceans, also called "lag parameter", key determinant of transient temperature change. While C_s has received a great deal of attention, its value is still highly uncertain (1; 3). Until recently, a range between 1.5°C and 4.5°C was commonly quoted (6). More recent studies have strongly argued for a wider range of 0.5°C to 9°C or even 10°C (1; 8). Regarding the lag parameter, its value may either be considered to be approximately independent of C_s, or it may be assumed to vary inversely with C_s.[1] The latter case results in higher transient temperature increases than with a fixed value of the lag parameter (for example, in our results, we observed that, when using a fixed lag parameter, the smallest achievable temperature increase is 0.5°C lower that when assuming a variable lag value). In the main analyses presented here, we make the prudent assumption of a variable lag, and we adopt the values adopted by the EMF-22 group[2] for the purpose of conducting comparative analyses of climate stabilization strategies with different models (Table 2.1). It is also assumed that the uncertainty on C_s and on the lag parameter will be fully resolved in 2040,[3] and that no significant additional knowledge will be obtained before that resolution date. In addition, sensitivity analysis on the date of resolution is presented in Section 2.5.

Table 2.1 Uncertain values of the climate sensitivity and the lag parameter

Climate Sensitivity	Likelihood	Corresponding Lag Parameter
1.5°C	0.25	0.065742
3°C	0.45	0.014614
5°C	0.15	0.010278
8°C	0.15	0.008863

Another potential source of uncertainty besides C_s is the annual rate at which the World economy develops, as this has a direct impact on economic demands and thus on GHG emissions. In this research, we also use the EMF-22 assumption that the base case global annual GDP growth rate is known until 2040. At that date, all future

[1] By linking C_s and σ_1, Yohe et al. (20) assume a deterministic relationship between the two parameters. Fussel (4) criticizes this relationship, since it results in values for the thermal capacity of the atmosphere and the upper oceans that are outside the physically plausible range. Moreover, the probabilistic relationship underestimates the true uncertainty about the transient climate response.

[2] The Energy Modelling Forum is an international forum on energy and environmental markets. The EMF-22 ongoing study, "Climate Policy Scenarios for Stabilization and in Transition", focuses on comprehensive analyses of long-run climate stabilization policies under uncertainty as well as intermediate-term transition policies.

[3] A recent paper by Weizmann (18) argues the case for the impossibility to fully resolve the uncertainty on Cs in finite time. If this were the case, the Stochastic Programming approach would be much simplified.

annual global growth rates until 2100 are assumed to be revealed and may have one of two equally probable values: a high value (equal to 4/3 of the base case rate), and a Low value (equal to 2/3 of the base case rate). The same simple-to-double growth rate assumption is used for the GDP growth rate of each region of ETSAP-TIAM. Regional GDP growth rates affect the growth rates of each energy service demand having GDP as a driver. World GDP starts from 32 trillion $ in 2000 and reaches 260 trillion $ (Base), 181 trillion $ (Low) or 385 trillion $ (High) in 2100.

Year 2040 corresponds to the beginning of the period 2040–2060 of the TIMES model. This period is called "2050" in results provided by TIMES. Therefore, all the results presented for years 2050 and after correspond to the part of the event tree after uncertainty is resolved, while results presented for years 2030 and before correspond to the part of the event tree before uncertainty is resolved.

2.3 Modeling Uncertainty in a Large Scale Integrated Energy-Climate Model

2.3.1 The TIMES Integrated Assessment Model (ETSAP-TIAM)

ETSAP-TIAM (TIMES Integrated Assessment Model) is a detailed, technology-rich Global TIMES model, where the World is divided in 15 regions in the version of the model (version 2007-2008) used for the work presented here. The 15 regional models integrated in TIAM are: Africa, Australia-New Zealand, Canada, Central and South America, China, Eastern Europe, Former Soviet Union, India, Japan, Mexico, Middle-East, Other Developing Asia, South Korea, United States, and Western Europe. In addition, the upstream and energy trade sectors in each country are split into OPEC/Non-OPEC. The regional modules are linked by trade variables of the main energy forms (coal, oil, gas) and of emission permits. Thus, impacts on trade (terms of trade) of environmental policies are taken into. ETSAP-TIAM's planning horizon extends from 2000 to 2100, divided into 7 periods of varying lengths, suitably chosen.

ETSAP-TIAM is a global instance of the TIMES model generator (full documentation is available from `www.etsap.org/documentation.asp`), where a bottom-up, detailed technological representation of each economic sector is combined with a key linkage to the rest of the economy via demands for energy services that are elastic to their own prices.

TIMES computes an inter-temporal dynamic partial equilibrium on energy markets, where demands for *energy services* are exogenously specified only in the reference case, and are sensitive to price changes (via a set of own-price elasticities) in all alternate scenarios. The equilibrium is driven by the maximization, via linear programming, of the total surplus (i.e. the sum of producers and suppliers surpluses), which acts as a proxy for welfare in each region of the model. Although TIMES does not encompass macroeconomic variables beyond the energy sector, accounting for price elasticity of demands captures a major element of feedback effects between the energy system and the economy. The surplus maximization is subject to many constraints, such as: supply bounds (in the form of detailed supply curves) for the primary resources, technical constraints governing the creation, operation, and abandonment of each technology, balance constraints for all energy forms and emissions, timing of investment payments and other cash flows, and the satisfaction of a set of demands for energy services in all sectors of the economy.

The construction of the base case demands for energy services is done by using the global General Equilibrium model GEM-E3 (`http://www.gem-e3.net/`), which provides a set of coherent *drivers* for each region and for the World as a whole, such as population, households, GDP, sectors outputs, and technical progress. These drivers are then transformed into growth rates for each of the 42 TIAM demands for energy services, via the generic relationship:

$$demand\_rate = driver\_rate\backslash times\ decoupling\_factor.$$

The decoupling factors account for phenomena such as saturation (factor is then less than 1) and suppressed demands (factor is then larger than 1), and are in part empirically based. Most demands have economic growth as their driver. As already mentioned, the demands of ETSAP-TIAM are user-specified only for the reference scenario, and are prone to endogenous changes in all alternate scenarios, in response to endogenously changing demand prices. The elasticities of demands to their own price range from 0 to -0.6, with a majority in the range -0.2 to -0.3.

ETSAP-TIAM comprises several thousand technologies in all sectors of the energy system (see sketch in Figure 2.1). A technology may represent any process that produces, transforms, conveys, and/or consumes energy and/or emissions (and some materials). It is characterized by several technical and economic parameters and by emission coefficients for the three main GHG's: CO_2, CH_4, and N_2O. Energy and industrial emissions of these three gases are all modeled. In addition, net CO_2 emissions from land use are represented, and non-energy CH_4 and N_2O emissions are also modeled (e.g. CH_4 and N_2O from landfills, manure, enteric fermentation, rice paddies, etc.). The model constructs a coherent image of the future energy system by choosing a mix of technologies to invest in and operate at each future period, with the objective of maximizing total surplus, while respecting the many constraints of the model. A complete description of ETSAP-TIAM's technological database is not possible within the limits of an article, but we wish to mention some options for GHG emission reductions available in the model: first, emission reductions may be done via the numerous fuel and technology switching options that are available in each sector, and via specific CH_4 and N_2O abatement options (e.g. suppression and/or combustion of fugitive methane from landfills, thermal destruction of N_2O in the adipic acid industry, etc.). CO_2 emissions may in some cases be captured and stored (CCS options) before their release into the atmosphere (e.g. CO_2 capture from the flue gas of fossil fueled power plants, from hydrogen production processes, and from oil extraction processes; storage in depleted oil fields, deep saline aquifers, deep oceans, etc.). Finally, atmospheric CO_2 may be partly absorbed and fixed by biological sinks such as forests; the model has six options for forestation and avoided

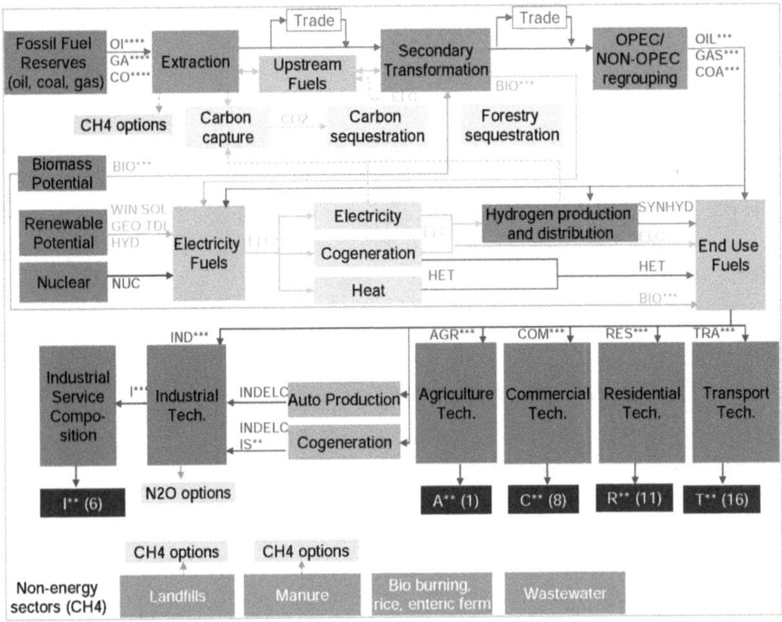

Fig. 2.1 ETSAP-TIAM Reference Energy System (version 2007-2008)

deforestation, as described in (17) and adopted by the EMF-22 group. Note also that methane emissions from the agriculture sector are fully accounted for, even if no abatement options are considered.

2.3.2 Using the Model

As noted before, most of climate equations from Nordhaus and Boyer (15) have been adapted and integrated into the model. One difference with these authors, we have adopted a separate representation of the N_2O and CH_4 atmospheric cycles, via simple one-box models where the concentration decays exponentially at constant rate.

ETSAP-TIAM may be used to evaluate different kinds of climate targets: emission limits directly, concentration bounds, bounds on radiative forcing, and finally, limits on global temperature change. However, the non-convexity of the radiative forcing expressions (see e.g. the forcing expression for CO_2 as equation (2.1)) precludes using the temperature equations as regular constraints of the ETSAP-TIAM model. Therefore, we have linearized the forcing expressions for CO_2, CH_4, and N_2O, within the useful concentration ranges. The result is an approximation that remains within 1% of the true forcing value within the range of interest for forcing values (i.e. between 300 and 600 ppmv CO_2-eq).

$$\Delta F(t) = \gamma^{*} \frac{In(M_{atm}(t)/M_0}{In2} + FEX(T) \tag{2.1}$$

where:

- $\Delta F(t)$ is the increase of the radiative forcing at period t relative to pre-industrial level
- M_0 is the pre-industrial (circa 1750) reference atmospheric concentration of CO_2
- γ is the radiative forcing sensitivity to the doubling of atmospheric CO_2 concentration (3.7 W/m^2)

2.3.3 The Computation of Hedging Strategies

2.3.3.1 Stochastic Programming

The treatment of uncertainty is done via Stochastic Linear Programming in extensive form (2; 19). In this method, the model takes a single *hedging* strategy until resolution of uncertainty, so as to be best positioned to adapt to any of the possible long term futures (after the resolution date). In the application described in Section 2.4, the optimization criterion is the *expected value* of the total surplus. Other optimizing criteria may be preferred. Loulou and Kanudia (11) presents an application using the Minimax Regret criterion. Another approach is available in TIMES, in which the criterion to maximize is a combination of the expected surplus and of a risk term calculated as the linearized semi-variance. Section 2.6 discusses these alternate criteria.

A typical stochastic LP with the expected value criterion is written as follows, in the simpler two-stage case where all uncertainties are resolved at a single date θ:

$$\textbf{Maximize} \sum_{t} \beta(t) \sum_{s=1,toS} C(t,s) \cdot X(t,s) \cdot p(s) \tag{2.2}$$

Subject to:

$$A(t,s) \times X(t,s) \geq b(t,s) \tag{2.3}$$

and $X(t,1) = X(t,2) = \ldots X(t,S)$, if $t <$ resolution date θ
where

- s represent the possible *states of the world* (sow), $s = 1,2,\ldots,S$
- $p(s)$ is the probability that *sow s* realizes
- C and b are respectively the surplus and the RHS vectors of the LP
- A is the matrix of LP coefficients
- $X(t,s)$ is the vector of decision variables at time t, under state-of-the-world s
- $\beta(t)$ is the discounting factor that converts 1$ from time t to time O.

Remark 1. We insist on the fact that the main interest of a hedging strategy resides in its description of what to do prior to the resolution date. In contrast, traditional deterministic scenario analysis computes multiple strategies even prior to the resolution date, leaving the decision maker in a quandary as to which one should be followed. Once uncertainty is resolved, the decision maker no longer faces uncertainty, and her decisions result from optimizing a deterministic problem from θ onward. Nevertheless, the computation of the hedging strategy must also take into account *all possible outcomes after the resolution date*. In other words, short term hedging decisions are devised while taking the uncertain long term into consideration. This is the essence of decision under risk, and in particular of stochastic programming.

The above discussion leads to an interesting classification of uncertain events that can be very useful in deciding which event(s) to model explicitly via stochastic programming, and which not: suppose it may be established that a certain event does NOT induce anticipatory actions. By this we mean that the hedging strategy is insensitive to the various possible outcomes of the event. In such a case, it is clear that there is no advantage in explicitly modeling the event as part of the S.P. event tree. Hence, only those events that are likely (or proven) to influence the Hedging strategy should be an explicit part of the event tree. Of course, the question now is how to detect such events, and the answer to that question is quite empirical. Some experimentation, and some judgment is required in order to sort out the two types of event. We shall come back to this question in the next paragraph and also in Section 2.5.

2.3.3.2 The Choice of the Two Uncertain Parameters

For a given temperature target, the two selected uncertain parameters are, as illustrated by Figure 2.2: i) the climate sensitivity C_s (four possible values), and ii) the vector of energy service demands resulting from the future economic growth (two possible values). The combination of these two uncertainties leads to 8 possible States of the World (SoW). However, after conducting stochastic optimizations with the 8 SoW's, it was *observed* that the impact of economic uncertainty on the hedging strategy before 2040 was quite negligible. In other words, the hedging decisions taken *before* 2040 are quite insensitive to the values of economic demands *after* 2040. In still other words, there is no anticipatory effect for the economic growth. Therefore, we decided to eliminate economic growth as an explicit uncertainty in our main runs reported in Section 2.4, and to assess the impact of uncertain economic growth on the hedging strategy as one kind of sensitivity analysis in Section 2.5. The resulting event tree, with only *Cs* as the uncertain parameter, has 4 branches, as shown in Figure 2.3.[4]

---

[4] Reducing the number of sow's has a direct impact on the computational time to resolve the LP. Typical time for solving the 8 sow problem was 440 minutes versus only 80 minutes for the 4 sow problem

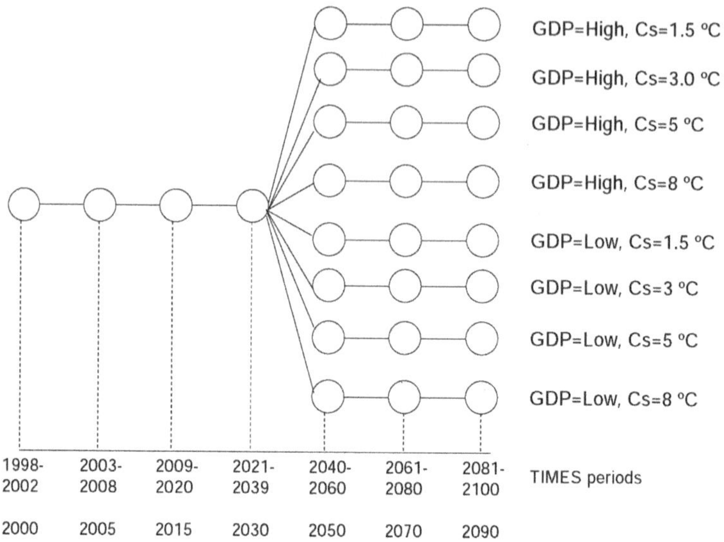

Fig. 2.2 The event tree

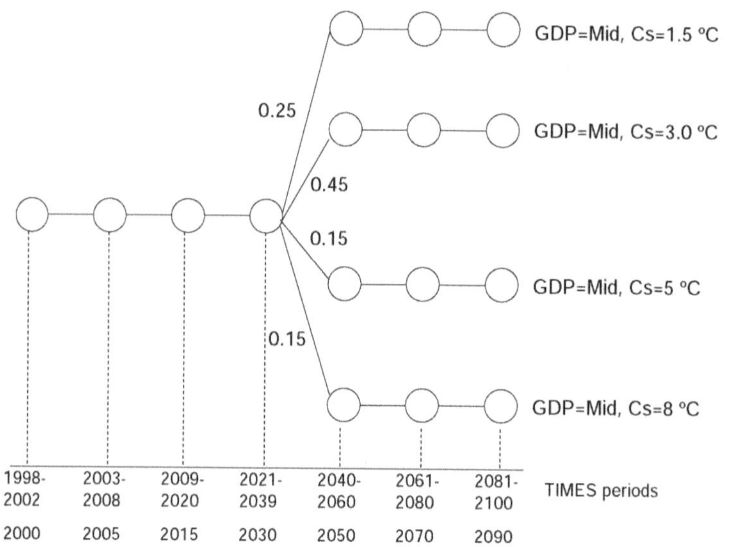

Fig. 2.3 The reduced event tree

2.4 A Treated Example: Hedging Strategy and Perfect Forecast Strategies for a 2.5°C Target on Global Temperature Change

2.4.1 Hedging and Perfect Forecast Strategies

Our initial objective was to calculate hedging strategies for two alternative scenarios, where the alternative targets for temperature change[5] are 2°C and 3°C. As it

[5] The targets actually set in the model are in year 2100. However, we are able to calculate the evolution of the global temperature after 2100, by assuming that emissions decline linearly to 0 from 2100 to 2200.

turned out, with the options present in the model, the 3°C target is achievable at very moderate cost, while the more severe 2°C target is only achievable at very high cost. Therefore, only the intermediate 2.5°C scenario will be discussed in detail in this paper. Moreover, additional model runs revealed that the smallest achievable temperature increase is close to 1.9°C, albeit at extremely large cost, given the options for GHG control present in the model and the GDP growth assumptions. This means that more severe temperature targets would require additional CO_2 abatement potential that is currently not yet seen as realistic.[6] Figure 2.5 shows that in order to keep global temperature change below the 1.9°C upper bound, the trajectory of CO_2-eq concentration must remain almost constant throughout the 21st century.

In addition to the hedging strategy, we also computed four (deterministic) perfect forecast strategies (noted PF), each assuming that the value of C_s is known as early as the first period. The theoretical interpretation of the four PF strategies is that of an optimal strategy if one assumes that the uncertainty is resolved at the beginning of the horizon. The PF's may be used to compute the Expected Value of Perfect Information (EVPI), which is the expected gain in welfare accrued if perfect information if available, i.e.:

$$EVPI = \sum_{s=1toS} p(s) \cdot \left[O_{PF(s)} - O_{HEDG}\right] \tag{2.4}$$

where

- $O_{PF(s)}$ is the surplus of the PFs strategy ($s = 1$ to S)
- O_{HEDG} is the expected surplus of the hedging strategy.

Another finding of the research is that when C_s = 1.5°C, the Base case satisfies the 2.5°C temperature constraint at all times, (provided emissions after 2100 decline linearly to 0 by 2200 as assumed here). Therefore, the $PF_{Cs=1.5°C}$ strategy is not different from the Base case.

2.4.2 Another Interpretation of the Perfect Forecast Strategies

In the detailed analysis of results for 2.5°C, we compare the results of the hedging strategy with those of a Base case where no climate target is imposed, but we also compare them with those of the Perfect Forecast strategies defined above. The rationale for this comparison stems from the following important remark: apart from its theoretical meaning as a perfect forecast strategy, any given PF strategy may be re-interpreted as a *heuristic* strategy *until the resolution date*. Indeed, a policy maker may wish to calculate a heuristic strategy that simply assumes a "guestimate" of Cs. This is easily done by optimizing a deterministic instance of the problem. Such a PF strategy is perfectly feasible (albeit not optimal) *until the resolution date*. After that

[6] No abatement options are available for rice production, enteric fermentation and biomass burning, whose CH_4 emissions are included in the model. This contributes to the infeasibility of any target smaller than 1.9°C.

date, decisions taken in the PF strategy are not realistic, since they ignore the fact that the value of C_s has indeed been revealed.

In this light, we shall discuss PF results only before 2040.[7] One finding is that $PF_{Cs=5°C}$ is the deterministic strategy that is closest to the optimal hedging one, although some significant differences between $PF_{Cs=5°C}$ and hedging exist in some areas, as we shall see. Therefore, when comparing Hedging with deterministic strategies, we shall always use $PF_{Cs=5°C}$ (and then only before 2040).

2.4.3 *Cost Analysis as a Proxy for the Target's Attainability*

We define the expected *cost* of a strategy as the Net Present value of the loss of expected surplus, relative to that of the base case. This provides a convenient indicator of the overall difficulty of reaching a particular target, and therefore a convenient way to compare various strategies. In addition to the NPV, we are interested in the marginal cost of one tonne of GHG.

2.4.3.1 Loss of Surplus and Expected Value of Perfect Information

The global net present value of the surplus attached to a climate strategy represents a compact measure of the social welfare associated with that strategy. Table 2.2 shows the expected loss of total surplus of the hedging strategy and of the perfect forecast strategy, relative to that of Base taken as reference. The loss of surplus when following Hedging is 35% higher than the expected loss for the perfect information strategy.[8] This difference represents the expected value of perfect information (210 B$ in NPV).

2.4.3.2 Marginal Cost of GHG

We first recall that the environmental constraint is defined in terms of global atmospheric CO_2-equivalent concentration. Thus, CO_2, CH_4 and N_2O have the same marginal cost in all regions and all sectors of the model.

Before 2040, the marginal cost of GHG in the hedging strategy remains low (Table 2.3). The analysis of hedging abatement options before 2040 shows that rel-

[7] With additional work, each PF strategy may also become a complete strategy as follows: freeze all PF decisions until 2040 at their observed values in the solution, and then re-optimize the system over periods post-2040 periods with each of the Cs values. In this way, each PF strategy gives birth to four post-2040 trajectories, which, taken together, constitute a bona fide complete strategy. This was not implemented in this research, but is illustrated (11).

[8] The corresponding annuities represent less than 0.1% of the World GDP (33000 B$ in year 2000). However, the stream of expenditures would clearly be lower in early years and higher in later years. Furthermore, equity issues might dictate an uneven imputation of the overall cost among the regions.

Table 2.2 Loss of surplus and expected value of perfect information

Strategy	**Loss of surplus** ($NPV_{5\%}$ in B$)	**Probability**	**Expected loss** (NPV in B$ and annuity in B$/year)	**EVPI** (NPV in B$ and annuity in B$/year)
BASE	0	1	–	–
PF Cs=1.5°C	0	0.25	–	–
PF Cs=3°C	43	0.45	–	–
PF Cs=5°C	580	0.15	–	–
PF Cs=8°C	3353	0.15	–	–
Total PF			610 (31)	–
HEDGING	820		820 (41)	210 (11)

EPVI = Expected loss $_{\text{HEDGING}}$ − Expected loss $_{\text{PERFECT FORECAST}}$

Table 2.3 Marginal cost of GHG ($/t$CO_2$)

Year	2000	2005	2015	2030	2050	2070	2090
TIMES periods	*1998–2002*	*2003–2008*	*2009–2020*	*2021–2039*	*2040–2060*	*2061–2080*	*2081–2100*
HEDGING Cs=1.5°C	1	2	4	10	0	0	0
HEDGING Cs=3°C					0	2	3
HEDGING Cs=5°C					11	40	80
HEDGING Cs=8°C					176	620	1236
PF Cs=3°C	0	0	0	1	2	7	14
PF Cs=5°C	0	1	2	4	12	43	86
PF Cs=8°C	3	7	12	28	84	296	589

atively inexpensive forestry measures contribute to this low price. The fact that no abatement option is available for methane from rice production, enteric fermentation and biomass burning, contributes to the observed high GHG price in the late horizon (up to more than 1200$/t$CO_2$), when methane represents the most important remaining GHG due to the lack of reduction options for agricultural CH_4 emissions.

We also observe that none of the perfect forecast strategies is able to provide a good approximation of the expected GHG price under uncertainty, although $PF_{Cs=5°C}$ is the closest to hedging in that respect.

2.4.4 Global Emissions and Climate Results

2.4.4.1 Base Case Emissions

The base case GHG emission trajectory (Figure 2.4) as well as the atmospheric GHG concentration reached in 2090 (Figure 2.5) are fairly close to the B2 Emission Scenario proposed by the Intergovernmental Panel on Climate Change (6; 14). CO_2 remains the most important GHG (around 79%), followed by CH_4 (around 19%) and N_2O (less than 2%). As for sector emissions in the base case, the electricity and transportation sectors are the highest GHG contributors in 2000 (more than 40% of

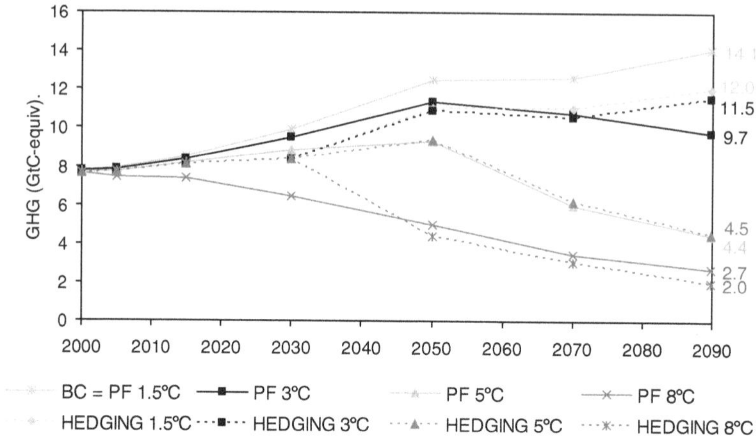

Fig. 2.4 GHG emissions of hedging and perfect forecast strategies

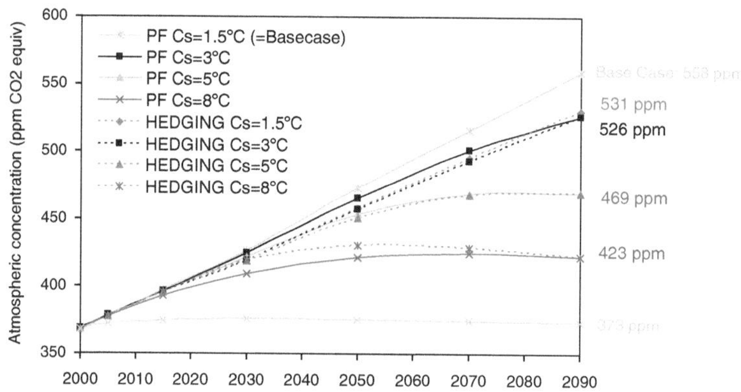

Fig. 2.5 Atmospheric concentration (CO_2-equiv) under hedging and perfect forecast strategies

total GHGs), and the electricity and industry sectors become the highest contributors at the end of the horizon (more than 48% of total GHG).

2.4.4.2 Emissions in the Optimal Hedging Strategy with 2.5°C Temperature Constraint

The situation is radically different under the 2.5°C temperature constraint, since both the electricity and industry sectors are able to reduce to almost zero (less than 3% of total GHG) their emissions in the most stringent branch, mainly thanks to CCS in the electricity sector, and switching to electricity in the industrial sector. In this most stringent branch, transport and agriculture are the highest remaining

GHG contributors (30% and 41% of total GHG). No such drastic decrease of CH_4 emissions is possible because some non-energy agriculture-related sources have no abatement options in the model. Based on emissions, the $PF_{Cs=5°C}$ strategy is also the deterministic strategy that is closest to the optimal hedging strategy before 2040.

Atmospheric concentration obtained with the lowest value of C_s (Figure 2.5) is lower in Hedging than in Base although no target was imposed on this branch of the Hedging. This is because hedging actions taken pre-2040 push concentration downward. Again, $PF_{Cs=5°C}$ is the PF strategy that is closest to Hedging before 2040.

In the base case, the temperature increase in 2090 is in the range from 1.4°C to 2.4°C, depending on C_s (Figure 2.6). In all hedging branches, temperature peaks within he 22nd century, and then declines, so that the equilibrium temperature is always lower than the maximum observed temperature from 2000 to 2200. This might not necessarily be the case for other temperature scenarios, or if a slower emission decline was assumed after 2100.

2.4.5 Robust (Hedging) Actions

2.4.5.1 CO_2 Sequestration

Sequestration by forests appears to be a very robust abatement option, since it penetrates in the hedging strategy as early as 2005 (Table 2.4) and uses its full potential. In fact, it plays a transitional role until less expensive energy options become available. As regards CCS options (with sequestration in deep oceans, saline aquifers, coal bed methane recovery, depleted oil and gas fields, enhanced oil recovery), they are much less robust, as they penetrate only slightly in 2030 in the hedging strategy, while they are used much earlier (in 2005) and at a higher level in $PF_{Cs=8°C}$, and used only after 2040 in the other PF strategies. In other words, no perfect forecast strategiy is able to reproduce the hedging strategy.

2.4.5.2 Electricity Sector's Actions

Electricity production is shown in Table 2.5. The first observation is that, as expected, electricity production up to 2030 takes a middle-of-the-road course up to 2030, compared to the PF strategies.

In the pre-2040 periods, we note significant differences in the Hedging and $PF_{Cs=5°C}$ strategies mainly in two categories: first, the PF strategy widely overestimates the amount of coal based electricity production (with CCS) compared to Hedging. In contrast, it underestimates the optimal amount of biomass fueled electricity and also the amount of hydroelectricity, compared to Hedging,. For other types of electricity (from gas and nuclear), $PF_{Cs=5°C}$ production levels are quite close to the optimal hedging amounts over the entire pre-2040 period.

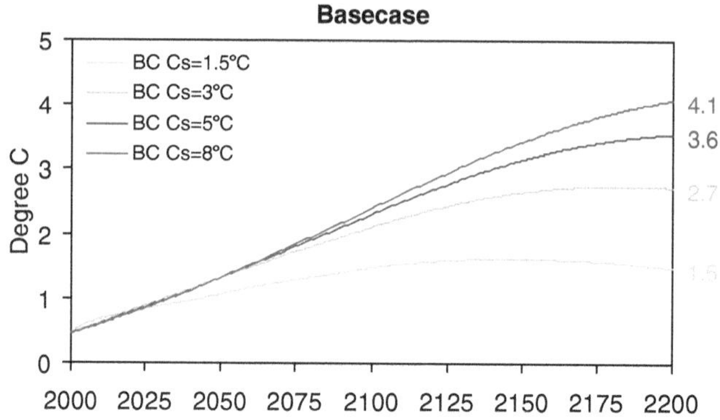

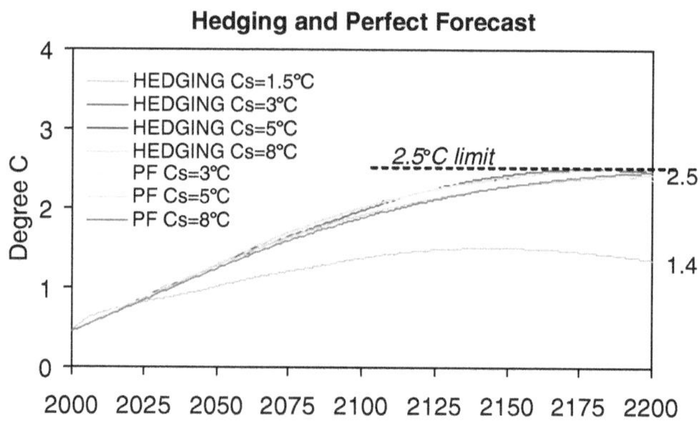

Fig. 2.6 Temperature increase 2000–2200*

Moreover, hydroelectricity (and, to a lesser extent, wind power too) and the shutdown of coal plants without CCS might qualify as hedging actions, since they appear before 2040. This is not the case of either power plants with CCS or nuclear plants.

In conclusion, the hedging strategy is significantly different from any of the PF strategies, which confirms the relevance of using stochastic programming.

Table 2.4 Contribution of CCS and forestry to the total GHG reduction

<table>
<tr><td colspan="7">Contribution of CCS to GHG (CO_2 equiv) reduction</td></tr>
<tr><td>Year</td><td>2005</td><td>2015</td><td>2030</td><td>2050</td><td>2070</td><td>2090</td></tr>
<tr><td>TIMES periods</td><td>2003–2008</td><td>2009–2020</td><td>2021–2039</td><td>2040–2060</td><td>2061–2080</td><td>2081–2100</td></tr>
<tr><td>HEDGING Cs=1.5°C</td><td rowspan="4">0.0%</td><td rowspan="4">0.0%</td><td rowspan="4">2.9%</td><td>0.0%</td><td>0.0%</td><td>0.0%</td></tr>
<tr><td>HEDGING Cs=3°C</td><td>0.0%</td><td>0.0%</td><td>0.0%</td></tr>
<tr><td>HEDGING Cs=5°C</td><td>1.0%</td><td>5.3%</td><td>10.8%</td></tr>
<tr><td>HEDGING Cs=8°C</td><td>17.0%</td><td>10.6%</td><td>10.7%</td></tr>
<tr><td>PF Cs=1.5°C</td><td>0.0%</td><td>0.0%</td><td>0.0%</td><td>0.0%</td><td>0.0%</td><td>0.0%</td></tr>
<tr><td>PF Cs=3°C</td><td>0.0%</td><td>0.0%</td><td>0.0%</td><td>0.0%</td><td>1.9%</td><td>3.7%</td></tr>
<tr><td>PF Cs=5°C</td><td>0.0%</td><td>0.0%</td><td>0.0%</td><td>1.1%</td><td>6.6%</td><td>10.9%</td></tr>
<tr><td>PF Cs=8°C</td><td>7.3%</td><td>3.7%</td><td>5.6%</td><td>17.7%</td><td>13.3%</td><td>11.9%</td></tr>
<tr><td colspan="7">Contribution of forestry sequestration to GHG (CO_2 equiv) reduction</td></tr>
<tr><td>Year</td><td>2005</td><td>2015</td><td>2030</td><td>2050</td><td>2070</td><td>2090</td></tr>
<tr><td>TIMES periods</td><td>2003–2008</td><td>2009–2020</td><td>2021–2039</td><td>2040–2060</td><td>2061–2080</td><td>2081–2100</td></tr>
<tr><td>HEDGING Cs=1.5°C</td><td rowspan="4">35%</td><td rowspan="4">53%</td><td rowspan="4">29%</td><td>65%</td><td>99%</td><td>97%</td></tr>
<tr><td>HEDGING Cs=3°C</td><td>61%</td><td>85%</td><td>78%</td></tr>
<tr><td>HEDGING Cs=5°C</td><td>31%</td><td>27%</td><td>21%</td></tr>
<tr><td>HEDGING Cs=8°C</td><td>12%</td><td>18%</td><td>16%</td></tr>
<tr><td>PF Cs=1.5°C</td><td>0%</td><td>0%</td><td>0%</td><td>0%</td><td>0%</td><td>0%</td></tr>
<tr><td>PF Cs=3°C</td><td>85%</td><td>86%</td><td>77%</td><td>53%</td><td>61%</td><td>41%</td></tr>
<tr><td>PF Cs=5°C</td><td>44%</td><td>65%</td><td>43%</td><td>29%</td><td>26%</td><td>20%</td></tr>
<tr><td>PF Cs=8°C</td><td>25%</td><td>27%</td><td>16%</td><td>13%</td><td>19%</td><td>17%</td></tr>
</table>

2.4.5.3 End-Use Sectors

In *transportation*, Hedging stays close to the PF strategies and even to the Base case before 2040. This is due to two causes: first, vehicles have a rather limited technical life, so that pre-2040 decisions do not have a lasting effect after resolution time. The other important cause of the observed insensitivity of this sector is that the CO_2 price signal is simply not strong enough before 2040 to warrant a large departure from traditional fuels. After resolution time, of course, the strategies do differ, and they do so in a fairly predictable way: the larger C_s values entail smaller market shares for RPP's and larger for alcohols and natural gas. Electricity keeps a very limited market share, and hydrogen (mainly produced by plants with CCS) makes a belated and small appearance in 2090 only in the most extreme branch of the Hedging).

The hedging strategy in *residential and commercial* buildings is characterized by very few energy changes compared to base case before 2040, and by an increase of electricity after 2040 (replacing natural gas and RPPs) in the most severe branches of the Hedging, mainly for space heating purposes.

In *industry*, differences between Hedging and Base case actions are slight before 2040. The exception being that N_2O abatement options in adipic and nitric acid industries penetrate as early as 2005 in the hedging strategy and thus are are robust decisions. After 2040, natural gas and, to a lesser extent, electricity, replace coal after 2040 in the most stringent branches of the hedging, mainly in chemical and other industry sub-sectors.

Table 2.5 Electricity production (EJ/year)

Plant Type		2000	2005	2015	2030	2050	2070	2090
COAL FIRED	BASE, PF Cs=1.5°C	18	17	17	15	28	22	24
	PF Cs=3°C	18	17	17	15	25	18	15
	PF Cs=5°C	17	17	17	9	5	3	8
	PF Cs=8°C	16	17	11	0	12	12	14
	HEDGING Cs=1.5°C					25	24	23
	HEDGING Cs=3°C					25	21	20
	HEDGING Cs=5°C	16	17	17	5	5	3	8
	HEDGING Cs=8°C					14	10	12
OIL + GAS FIRED	BASE, PF Cs=1.5°C	5	10	18	34	54	57	61
	PF Cs=3°C	5	10	18	34	52	56	54
	PF Cs=5°C	6	10	18	34	56	39	29
	PF Cs=8°C	8	10	20	29	34	27	30
	HEDGING Cs=1.5°C					51	58	61
	HEDGING Cs=3°C					51	56	62
	HEDGING Cs=5°C	7	10	18	36	59	38	29
	HEDGING Cs=8°C					29	23	25
NUCLEAR	BASE, PF Cs=1.5°C	9	8	10	11	20	59	109
	PF Cs=3°C	9	8	10	11	20	59	109
	PF Cs=5°C	9	8	10	11	20	73	128
	PF Cs=8°C	9	8	10	13	28	74	136
	HEDGING Cs=1.5°C					20	59	109
	HEDGING Cs=3°C					20	59	109
	HEDGING Cs=5°C	9	8	10	11	20	73	128
	HEDGING Cs=8°C					28	74	138
HYDRO	BASE, PF Cs=1.5°C	9	9	10	11	13	22	26
	PF Cs=3°C	9	9	10	11	19	26	38
	PF Cs=5°C	9	9	10	15	30	39	44
	PF Cs=8°C	9	9	12	25	35	42	49
	HEDGING Cs=1.5°C					19	19	27
	HEDGING Cs=3°C					19	24	28
	HEDGING Cs=5°C	9	9	10	17	28	39	44
	HEDGING Cs=8°C					35	44	53
BIOMASS	BASE, PF Cs=1.5°C	0	0	0	0	1	1	1
	PF Cs=3°C	0	0	0	0	1	1	1
	PF Cs=5°C	0	0	0	0	1	1	1
	PF Cs=8°C	0	0	0	1	6	4	3
	HEDGING Cs=1.5°C					1	1	1
	HEDGING Cs=3°C					1	1	1
	HEDGING Cs=5°C	0	0	0	0	1	1	1
	HEDGING Cs=8°C					8	7	7
OTHER RENEWWABLES	BASE, PF Cs=1.5°C	0	0	0	0	1	1	1
	PF Cs=3°C	0	0	0	0	1	1	1
	PF Cs=5°C	0	0	0	0	1	1	2
	PF Cs=8°C	0	0	0	1	1	3	7
	HEDGING Cs=1.5°C					1	1	1
	HEDGING Cs=3°C					1	1	1
	HEDGING Cs=5°C	0	0	0	0	1	1	2
	HEDGING Cs=8°C					3	7	22
TOTAL	BASE, PF Cs=1.5°C	41	45	56	72	118	162	222
	PF Cs=3°C	41	45	56	72	118	161	219
	PF Cs=5°C	42	45	56	70	114	155	214
	PF Cs=8°C	42	44	54	68	115	162	239
	HEDGING Cs=1.5°C					117	162	222
	HEDGING Cs=3°C					117	162	222
	HEDGING Cs=5°C	42	45	56	70	114	155	214
	HEDGING Cs=8°C					117	166	257

Demands (and thus economic activity) are affected by the introduction of the climate target, since the rising GHG price induces a rise in demand prices, and thus a decrease in demand levels. Their reduction starts as soon as 2005, remaining small until 2030 and reaching up to 5% in buildings and 6% in industry in the longer term.

The reduction of *upstream* emissions until 2030 is the result of both changes in the primary energy structure driven by final energy changes (for example, CO_2 and CH_4 reduction in coal extraction), and of specific GHG abatement measures (for example, degasification and pipeline injection of CH_4 in coal sector, inspection and maintenance of gas distribution facilities, flaring instead of venting, etc.) In fact, a few CH_4 reduction options appear to be non-regret measures and penetrate even in the base case.

Finally, CH_4 capture options in *landfill* and, to a lesser extent, *manure* emission abatement measures also appear to be either non-regret or robust (penetration before 2040 in the hedging strategy). In fact, we observe that the relative CH_4 reduction is more important than the CO_2 reduction in the short term, due to the availability of these low-cost CH_4 capture options in upstream and landfills. This result is in line with the literature (e.g. (7)).

2.4.6 Super-Hedging Actions

A "super-hedging" action is an action that penetrates more in the hedging strategy than in *any* of the perfect forecast strategies. The existence of such an action seems counter-intuitive, since it lies outside limits defined by the perfect forecast strategies, but such actions nevertheless do exist, which confirms that stochastic analysis of future climate strategies may propose decisions that are beyond any combination of the deterministic strategies (9).

Electricity production from renewables, fuel switches in industry (to biomass and gas), consumption of ethanol in several subsectors, consumption of geothermal in commercial buildings and biomass in residential buildings, and finally CH_4 abatement actions are all super-hedging actions.

2.5 Sensitivity Analyses

Sensitivity analyses were undertaken on: the exogenous radiative forcing, the very long term emission profile, the date of resolution of uncertainties, the pace of development of nuclear power plants, and the own price elasticities of demands.

In our main experiment, the assumed value of *exogenous forcing* in ETSAP-TIAM is 0.4 W/m^2 indefinitely, a fairly large value. Such forcing accounts for substances in the atmosphere that are not explicitly modeled in TIAM, and also for a possible increase in solar activity. We conducted a simple calculation: keeping the Hedging strategy calculated in Section 2.4, we simply varied the value of the residual exogenous forcing from 0 to 0.8 W/m^2, and we re-calculated (outside the model) the resulting temperature changes. We found that the resulting equilibrium temperature of the Hedging strategy remains less than 2.5°C across most of the range, reaching 2.8°C for the highest value (0.8) of the exogenous forcing. Although these temperature shifts are not negligible, they do not drastically depart from the temperature changes observed in the main hedging strategy.

Changing the assumption about the *post-2100 emission curve* (for instance extending the period of emission decrease to 200 years instead of 100 years), has of course no impact on the equilibrium temperature, but has an impact on the peak temperature. However, this impact remains very small. This analysis is most reassuring, as it tends to confirm that emission policies beyond 2100 have a small impact on temperature increase, as long as a policy of eradicating all emissions is followed, irrespective of the speed of that eradication.

Advancing the date at which the climate uncertainty is resolved to 2020 (instead of 2040) results in welfare savings of 159 B$, i.e. a full 3/4 of the EVPI. Such an analysis may provide a useful guide in deciding research expenditures in the climate change domain.

There may be societal and political reasons that may warrant limiting the degree of *penetration of nuclear power*. Therefore, we have undertaken sensitivity analyses on both the level of nuclear power in the base case and on the maximum allowed level of nuclear energy. In both cases, other reduction options (wind, solar, biomass etc.) penetrate to replace the nuclear loss, and the loss of surplus of the new hedging strategy is moderately increased. This confirms that *nuclear per se does not qualify as a robust abatement option* but also that the limitation of nuclear penetration does not seriously compromise the possibilities to satisfy a 2.5°C target at an "acceptable" cost.

Finally, if *demand elasticities* are set to 0, the expected loss of total surplus of the hedging strategy increases by almost 15%, and the marginal cost of GHG reduction is around 19% higher compared to the hedging strategy with elastic demands (higher electricity consumption, higher penetration of hydrogen and natural gas in the transportation sector, higher penetration of low emitting power plants etc.). Moreover, the reduction of emissions starts earlier, so that emissions are smaller before 2040 and higher in the long term compared to the hedging strategy with elastic demands.

2.6 Alternate Criteria for Stochastic Programming

The Expected Cost criterion, although widely used, is justified only if the policy maker is risk neutral. Risk neutrality usually applies when the payoffs attached to each outcome stay within a reasonably narrow range, and when they do not represent very large losses or gains as compared to the mean payoff. In all other cases, considerations of risk aversion or risk preference should be introduced in the criterion to be optimized. In the most general terms, the policy maker should choose a criterion that represents her utility function. We review here two such utility functions, both well-known, although not often utilized in large scale models.

2.6.1 *Expected Cost Criterion with Risk Aversion*

2.6.1.1 E-V Approach

The E-V model (an abbreviation for Expected Value-Variance) was pioneered by Harry Markowitz (13) for applications in financial portfolios.

In the E-V approach, it is assumed that the variance of the cost is an acceptable measure of the risk attached to a strategy in the presence of uncertainty. The variance of the cost $\boldsymbol{C_s}$ of a given strategy $\boldsymbol{s}$ is computed as follows:

$$Var(C_s) = \sum_j p_j \cdot (Cost_{j|s} - EC_s)^2$$

where $\boldsymbol{Cost_{j|s}}$ is the cost when strategy $\boldsymbol{s}$ is followed and the $\boldsymbol{j^{th}}$ state of nature prevails, and $\boldsymbol{EC_s}$ is the expected cost of strategy $\boldsymbol{s}$, defined as usual by:

$$EC_s = \sum_j p_j \cdot Cost_{j|s}$$

The E-V approach thus replaces the expected cost criterion by the following utility function to minimize:

$$U = EC + \lambda \cdot \sqrt{Var(C)}$$

where $\lambda > 0$ is a measure of the risk aversion of the decision maker. For $\lambda = 0$, the usual expected cost criterion is obtained. Larger values of λ indicate increasing risk aversion.

Taking risk aversion into account by this formulation would lead to a non-linear, non-convex model, with all its ensuing computational restrictions. These would impose serious limitations on model size. We therefore propose a linearized version that is more suitable for large scale applications.

2.6.1.2 Utility Function with Linearized Risk Aversion

This criterion mimics the E-V approach while eliminating two drawbacks. First, it eliminates the lower part of the range of variation of the cost, which indeed should not be considered as a risk. Second, it is linear and thus compatible with the overall LP model. To avoid non-linearity, it is possible to replace the semi-variance by the Upper-absolute-deviation, defined by:

$$UpAbsDev\,(Cost_s) = \sum_j p_j \cdot \{Cost_{j|s} - EC_s\}^+$$

where $y = \{x\}^+$ is defined by the following two *linear* constraints:

$$y \geq x, y \geq 0$$

and the utility function is now written as the following *linear* expression:

$$U = EC + \lambda \cdot UpsAbsDev\,(C)$$

This is the expected utility formulation implemented into the TIMES model generator.

2.6.2 The Savage Criterion (Minimax Regret)

The Minimax Regret Criterion, also known as Savage Criterion (16), is one of the more credible criteria for selecting public policies when the likelihoods of the various possible outcomes are not known with sufficient precision to use the classical expected value or expected utility criteria. In order to fix ideas, we shall assume in what follows that a certain decision problem is couched in terms of a cost to minimize (a symmetric formulation is obtained in the case of a payoff to maximize). We may thus denote by $C(z,s)$ the cost incurred when strategy s is used, and outcome z occurs. The Regret $R(z,s)$ is defined as the difference between the cost incurred with the pair (z,s) and the least cost achievable under outcome perfect information on z, i.e.:

$$R(z,s) = C(z,s) - \underset{t \in S}{Min}\, C(z,t), \qquad \forall z \in Z, \quad s \in S$$

where Z is the set of possible outcomes and S is the set of feasible strategies. Note that, by construction, a regret $R(z,s)$ is always non negative, and that it has value 0 for one outcome at least. Note also that the term 'regret' is particularly well chosen as it does quantify how much the policy maker would regret having chosen strategy s when outcome z occurs.

A *Minimax Regret (MMR) strategy* is any s^* that minimizes the worst regret:

$$s^* \in \underset{s \in S}{ArgMin}\{\underset{z \in Z}{Max}\, R(z,s)\}$$

and the corresponding Minimax Regret is equal to:

$$MMR = \underset{s \in S}{Min}\{\underset{z \in Z}{Max}\, R(z,s)\}$$

2.6.3 Application of Minimax Regret in Large Scale Linear Programs

We now turn to the application of the above definition to the case when the cost $C(z,s)$ is not an explicit expression. Rather, it is implicitly computed via an optimization program. This is the case in particular when using a model such as TIMES, where the long term energy system least cost (i.e. max surplus) is computed by solving a large scale linear program. The notation of Section 2.3 is used again, with appropriate changes. In the formulation 2.5 below, the A matrix defines the very large number of techno-economic constraints of the model, and the last group of constraints have some uncertain parameters as coefficients or as right hand sides. We assume for simplicity of notation, that all uncertainties are in the RHS, D, and that they are all resolved at the same date denoted by t^*. In the absence of uncertainty, the TIMES linear program has the following structure:

$$\begin{aligned}
&\textbf{Maximize } C \times X \\
&\textbf{Subject to:} \\
&A \times X \geq b \\
&E \times X \leq D
\end{aligned} \tag{2.5}$$

Assuming now that the RHS, D, is uncertain, and that the uncertainty is resolved at time t^*, we observe that prior to t^*, all decisions are taken under uncertainty, whereas at t^* and later, decisions are taken under perfect knowledge of the value of D. It is convenient to decompose the vector X of decision variables into two vectors (X_1 and X_2), X_1 representing the decisions to be taken prior to t^*, and X_2 those decisions at t^* and later. We shall assume that the uncertain vector D may take an arbitrary but finite number n of distinct values: $D_1, D_2, \ldots, D_n$.

We denote by $M(D)$ the minimum value obtained from the minimization of the above LP when vector D is known. This is the same as calculating the minimum cost under perfect information on D. We therefore may now theoretically calculate the *Regret* of strategy X as follows:

$$R(X, D_i) = C^t X - M(D_i)$$

And the maximum regret of strategy X as:

$$Max_i \left\{ C^t X - M(D_i) \right\}$$

Finally, the Minimax Regret strategy is an optimal solution to the following optimization program:

$$\begin{aligned}
MMR = &\underset{X_1, X_2}{Min} \, \underset{i}{Max} \left[C_1^t X_1 + C_2^t X_2^i - M(D_i) \right] \\
s.t. \quad &A_1 X_1 + A_2 X_2^i \geq b, \qquad i = 1, 2, \ldots, n \\
&E_1^t X_1 + E_2^t X_2^i \leq D_i, \qquad i = 1, 2, \ldots, n
\end{aligned} \tag{2.6}$$

The above program is not quite an L.P., but may be converted into one by introducing a new variable:

$$
\begin{aligned}
MMR = \; & \underset{X_1,X_2,\phi}{Min} \; [\phi] \\
s.t. \quad & \phi \geq C_1^t X_1 + C_2^t X_2^i - M(D_i), \quad \forall i \\
& A_1 X_1 + A_2 X_2^i \geq b, \quad \forall i \\
& E_1 X_1 + E_2^i \leq D_i, \quad \forall i
\end{aligned}
\tag{2.7}
$$

Note carefully that a *bona fide* strategy X is such that X_1 is common to all outcomes D, whereas there is a different vector X_2^i for each outcome D_i. This is so because decisions made at t^* and later take into account the knowledge of the true value of D that realizes at t^*. Hence, the LP (2.8) has up to n replications of the constraints, and of the X_2 variables (to be more precise, all constraints which do not involve X_2 variables are not replicated, and therefore, the size of (2.7) may be significantly smaller than n times the size of LP (2.5).

Important remark: An unfortunate phenomenon occurs when (2.7) is solved: since all that matters when computing the *MMR* strategy is indeed the value of the Minimax Regret, all other regrets are left free to take any values, as long as these values remain below the *MMR*. This remark is equivalent to saying that (2.7) is highly degenerate. For example, in one instance reported in (11), the *MMR* is equal to 3,311 M$, but when (2.7) is solved, it turns out that *each* of the n regrets (i.e. each right-hand-side of the first constraints of (2.7)), is found to be also equal to 3,311 M$. This is undesirable, as in practice, depending upon the actual value of z which realizes, the regret can be quite much lower than *MMR*. In order to remove the dual degeneracy, it is useful to proceed in two phases: first, consider (2.8) as essentially only a way of computing the partial strategy up to the resolution date, i.e. X_1. Next, when this is done, each X_2^i may be computed independently by (a) fixing X_1 at its optimal value (call it X_1^*), and (b) for each i, solving the following linear program:

$$
\begin{aligned}
& \underset{x_2}{Min} \left[c_1^t X_1^* + c_2^t X_2 - M(D_i) \right] \\
s.t. \quad & A_1 X_1^* + A_2 X_2 \geq b \\
& E_1^t X_1^* + E_2^t X_2 \leq D_i
\end{aligned}
\tag{2.8}
$$

The largest LP to solve is clearly (2.7), which has the same approximate size as a classical stochastic LP defined on the same problem instance, and using the expected value criterion. In addition, $n-1$ smaller problems (2.4) must be solved, in order to compute the $n-1$ non degenerate strategies after date t.

2.7 Conclusion

In this article, the classical Stochastic Programming technique is presented and applied to large scale instances of the Integrated Assessment Model (ETSAP-TIAM). The instances solved and discussed lead to the long term analysis of climate stabilization strategies under high uncertainty of climate sensitivity Cs (in the range 1.5 to 8°C) and of economic growth (simple-to-double GDP growth rates from 2040). Both uncertainties are assumed to be resolved in 2040. The methodology relies on the computation of a hedging strategy based on the maximization under uncertainty (via Stochastic Programming) of total World surplus over the 21st century. The properties of the resulting strategies are stressed, and a class of hedging and super-hedging actions is identified.

Amongst the most noticeable results, the model reveals that the smallest achievable temperature increase is close to 1.9°C, albeit at a very large cost, by a combination of energy switching, capture and storage of CO_2, CO_2 sequestration by forests and non-CO_2 emission reduction options. This means that more severe temperature targets would require additional GHG abatement potential that is currently not yet seen as realistic. Moreover, the impact of uncertainty of the climate sensitivity parameter C_s is major, requiring the implementation of early actions (before 2040) in order to reach the temperature target. In other words, the "wait and see" approach is not recommended. Robust abatement options include: substitution of coal power plants by hydroelectricity, sequestration by forests, CH_4 and N_2O reduction. Nuclear power plants, electricity production with CCS, and end-use fuel substitution do not belong to early actions. Among them, several options appear also to be super-hedging actions i.e. they penetrate more in the hedging strategy than in any of the perfect forecast strategies (e.g. hydroelectricity, CH_4 reduction), proving that stochastic analyze of future climate strategies might give insights that are beyond any combination of the deterministic strategies. In contrast, the uncertainty of the GDP growth rates has very little impact on pre-2040 decisions. This insensitivity is a pleasant surprise, as it shows that the hedging strategy for only one random parameter (C_s) is also a quasi-optimal strategy when the two types of uncertainty are present.

The comparison of hedging with perfect forecast strategies shows that a deterministic strategy with Cs=5°C is closest to the hedging strategy. However, the two differ in several key aspects, and this confirms the relevance of using stochastic programming in order to analyze preferred climate policies in an uncertain world where the correct climate response is known only far into the future. In particular, the perfect forecast strategy provides a poor approximation of the optimal electricity production mix, of the price of carbon, and of the penetration of several sequestration options.

Among the more sensitive parameters of the problem, resolving the uncertainties in 2020 rather than 2040 induces a 19% reduction in the loss of expected surplus, and keeping the same hedging strategy while assuming a doubling of the exogenous forcing has a non negligible (although moderate) raises global temperature by 0.3°C.

Two modifications of the criterion expected cost used in Stochastic Programming are explicited: the first one is the Expected value-semi-variance criterion, linearized in order to be integrated into the ETSAP-TIAM model. The second is the Savage criterion consisting in minimizing the maximum regret. This criterion requires a major modification of the original LP underlying the TIAM model.

Future work will operationalize the Minimax Regret criterion as an integral option of the TIAM model.

Acknowledgements This work is the main contribution of the Energy Technology Systems Analysis Programme (ETSAP) to Work Group 1 of the EMF-22 program of research. ETSAP is the principal sponsor of the development of the TIMES Integrated Assessment Model (ETSAP-TIAM) used to conduct our analysis.

References

[1] Andronova N, Schlesinger ME (2001) Objective estimation of the probability distribution for climate sensitivity. *Journal of Geophysical Research (Atmospheres)* 106(D19):22605–22612

[2] Dantzig GB (1955) Linear Programming Under Uncertainty. *Management Science* 1:197–206

[3] Forest CE, Stone PH, Sokolov AP, Allen MR, Webster MD (2002) Quantifying Uncertainties in Climate System Properties with the Use of Recent Climate Observations. *Science* 205:113–117

[4] Fussel H-M (2006) Empirical and Methodological Flaws in Applications of the DICE Model. Poster, Center for Environmental Science and Policy, Stanford University

[5] GEM-E3. Computable General Equilibrium Model for studying Economy-Energy-Environment Interactions for Europe and the World, available at: `http://www.gem-e3.net/`

[6] Houghton JT, Ding Y, Griggs DJ, Noguer M, van der Linden PJ, Xiaosu D (2001) *Climate Change 2001: The Scientific Basis. Contribution of Working Group I to the Third Assessment Report of the Intergovernmental Panel on Climate Change*, Cambridge University Press (UK), 944

[7] Hyman R, Reilly J, Babiker M, De Masin A, Jacoby H (2003) Modeling Non-CO_2 Greenhouse Gas Abatement. *Environmental Modeling and Assessment* 8(3):175–186

[8] IPCC (2007) *Climate Change 2007: The Physical Science Basis. Contribution of Working Group I to the Fourth Assessment Report of the Intergovernmental Panel on Climate Change*. In: Solomon S, Qin D, Manning M, Chen A, Marquis M, Averyt KB, Tignor M, Miller HL (eds) Cambridge University Press, Cambridge, UK and New York, USA

[9] Kanudia A, Loulou R (1998) Robust Responses to Climate Change via Stochastic MARKAL: The Case of Québec. *European Journal of Operational Research* 106(1):15–30
[10] Loulou R (2007) ETSAP-TIAM: The TIMES Integrated Assessment Model – Part II: Mathematical Formulation. Computational Management Science special issue on Energy and Environment 5(1–2): 7–40.
[11] Loulou R, Kanudia A (1999) Minimax Regret Strategies for Greenhouse Gas Abatement: Methodology and Application. *Operations Research Letters* 25(5):219–230
[12] Loulou R, Labriet M (2007) ETSAP-TIAM: The TIMES Integrated Assessment Model – Part I: Model Structure. *Computational Management Science* special issue on Energy and Environment 5(1–2): 41-66.
[13] Markowitz HM (1952) Portfolio Selection. *Journal of Finance* 7(1):77–91
[14] Nakicenovic N (2000) *Special Report on Emissions Scenarios. A Special Report of Working III of the Intergovernmental Panel on Climate Change*, Cambridge University Press (UK), 599
[15] Nordhaus WD, Boyer J (1999) *Roll the DICE Again: Economic Models of Global Warming*. Manuscript Edition, Yale University
[16] Raiffa H (1968) *Decision Analysis*. Addison-Wesley, Reading, Mass.
[17] Sathaye J, Makundi W, Dale L, Chan P, Andrasko K (2005) GHG Mitigation Potential, Costs and Benefits in *Global Forests: A Dynamic Partial Equilibrium Approach*. LBNL Report 58291, Lawrence Berkeley National Laboratory
[18] Weitzman ML (2008) On Modeling and Interpreting the Economics of Catastrophic Climate Change. Manuscript, REStat FINAL Version July 7, 2008
[19] Wets RJB (1989) Stochastic Programming. In: Nemhauser GL, Rinnooy K, Alexander HG, Todd Michael J. (eds) *Handbooks in OR and MS*, Vol. 1, Elsevier Science Publishers, Amsterdam
[20] Yohe G, Andronova N, Schlesinger M (2004) To Hedge or Not Against an Uncertain Climate Future? *Science* 306:416–417

Chapter 3
Robust Optimization for Environmental and Energy Planning

F. Babonneau, J.-P. Vial, and R. Apparigliato

Abstract Uncertainty is often present in environmental and energy economics. Traditional approaches to optimization under uncertainty, e.g., *stochastic programming*, *chance-constrained programming* or *stochastic dynamic programming*, encounter the most severe numerical difficulties because models in this area are large and complex, already in their deterministic formulation. The goal of the present chapter is to introduce a relatively new field, known as *robust optimization*, as an alternative to traditional methods and formulations. Through an illustrative example, we suggest ways of putting robust optimization at work in environmental and energy optimization models.

3.1 Robust Optimization in short

Uncertainty is often present in environmental and energy economics. As models in this area are often large and complex, introducing uncertainty with traditional approaches, e.g., *stochastic programming* (26; 25; 18; 33), *chance-constrained programming* (23; 22; 34; 36) or *stochastic dynamic programming* (13), generally leads to numerical intractable model as soon as a relevant representation of uncertainty is sought.

The goal of the present contribution is to introduce a relatively new field, known as *robust optimization* (37), which is an alternative to traditional methods and for-

F. Babonneau
ORDECSYS, 4 place de l'Etrier, 1224, Chêne-Bougeries, Suisse, e-mail: fbabonneau@ordecsys.com

J.-P. Vial
ORDECSYS, 4 place de l'Etrier, 1224, Chêne-Bougeries, Suisse, e-mail: jpvial@ordecsys.com

R. Apparigliato
GdF-Suez, Paris, France, e-mail: romain.apparigliato@gmail.com

J.A. Filar and A. Haurie (eds.), *Uncertainty and Environmental Decision Making*, International Series in Operations Research & Management Science 138,
DOI 10.1007/978-1-4419-1129-2_3,

mulations. The main feature of this approach is that it is does not resort to the calculus of probability, which makes it immune against the curse of dimensionality and computational intractability. In the meantime, robust optimization turns out to be a safe approximation of chance constrained programming (23). We shall illustrate our presentation through an example, a simplified version in power supply subject to constraints on admissible concentration of pollutants. This will provide a support to the presentation of the main ideas, to the methodology and to the value of the solutions.

Dealing with uncertainty raises formidable theoretical and practical modeling problems. On the other hand, almost all solution methods stumble on intractability issues. In the last decade, a new approach has emerged, which provides new ways to attack the problem. It is named Robust Optimization, and works on a new paradigm. To present it in few words only, we shall contrast it with traditional methods, such as stochastic programming, chance-constrained programming and/or dynamic programming. Roughly speaking, traditional methods posit the prerequisite of a well-defined probability model of the uncertainties involved in the problem; they next expand the mathematical programming model of the deterministic version of the problem to incorporate the uncertainty. This last operation usually goes with an increase of complexity that, most of the time, puts the computation of the solution out of reach of the current optimization methods. This phenomenon is sometimes described as the *curse of dimensionality* and/or *computational intractability* of the model[1]. In contrast, the primary concern of Robust Optimization is to overcome the complexity issue in adopting a non-probabilistic formulation of the uncertainty. The main underlying idea is to start with a much simplified, if not simplistic, description of uncertainty and look for solutions that remain satisfactory for all realizations of the uncertain parameters that are allowed by the uncertainty model. Solutions having this property are named *robust*. In Robust Optimization, no probability model is assigned to the uncertainty, which makes it possible to avoid expensive computations of multi-dimensional integrals associated with probabilities and expectations. Computing robust solution becomes a numerically tractable operation.

At first sight Robust Optimization can be assimilated to a worst case approach on a selected subset of possible realizations of the uncertainty parameters. This may give mixed feelings to people attached to the probabilistic aspect of the problem of interest. Of course, one may object that Robust Optimization has the definite advantage of avoiding the dramatic computational shortcomings of traditional approaches, in particular in multistage problems. Indeed, representing with a minimum of accuracy probabilities as well as computing expectations with multivariate distributions are formidable handicaps in the framework of optimization problems. But, recent results in Robust Optimization (5; 22; 23; 20) offer a more positive view. In few words, these results consist in a lower bound on the probability that the computed solution remains robust when the whole set of possible realizations—those in the uncertainty set and those outside of it— is considered. These strong results are obtained at the cost of rather mild assumptions on the probability model of uncertainty

[1] In very loose terms, a model is numerically intractable if no method can guarantee that a solution can be obtained in polynomial time.

(independence, a bounded range and an average value at the middle of the range). Probabilities, which are discarded in the initial model, reappear unexpectedly and give stronger confidence in Robust Optimization.

The first robust formulation for an optimization problem with uncertainty parameters has been proposed by Soyster (37) at the beginning of the seventies. The concept was taken over in the nineties by El-Gahoui and Lebret (28) and by Ben-Tal and Nemirovski (9). Since then, robust optimization has been intensively studied in the literature. The main theoretical contributions are (28; 10; 8; 4; 12; 15; 14).

Robust optimization is operational and useful for a large set of decision problems with uncertainty. One can mention decision problems in finance and risk measure (17; 19; 21; 30), in supply chain and inventory management (1; 6; 16), in telecommunications (35) and in management of electricity production in hydraulic valley (2; 3). This approach is also adapted in complement of optimization techniques such as constraints in probability (20), dynamic optimization (32), or stochastic optimization (24).

The chapter is organized as follows. In Section 2, we present our illustrative example and show that its deterministic solution performs poorly in an uncertainty context. Section 3 is concerned with an application of the robust optimization concepts developed in the next sections to the illustrative example with static uncertainty. In Section 4, we introduce those basic concepts of robust optimization in a static framework. In Section 5, we examine the dynamic case and propose the concept of Linear Decision Rules to cope with the adaptive nature of decisions. Section 6 is devoted to applications to the illustrative example, when the demand over time is the source of uncertainty. We compare the robust optimization solution with a stochastic programming approach. In Section 7 one explores a hybrid approach mixing stochastic programming and robust optimization. In Section 8 we show that bounds on the probability that the robust solution satisfies the constraint with uncertain coefficients can be obtained at the cost of a mild assumption on the probability distribution of the uncertain coefficients. This result shows an obvious relationship between chance constrained programming and robust optimization. Finally, in Section 9 we present an extension of the robust optimization concept that covers both the cases when the uncertainty lies in the uncertainty set and when it lies out of it. A short conclusion discusses possible issues in dealing with practical models in environmental and energy planning that are necessarily of much larger size than the illustrative example of this note. It also gives hints on new developments in the field of robust optimization.

3.2 An example in power supply under pollution constraint

To help the reader on the concepts of robust optimization that will be presented in this chapter, we illustrate them on a simple environmental and energy planning problem, inspired by the MARKA-Geneva model (27). The energy planning side concerns the simulation of the evolution of the energy system for a region. The model is based on the assumption of an optimal use of the resources and the simulated planning is obtained by minimizing a welfare function. The energy planning model is enriched to account for a cap on pollutant level in the various subregions. The pollutant level in a given subregion results from local emissions by energy producing sources and from transfers from other subregions. The model is dynamic with three periods.

In this model, we consider two types of uncertainties. One is related to the pollutant transfer rates from one subregion to another; the other one concerns the demand. Those two sources of uncertainties are of different natures. We model the demand as an auto-regressive process: a high demand at a time period is an indication that the demand will also be relatively high in the next time period. The decision process must make use of this information and be adaptive. The pollution effect does not enjoy this property. The transfer rates are not stable, but they do not follow any evolution pattern over time. The process is described by an i.i.d. process. In the rest of the chapter, we shall use the terminology region for subregion.

3.2.1 *Deterministic formulation*

The problem we consider here is the planning of energy production under environmental constraints. The model has four regions, three periods, and three technologies TEM. We denote $\mathscr{R} = \{r_1, r_2, r_3, r_4\}$ the set of regions, $\mathscr{T} = \{t_1, t_2, t_3\}$ the set of time periods and $\mathscr{P} = \{p_1, p_2, p_3\}$ the set of production technologies. The model assumes a single output with production target 10, 12 and 14 for the periods t_1, t_2 and t_3 respectively. The company is committed to serve the demand at all times. Nevertheless, delivery failures may occur if the demand is higher than the production capacity, but the shortage costs, presumably very high, are not known. Thus violations on the demands should be avoided by all means.

Three technologies can be used to produce the output. The admissible production level for each technology depends on the installed capacity. This installed capacity depends on installations performed prior to the planning horizon, thereafter named *residual capacity*, and on investments in the successive periods. These technologies can be installed in four different regions. The residual capacities are given in Table 3.2.

The total system cost is the sum over the three periods and the four regions of the capacity investment, maintenance and operation costs for the three technologies. The costs are reported in Table 3.1.

Table 3.1 Maintenance, investment and operation costs for the three technologies.

Cost	r_1	r_2	r_3	r_4
Maintenance (M_r)	1	1	1	1
Investment (I_r)	5	3	4	5
Operation (O_r)	2	2	3	2

Table 3.2 Residual capacities ($res_{p,t,r}$).

	Technology p_1			Technology p_2			Technology p_3		
	t_1	t_2	t_3	t_1	t_2	t_3	t_1	t_2	t_3
r_1	1	1	-	0.5	-	-	-	-	-
r_2	-	-	-	-	-	-	-	-	-
r_3	-	-	-	-	-	-	-	-	-
r_4	-	-	-	-	-	-	1	1	-

The technologies have different pollutant emission rates, given in Table 3.3, in each region. A pollution transport and dispersion process takes place. A source-receptor matrix specifies, for each source location, the proportion of the emitted pollutant that is deposited in the different receptor locations (see Table 3.4). These data are used to determine the level of concentration (immission) in each region. A standard environmental quality, $Q = 1.5$, is imposed on this concentration for all regions and for all periods.

Table 3.3 Emission rates ($E_{p,r}$) by production technology and by region.

	r_1	r_2	r_3	r_4
p_1	0.7	0.8	0.8	0.6
p_2	0.5	0.4	0.7	0.7
p_3	0.8	0.9	0.6	0.7

Table 3.4 Source-receptor transfer matrix (G_{r_i,r_j}).

	r_1	r_2	r_3	r_4
r_1	0.5	0.1	0.1	0.05
r_2	0.1	0.4	0.04	0.1
r_3	0.09	0.05	0.5	0.1
r_4	0.05	0.1	0.1	0.6

The model gives conditions under which a joint investment and production plan meets the demand needs and the environmental standards at minimal cost.

For the model, we define the following variables:

- $x_{p,t,r}$: production of technology p, at period t and in region r.
- $y_{p,t,r}$: capacity investment of technology p, at period t and in region r.
- $z_{p,t,r}$: installed capacity of technology p, at period t and in region r.
- $e_{t,r}$: emission at period t and in region r.

The optimization problem is given by

$$\min_{x\geq 0, y\geq 0, z, e} \sum_r (O_r \sum_{p,t} x_{p,t,r} + I_r \sum_{p,t} y_{p,t,r} + M_r \sum_{p,t} z_{p,t,r}) \tag{1a}$$

$$z_{p,t,r} = res_{p,t,r} + \sum_{\tau \leq t} y_{p,\tau,r} \quad \forall t \in \mathscr{T}, \forall p \in \mathscr{P}, \forall r \in \mathscr{R} \tag{1b}$$

$$x_{p,t,r} \leq z_{p,t,r} \quad \forall t \in \mathscr{T}, \forall p \in \mathscr{P}, \forall r \in \mathscr{R} \tag{1c}$$

$$\sum_{p,r} x_{p,t,r} = d_t \quad \forall t \in \mathscr{T} \tag{1d}$$

$$\sum_p E_{p,r} x_{p,t,r} = e_{t,r} \quad \forall t \in \mathscr{T}, \forall r \in \mathscr{R} \tag{1e}$$

$$\sum_{\rho \in \mathscr{R}} e_{t,\rho} G_{\rho,r} \leq Q \quad \forall t \in \mathscr{T}, \forall r \in \mathscr{R}. \tag{1f}$$

In that formulation, the equality constraints (1b) express the total installed capacities, $z_{p,t,r}$, as the sum of the residual capacities and the capacities that are installed in the previous periods. Inequality constraints (1c) limit production to the installed capacities. Equations (1d) are demand constraints. The last two sets of constraints firstly compute the pollutant emissions resulting from all technologies (1e), and secondly impose bounds on the final emissions resulting from the dispersion process (1f).

Problem (3.1) has 108 variables and 99 constraints. Note that (3.1) can be formulated in a more compact way by removing the z and e variables and the equality constraints (1b) and (1e) (see Subsection 3.6 for more details). Solving problem (3.1) gives the optimal objective value 162.056. The optimal investments, productions and emissions are reported in Table 3.5, 3.6 and 3.7, respectively.

Table 3.5 Optimal investment schedule.

	t_1			t_2			t_3			
	p_1	p_2	p_3	p_1	p_2	p_3	p_1	p_2	p_3	Total
r_1	-	1.41	-	-	0.77	-	-	-	-	2.18
r_2	-	6.09	-	-	1.73	-	-	0.15	-	7.97
r_3	-	-	-	-	-	-	-	-	3.85	3.85
r_4	-	-	-	-	-	-	-	-	-	0
Sum	-	7.50	-	-	2.50	-	-	0.15	3.85	
Total per period		7.50			2.50			4.00		

We observe in Table 3.5 and Table 3.6 that the optimal solution is to invest in technology p_2 only, except in period t_3 where technology p_3 is also selected. As far as production is concerned, the three technologies are used in the first two periods, but the first production technology is abandoned in the last period. Table 3.7 shows that only two emission constraints are active at the optimum (in region r_2 at periods t_2 and t_3), and one is close to be active (in region r_3 at period t_3). Should the optimal investment/production plan be implemented within an uncertain environment, then,

Table 3.6 Optimal production schedule.

	t_1			t_2			t_3			
	p_1	p_2	p_3	p_1	p_2	p_3	p_1	p_2	p_3	Total
r_1	1.00	1.91	-	1.00	2.18	-	-	2.18	-	8.27
r_2	-	6.09	-	-	7.82	-	-	7.97	-	21.88
r_3	-	-	-	-	-	-	-	-	3.85	3.85
r_4	-	-	1.00	-	-	1.00	-	-	-	2.00
Sum	1.00	8.00	1.00	1.00	10.00	1.00	-	10.15	3.85	
Total per period		10.00			12.00			14.00		

Table 3.7 Emission associated with the optimal solution.

	t_1	t_2	t_3
r_1	1.16	1.24	1.07
r_2	1.21	**1.50**	**1.50**
r_3	0.33	0.37	1.39
r_4	0.75	0.82	0.60

those constraints would potentially be critical and one should expect violations for some variations in the entries of the source-receptor transfer matrix.

3.2.2 Uncertainties in the power supply model

In the power supply model, we consider two sources of uncertainties, one on the pollutant diffusion coefficients and the second one on the demand.

First we define the uncertainty on the coefficients of the source-receptor matrix G. Let ξ be a random variable with values in the interval $[-1,1]$, each coefficient of the matrix G is given by

$$G_{i,j}(\xi) = \bar{G}_{i,j} + \hat{G}_{i,j}\xi_{i,j}, \tag{3.2}$$

where $\bar{G}_{i,j}$ is the average coefficient reported in Table 3.4 and $\hat{G}_{i,j}$ corresponds to the coefficient variability. Here we set $\hat{G}_{i,j} = 0.1\bar{G}_{i,j}$.

We now focus on the demand uncertainty. We adopt the following autoregressive model

$$d_{t+1} = d_t + \alpha_{t+1} + \hat{d}_{t+1}\eta_{t+1},$$

where η_{t+1} is a random variable with values in the interval $[-1,1]$ and α_{t+1} is a deterministic trend factor. This formula can be explicited using backward substitution. Each demand d_t appears then as a function of the past random factor η_τ for $\tau = 1,\ldots,t$. In our case with horizon of length 3, we have

$$d_1 = \bar{d_1} + \hat{d_1}\eta_1 \quad (3a)$$
$$d_2 = \bar{d_2} + \hat{d_1}\eta_1 + \hat{d_2}\eta_2 \quad (3b)$$
$$d_3 = \bar{d_3} + \hat{d_1}\eta_1 + \hat{d_2}\eta_2 + \hat{d_3}\eta_3 \quad (3c)$$

where $\bar{d_i}$ and $\hat{d_i}$ are the average demand and the variability of the demand, respectively. We have $\bar{d_1} = 10$, $\bar{d_2} = 12$, $\bar{d_3} = 14$ and $\hat{d_i} = 0.1\bar{d_i}$, $\forall i$.

3.2.3 Validation process

The validation process is an empirical study of the behavior of the solution under a set of simulated values for the uncertain parameters. In our study, an item of the simulation process is a set of independent realizations of the random factors (demand and/or emission transfer parameters) over the three periods. We shall use the terminology "scenario" to name one such set of realizations. To generate a scenario, one needs a probabilistic model of the random factors. In our illustrative example, we choose to have the underlying factor to be i.i.d. with a uniform distribution on the range $[-1,1]$. The validation is performed on a sample of scenarios of size 1000; each scenario in the sample is a multidimensional vector of realizations of the random factors η and ξ. The demands and the pollutant diffusion coefficients are computed via (3.3) and (3.2), respectively.

In the evaluation of performance, we must differentiate among the two types of uncertainties. Let us start with the emission transfer coefficients. Since there is no recourse associated with this type of uncertainty, the investment/production can be implemented as such, without a risk of violating the demand constraint. Of course, the constraint on the air quality may not be satisfied, but this is just recorded without any modification of the solution itself.

The case of an uncertain demand raises a new issue. One could just record the violations of the demand constraint from above or from below as argued in the previous case, but this would not be realistic. Since the production is adapted to the manifested demand, it is difficult to stick to the view that the production should be maintained at its scheduled value if the demand turns out to be smaller than the production. In the validation process we adopt the following strategy. Let $x_{p,t,r}$ and $y_{p,t,r}$ be the production and the investment in a particular solution. If the total planned production $\sum_{p,r} x_{p,t,r}$ at period t is less than the demand d_t, the production is kept as such, and we record a shortage. The ensuing constraint violation is measured in relative value $(\max\{d_t - \sum_{p,r} x_{p,t,r}, 0\})/d_t$. If the demand is less than the planned production, production in each region with each technology is uniformly downsized by the common factor $d_t / \sum_{p,r} x_{p,t,r}$. The treatment of the constraint on the pollutant level in each region is simpler. It measures the amount of relative violation.

3.2.4 *Evaluation of the deterministic solution in the uncertain environment*

We now subject the deterministic optimal solution computed in the previous section (see Tables 3.5, 3.6 and 3.7) to the simulation process.

Impact of uncertainty on pollutant diffusion

Table 3.8 gives the simulation results with uncertainty on the pollutant diffusion coefficients. We observe that for only 46.5% of the simulations all the air quality constraints are satisfied. For 4.2%, 47.8%, and 1.5% of the cases, one, two and three air quality constraints are violated, respectively. The average and maximum relative violations are about 4.3% and 9.3%, respectively. An analysis of the constraint violations has revealed that the phenomenon occurred in the three pairs (region and periods) that were critical or near critical in the deterministic study. In the other pairs there is enough slack to absorb variations in the transfer rates.

Table 3.8 Simulation results with uncertain source-receptor matrix.

Average relative violation (%)	4.3
Maximum relative violation (%)	9.3
% of satisfaction	46.5
% of one violation	4.2
% of two violations	47.8
% of three violations	1.5

Impact of uncertainty on the demand

In Table 3.9, we report the simulation results conducted on the optimal deterministic solution with a demand uncertainty.

Table 3.9 Simulation results with uncertain demands.

Predicted cost performance	162.056
Observed cost performance	160.871
Scenarios with demand violation(s) in %	62.0
Conditional average relative violation in %	2.5
Average number of violations per scenario	2.0

Contrary to the previous experiment, the uncertainty on the demand has an impact on the cost performance. As a result we distinguish between the predicted performance (the optimal cost in the deterministic model) and the observed performance (the average cost in the simulation). Surprisingly enough, the observed performance is better than the predicted one. The paradox is apparent and the explanation is straightforward. When the demand is lower than the average value used in the deterministic model, the production levels are scaled down to match the actual demand and the production costs less than predicted by the deterministic plan. But when the demand is higher than this average, the production cannot be increased, because the capacity constraints in the deterministic optimal solutions are tight; the production costs remain as computed in the deterministic model, but part of the demand is not satisfied. Thus, the seemingly improved cost performance must be balanced with demand violations. In that respect, we observe a demand violation in 62% of the scenarios and a conditional average of unsatisfied demand of 2.5%. Finally, when a violation occurs in a scenario the average number of periods with violation is approximately 2.0 (over the total of 3). The conditional average demand violation is small with respect to the average demand, but one should recall that the range of variation of the demand around the average is $\pm 5\%$, $\pm 10\%$ and $\pm 15\%$, in periods $1, 2$ and 3, respectively.

3.3 Case study: robust solution to a problem with uncertain pollutant transfer coefficients

To motivate robust optimization, we propose a simple fix to handle uncertainties in the power supply problem. We limit this illustration to uncertainties on the pollutant diffusion coefficients, and consider that the demands are fixed. We shall use heuristic arguments only and postpone theoretical justifications to later sections.

Let us start with a general formulation of a linear constraint with uncertain coefficients

$$\sum_{j=1}^{n} \tilde{a}_j x_j \leq b, \tag{3.4}$$

where $\tilde{a}_j$ are uncertain. For the sake of a simpler presentation, we assume b to be certain. We further describe the uncertain coefficients as linear functions of an underlying random factor ξ

$$\tilde{a} = \bar{a} + P\xi,$$

where $\xi \in \mathbb{R}^m$ and P is an $n \times m$ matrix. We further assume that the random factor has a symmetric distribution with mean 0. The certain vector $\bar{a}$ is usually named the normal factor. We can thus focus on the uncertain component of the constraint

$$\underbrace{\bar{a}^T x}_{\text{certain}} + \underbrace{(P^T x)^T \xi}_{\text{uncertain}} \leq b.$$

We now evoke a common sense engineering approach that consists in replacing the uncertain term by a safety term $\kappa > 0$. By taking a large enough safety term, we give a sufficient guarantee that the solutions of $\bar{a}^T x + \kappa \leq b$ will almost always remain feasible to (3.4). This says nothing on the critical way to choose the safety factor, but many practitioners will be receptive to a so-called 2σ, 3σ, possibly 6σ approach. This industrial practice is often justified through the following probabilistic and statistical argument. Assuming that the underlying random factor ξ is a vector with probabilistically independent components with mean 0 and standard deviation σ, we can state that the uncertain component $(P^T x)^T \xi$ is a random variable with mean 0 and standard deviation $||P^T x||_2$. In the $k \times \sigma$ approach, the safety term is $\kappa = k||P^T x||_2$, with the implicit argument that for most distributions a large enough k, say $k = 3$, will guarantee that a solution to $\bar{a}^T x + k||P^T x||_2 \leq b$ will satisfy the probabilistic constraint $\sum_{j=1}^{n} \tilde{a}_j x_j \leq b$ with high probability.

The two salient features of this "engineer-like" approach can be summarized as follows. An uncertain constraint is replaced by a deterministic constraint with a safety term, and the safety term depends on the decision variables x. Applying this idea to our problem of interest, we obtain that the robust equivalent of the air quality constraints (1f) is

$$\sum_{\rho \in \mathscr{R}} e_{t,\rho} \bar{G}_{r,\rho} + k|| \sum_{\rho \in \mathscr{R}} e_{t,\rho} \hat{G}_{r,\rho} ||_2 \leq Q \quad \forall t \in \mathscr{T}, r \in \mathscr{R}. \tag{3.5}$$

This formulation does not increase the problem size, but the linear air quality constraints are replaced by nonlinear ones. Those constraints are convex conic quadratic, and problems with such constraints can be solved very efficiently by modern solvers (e.g., the open source code (38)).

If we wish to remain in the realm of linear programming, we have to replace the safety factor by terms that are amenable to linear inequalities. The following bound will be established in the next section

$$||\alpha||_2 \leq \min_{\beta} \{\sqrt{m}\, ||\alpha - \beta||_\infty + ||\beta||_1\},$$

where $||\alpha||_1 = \sum_{i=1}^{m} |\alpha_i|$ and $||\alpha||_\infty = \max_i |\alpha_i|$. Hence, the alternative, linear but more restrictive formulation,

$$\sum_{\rho \in \mathscr{R}} e_{p,\rho} \bar{G}_{r,\rho} + k\sqrt{m}|| \sum_{\rho \in \mathscr{R}} e_{p,\rho} \hat{G}_{r,\rho} - \beta ||_\infty + k||\beta||_1 \leq Q \quad \forall p \in \mathscr{P}, r \in \mathscr{R}. \tag{3.6}$$

In this experiment, we solve the robust equivalent problem with the two formulations, the first one (3.5) with the conic quadratic constraint, and the other one (3.6) with linear constraints. We use the same k in the two formulations, but make this k vary to experiment with different degrees of safety. We shall compare the behavior of the robust solutions with the deterministic solution. The results are reported in Table 3.10.

In the second set of experiments we use the formulation with linear constraints (3.6). We shall prove in the next section that (3.6) is equivalent to a system of linear inequalities. It turns out that the robust formulation of a typical air quality constraint (1f) corresponding to the pair $(t \in \mathscr{T}, r \in \mathscr{R})$ is

Table 3.10 Variable transfer coefficients: behavior on the sample of 1000 scenarios of a robust solution with the conic quadratic formulation (3.5).

	Deterministic	Robust		
		$k=0.8$	$k=1$	$k=1.2$
Cost performance	162.06	163.76	164.21	164.64
Conditional average of relative violation (%)	4.3	1.0	0.5	-
Maxi relative violation (%)	9.3	3.0	1.5	-
proportion of scenarios with violation (%)	53.5	25.4	6	0
proportion of scenarios with one violation (%)	4.2	17.2	5	-
proportion of scenarios with two violations (%)	47.8	6.4	1	-
proportion of scenarios with three violations (%)	1.5	1.8	-	-

$$\sum_{i\in\mathscr{R}} e_{t,i}\bar{G}_{r,i} + k\Big(\sum_{i\in\mathscr{R}} u_{t,r,i} + 2v_{t,r}\Big) \leq Q \tag{7a}$$

$$u_{t,r,\rho'} + v_{t,r} \geq \sum_{\rho\in\mathscr{R}} e_{t,\rho}\hat{G}_{\rho',\rho} \quad \forall \rho' \in \mathscr{R} \tag{7b}$$

$$u \geq 0, v \geq 0 \tag{7c}$$

The above robust counterpart is obtained from the set of inequalities (3.10) in Proposition 1 (Section 3.4). The reader will notice that (3.10) also includes constraints like (7b), but with right-hand side $-\sum_{\rho\in\mathscr{R}} e_{t,\rho}\hat{G}_{\rho',\rho}$. In this particular problem, we know that $\sum_{\rho\in\mathscr{R}} e_{t,\rho}\hat{G}_{\rho',\rho} \geq 0$, which makes one half of the constraints redundant. So, we eliminate them and obtain a robust equivalent model that is still linear but with 60 additional variables u's and v's and 48 additional constraints.

We solve the robust equivalent problem with different values of k and we report the results of the simulations for each robust optimal solution in Table 3.11.

Table 3.11 Variable transfer coefficients: behavior on the sample of 1000 scenarios of a robust solution with respect to the linear formulation (3.7).

	Deterministic	Robust		
		$k=0.6$	$k=0.8$	$k=1$
Cost performance	162.06	163.49	163.99	164.49
Conditional average of relative violation (%)	4.3	1.2	0.7	-
Max relative violation (%)	9.3	3.5	1.7	-
proportion of scenarios with violation (%)	53.5	29.1	12.4	0
proportion of scenarios with one violation (%)	4.2	14.9	9.4	-
proportion of scenarios with two violations (%)	47.8	12.7	2.9	-
proportion of scenarios with three violations	1.5	1.5	0.1	-

The results of Table 3.10 and Table 3.11 are very much alike. Nevertheless, we notice that with the same safety level k, say $k=0.8$ for a typical example, the robust solution with respect to the linear approximation (3.6) achieves a slightly worst cost objective (163.99 vs. 163.76) than the solution with respect to the conic quadratic formulation (3.5). In the meantime, the first solution achieves a better protection

Table 3.12 Robust investment schedule for $k = 0.8$ and the linear formulation (3.7).

	t_1			t_2			t_3			
	p_1	p_2	p_3	p_1	p_2	p_3	p_1	p_2	p_3	Total
r_1	-	1.39	-	-	1.51	-	-	0.10	-	3.00
r_2	-	5.94	-	-	0.91	-	-	0.16	-	7.01
r_3	-	-	-	-	-	-	-	-	3.74	3.74
r_4	0.17	-	-	0.08	-	-	-	-	-	0.25
Sum	0.17	7.33	-	0.08	2.42	-	-	0.26	3.74	
Total per period		7.50			2.50			4.00		

against constraint violation. These observations conform to the fact that for identical k the linear safety factor in (3.6) in Table 3.11 is an upper bound of the conic quadratic safety factor (3.5) in Table 3.10. The first one is more constrained is thus more constrained. The cost performance in not so good, but it ensures a lesser protection against violations. The chances are that in a simulation we observe a lesser number of constraint violations.

3.4 Robust Optimization for the static problem: theoretical developments

Let us recall the formulation of a linear constraint with uncertain coefficients

$$\sum_{j=1}^{n} \tilde{a}_j x_j \leq b.$$

The first basic assumption on the uncertain parameters is that they depend on some random factor ξ in a linear way.

Assumption 1 *The uncertain vector $\tilde{a}$ is written as*

$$\tilde{a} = \bar{a} + P\xi,$$

where $\xi \in \mathbb{R}^m$ and P is an $n \times m$ matrix.

The certain vector $\bar{a}$ is usually named the normal factor. We can thus focus on the uncertain component of the constraint

$$\underbrace{\bar{a}^T x}_{\text{certain}} + \underbrace{(P^T x)^T \xi}_{\text{uncertain}} \leq b. \tag{3.8}$$

Our present goal is to use information on ξ to build the safety factor introduced in the previous section (see inequalities (3.5) and (3.6)). The idea is to focus on a subset of all possible events that it is made of all realizations of the underlying uncertain factor ξ that the modeler deems necessary to protect against. This is the so-called *uncertainty set*. The robust version of the initial uncertain constraint $\tilde{a}^T x \leq b$ consists in enforcing the uncertain constraint (3.8), not for *all* possible realizations, but only by those in the uncertainty set; that is, the less restrictive constraint

$$\bar{a}^T x + (P^T x)^T \xi \leq b, \text{ for all } \xi \in \Xi,$$

where $\Xi \subset \mathbb{R}^m$ is the uncertainty set. A solution to this constraint is called *robust* with respect to Ξ. If Ξ is a continuous set, the robust constraint is a short-hand writing of an infinite number of simple linear constraints. This seems to put the whole approach into the realm of semi-infinite programming, making the computation of solutions a real issue. It turns out that this is not so, for a large variety of uncertainty sets.

We shall consider a few different types of uncertainty sets. Let us start with the ellipsoidal uncertainty set

$$\Xi = \{\xi \mid ||\xi||_2 \leq k\}.$$

In the rest of the chapter, we shall describe the ellipsoidal uncertainty set as the ball in the 2-norm, centered at the origin and with radius k. We denote it $B_2(0,k)$. We shall also use more general balls, such as

$$B_p(0,k) = \{\xi \mid ||\xi||_p \leq k\}$$

with $1 < p < \infty$ and $||\xi||_p = (\sum_{i=1}^m |\xi_i|^p)^{1/p}$. The following lemma gives the clue to the replacement of the robust constraint by a single finite-dimensional constraint which will be named the *robust equivalent*.

Lemma 1. *Let $1 < p < \infty$ and let q be such that $\frac{1}{p} + \frac{1}{q} = 1$. For any d*

$$\max\{d^T \xi \mid ||\xi||_p \leq k\} = k||d||_q.$$

Proof. The proof in the general case is simple but tedious: we omit it. We just illustrate the case of the 2-norm. (See Figure 3.1.) □

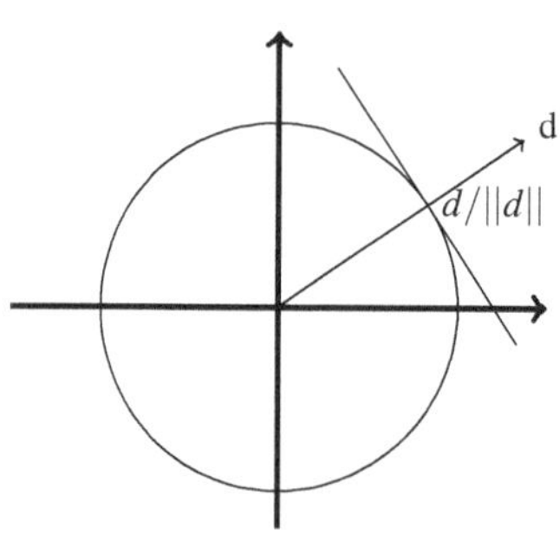

The maximum is achieved at $\xi^* = d/||d||$ and takes the value

$$d^T \xi^* = d^T \frac{d}{||d||} = ||d||.$$

Fig. 3.1 Maximum of a linear form over a ball in the 2-norm.

3.4.1 Robust equivalent: the case of the 2-norm

We can now state the robust equivalent of the robust constraint.

Theorem 1. *The robust equivalent of the constraint*

$$\bar{a}^T x + (P^T x)^T \xi \leq b, \text{ for all } \xi \in \Xi = \{\xi \mid ||\xi||_2 \leq k\},$$

is

$$\bar{a}^T x + k||P^T x||_2 \leq b.$$

Proof. The proof follows directly from the above lemma. It suffices to replace d by $P^T x$. □

The factor k plays a crucial role in Theorem 1. The larger its value, the greater the number of realizations ξ against which a solution of $\bar{a}^T x + k||P^T x||_2 \leq b$ is *immunized* in the constraint $\bar{a}^T x + (P^T x)^T \xi \leq b$ with uncertain coefficients . In the sequel we shall use the terminology *immunization factor k*, or *immunization level k*.

At this stage we should raise some fundamental issues:

1. Is the nonlinear formulation of the robust equivalent a potential source of complexity from a numerical point of view?
2. Should, or could, one consider alternative uncertainty sets?
3. What is the value of starting with the new concept of uncertainty set if one ends up with the same formulation of an engineering safety factor?

We shall answer the first two questions, leaving the answer of the last question to the final section of the chapter. Let us start with the first question. It is easy to show that the new constraint is convex (a property of the 2-norm). Moreover, it can be reformulated as

$$\bar{a}^T x + kz \leq b \tag{9a}$$

$$P^T x = u \tag{9b}$$

$$||u||_2 \leq z. \tag{9c}$$

The last constraint is conic quadratic, a feature that modern convex optimization codes handle about as efficiently as a linear constraint.

Taking the uncertainty set as the primary concept shifts the focus on numerically tractable robust equivalent. However, people concerned with a probabilistic approach may feel ill at ease with the apparent arbitrariness of the uncertainty set. However, there is a powerful theorem in probability theory that assets that under a very mild probabilistic assumption on the random factor ξ, one can provide a surprisingly strong lower bound on the probability that the initial uncertain constraint be satisfied by a robust solution (i.e., a solution to the robust equivalent).

The answer to the second question motivates the next section.

3.4.2 Robust equivalent: the case of the ℓ_1 and ℓ_∞ norms

For some reasons, e.g., a lack of access to a conic quadratic solver, one may want to remain in the realm of linear programming. This can be achieved by resorting to polyhedral uncertainty sets. To link the results with the previous study with ellipsoidal uncertainty sets, we resort to an approximation of the unit ball in the 2-norm by the intersection of balls in the 1-norm and the infinity norm.

The ℓ_1 and ℓ_∞ norms are natural extensions of the p-norm with $p=1$ and $p=\infty$. It generates the balls

$$B_1(0,k) = \{\xi \mid \sum_{i=1}^{m} |\xi_i| \leq k\}$$

and

$$B_\infty(0,k) = \{\xi \mid \max_{i=1,\ldots,m} |\xi_i| \leq k\}.$$

Both are polyhedral sets that can be represented by simple inequalities. The next lemma extends Lemma 1 to the two limit cases $p=\infty$ and $p=1$.

Lemma 2. *We have*

$$\max\{d^T\xi \mid ||\xi||_\infty \leq k\} = k||d||_1$$

and

$$\max\{d^T\xi \mid ||\xi||_1 \leq k\} = k||d||_\infty.$$

The $B_1(0,k\sqrt{m})$ and $B_\infty(0,k)$ balls can be jointly used to approximate the $B_2(0,k)$ ball in the m-dimensional space as it is illustrated in Figure 3.2. We can now give

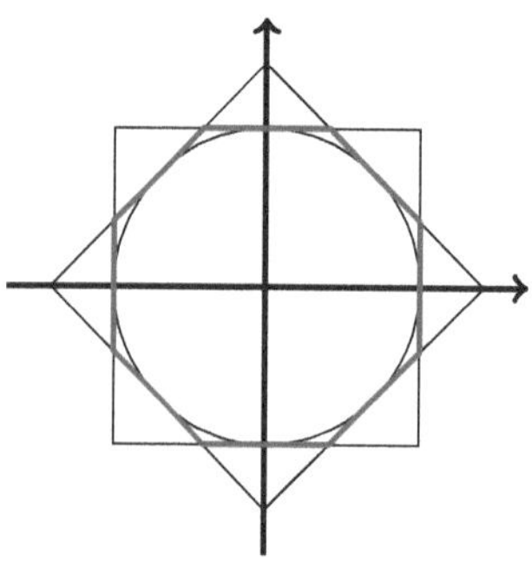

Fig. 3.2 The set $B_1(0,\sqrt{2}) \cap B_\infty(0,1)$

the robust equivalent relative to the uncertainty set $\Xi = B_1(0,k_1) \cap B_\infty(0,k_\infty)$.

Lemma 3. *We have*

$$\max_{\xi}\{d^T\xi \mid \xi \in B_1(0,k_1) \cap B_\infty(0,k_\infty)\} = \min_{w}\{k_1||d-w||_\infty + k_\infty||w||_1\}.$$

Proof. Let

$$z^* = \max_{\xi}\{d^T\xi \mid \xi \in B_1(0,k_1) \cap B_\infty(0,k_\infty)\}.$$

Let us replace $B_\infty(0,k_\infty)$ by linear inequalities in the above maximization problem. We obtain the alternative expression

$$\begin{aligned} \max_{\xi} \;\; & d^T\xi \\ & -k_\infty e \leq \xi \leq k_\infty e \\ & \xi \in B_1(0,k_1), \end{aligned}$$

where e is the vector of all ones of appropriate dimension. We form the partial Lagrangian

$$L(\xi,u,v) = d^T\xi + u^T(\xi + k_\infty e) + v^T(-\xi + k_\infty e) = k_\infty e^T(u+v) + (d+u-v)^T\xi.$$

Clearly

$$z^* = \max_{\xi \in B_1(0,k_1)} \; \min_{u \geq 0, v \geq 0} L(\xi,u,v)$$

and by linear programming duality

$$z^* = \min_{u \geq 0, v \geq 0} \; \max_{\xi \in B_1(0,k_1)} L(\xi,u,v).$$

The inner minimization problem is

$$\mathscr{L}(u,v) = \max_{\xi}\{k_\infty e^T(u+v) + (d+u-v)^T\xi \mid \xi \in B_1(0,k_1)\}$$

By Lemma 2 we have

$$\mathscr{L}(u,v) = k_\infty e^T(u+v) + k_1||d+u-v||_\infty.$$

One easily checks that an optimal solution of the minimization $\min_{u\geq 0, v\geq 0}\mathscr{L}(u,v)$ satisfies the complementary condition $u_i v_i = 0$ for $i = 1,\ldots,m$. Hence u and v can be viewed as the positive and negative parts of a real vector $w = u - v$. Therefore $e^T(u+v) = ||w||_1$ and one can write

$$z^* = \min_{w}\{k_1||d-w||_\infty + k_\infty||w||_1\}.$$

□

We can now state the robust equivalent for the new polyhedral uncertainty set.

Theorem 2. *The robust equivalent of the robust constraint*

$$\bar{a}^T x + (P^T x)^T\xi \leq b, \textit{ for all } \xi \in \Xi = \{\xi \mid B_1(0,k_1) \cap B_\infty(0,k_\infty)\},$$

is the constraint in x **and** *w*

$$\bar{a}^T x + k_1 ||P^T x - w||_\infty + k_\infty ||w||_1 \leq b.$$

Proof. Using Lemma 3, we may write the robust equivalent as $\bar{a}^T x + z^* \leq b$. Since for any w one has $z^* \leq k_1 ||P^T x - w||_\infty + k_\infty ||w||_1$, we have that the new inequality implies the robust inequality. By the strong linear duality theorem, the equivalence is achieved because there always exists a w^* such that $z^* = k_1 ||P^T x - w^*||_\infty + k_\infty ||w^*||_1$. □

Remark 1. The ball $B_1(0,k_1)$ in Theorem 2 can be replaced by a ball $B_p(0,k_p)$ in the norm ℓ_p, for any $1 < p < \infty$. One obtains the following robust equivalent

$$\bar{a}^T x + k_p ||P^T x - w||_q + k_\infty ||w||_1 \leq b$$

with q such that $1/p + 1/q = 1$.

For implementation purposes, it is convenient to replace the norm expressions into linear inequalities.

Proposition 1. *The robust counterpart*

$$\bar{a}^T x + k_1 ||P^T x - w||_\infty + k_\infty ||w||_1 \leq b$$

has the same set of solutions as the system of linear inequalities

$$\bar{a}^T x + k_1 t + k_\infty e^T w \leq b \tag{10a}$$

$$w + te \geq P^T x \tag{10b}$$

$$w + te \geq -P^T x \tag{10c}$$

$$w \geq 0,\ t \geq 0.$$

Proof. Let t be a scalar variable and write the minimization problem yielding $z^* = \min_w k_1 ||P^T x - w||_\infty + k_\infty ||w||_1$ as the linear program

$$\begin{aligned} \min\ & k_1 t + k_\infty e^T (u + v) \leq b \\ & -te \leq P^T x + u - v \leq te \\ & u \geq 0,\ v \geq 0,\ t \geq 0. \end{aligned}$$

This program can be simplified as follows. Consider first the case $(P^T x)_i > 0$. The minimization operation entails $u_i = 0$, and $v_i = \max\{(P^T x)_i - t, 0\}$. A similar reasoning yields $v_i = 0$ and $u_i = \max\{(-P^T x)_i - t, 0\}$, when $(P^T x)_i \leq 0$. Define $w \geq 0$ by $w_i = v_i$ if $(P^T x)_i > 0$ and $w_i = u_i$ if $(P^T x)_i \leq 0$. By construction, w is such that $w + te \geq P^T x$ and $w + te \geq -P^T x$.

Conversely, a solution (w,t) to (3.10) generates a solution (u,v,t) of the initial system. Indeed, set $u_i = 0$ and $v_i = w_i$ if $(P^T x)_i > 0$ and $u_i = w_i$ and $v_i = 0$ if $(P^T x)_i \leq 0$. We have thus that the minimization problem is equivalent to

$$\begin{aligned} \min\ & k_1 t + k_\infty e^T w \\ & w + te \geq P^T x \\ & w + te \geq -P^T x \\ & w \geq 0,\ t \geq 0. \end{aligned}$$

This concludes the proof. □

3.4.3 Robust equivalent: bounded polyhedral uncertainty set

The balls in ℓ_1 and ℓ_∞ are special cases of polyhedra. We extend the construction of the robust counterpart to the case of uncertainty sets defined by a bounded nonempty polyhedral set defined by $\{\xi \mid Q\xi \leq t\}$. In that case the robustness test for the constraint $(\bar{a}+P\xi)^T x \leq b$, $\forall \xi \in \Xi$ is given by

$$\max_{\xi}\{(P^T x)^T \xi \mid Q\xi \leq t\} \leq -\bar{a}^T x + b.$$

Proposition 2. *The condition*

$$\max_{\xi}\{(P^T x)^T \xi \mid Q\xi \leq t\} \leq -\bar{a}^T x + b$$

is satisfied if and only if there exists a solution (x,u) for the linear system

$$\begin{aligned} \bar{a}^T x + t^T u &\leq b \\ P^T x - Q^T u &= 0 \\ u &\geq 0. \end{aligned}$$

Proof. In the proof, x is a fixed parameter. A direct derivation is obtained using the theory of duality in linear programming. To define the worst case in the robust constraint, we consider the linear programming problem

$$\max_{\xi}\{(P^T x)^T \xi \mid Q\xi \leq t\}.$$

This problem is feasible and bounded. It has an optimal solution and its dual

$$\min_{u}\{t^T u \mid Q^T u - P^T x = 0,\ u \geq 0\}$$

also has an optimal value, which is equal to the optimal primal value. For all feasible primal-dual pair (ξ,u) we have the implication

$$\left.\begin{array}{r} Q\xi \leq t \\ Q^T u - P^T x = 0,\ u \geq 0 \end{array}\right\} \Rightarrow (P^T x)^T \xi \leq t^T u,$$

with equality if and only if the pair (ξ,u) is optimal. If we substitute the bounding value $t^T u$ to the maximization in the left-hand side of

$$\bar{a}^T x + \max_{\xi}\{(P^T x)^T \xi \mid Q\xi \leq t\} \leq b,$$

we obtain a stronger inequality, which coincides with the initial one when u is dual optimal. Consequently, the system

$$\begin{aligned} \bar{a}^T x + t^T u &\leq b \\ P^T x - Q^T u &= 0 \\ u &\geq 0, \end{aligned}$$

which contains the feasibility constraints of the dual is equivalent to the initial robust constraint. □

3.4.4 Uncertainties in the objective function

Let us consider the following linear problem with uncertainty data

$$\begin{aligned} \min_x \ & \sum_{j=1}^{n} c_j(\xi) x_j \\ & Ax \leq b. \end{aligned}$$

There is a big difference difference between the objective and the constraints. In the latter, the values achieved by the left-hand side are irrelevant, insofar as they are less than the right-hand side. In particular, the average value is not relevant. In contrast, the average value of the objective function value is by all means an important factor to be considered. Let us assume that the uncertain factor has a symmetric distribution around 0 and that $c(\xi)$ is linear. The average value of the objective is thus $\sum_{j=1}^{n} c_j(0) x_j$ and one can simply minimize it. But in so doing, one completely ignore possible bad outcomes with high values for the objective function.

A possible alternative is to focus on the worst case and consider the problem

$$\min_{x,z} \ z \tag{11a}$$

$$\sum_{j=1}^{n} c_j(\xi) x_j - z \leq 0, \ \forall \xi \in \Xi \tag{11b}$$

$$Ax \leq b. \tag{11c}$$

The z value in (3.11) is an upper bound for the objective function for all possible realizations in the uncertainty set. Minimizing this worst case performance is an acceptable decision criterion, but one may argue that shifting from one problem to the other amounts to pass from one extreme (no concern for the risk) to another (no concern for the average performance).

At minimal computational cost, one may consider a bi-criteria problem involving the two objectives. The Pareto frontier is obtained by solving the parametric problem

$$\min_x \sum_{j=1}^{n} c_j(0)x_j^*$$
$$\sum_{j=1}^{n} c_j(\xi)x_j \le z^* + \gamma, \ \forall \xi \in \Xi$$
$$Ax \le b.$$

In that formulation $\gamma > 0$ determines an arbitrage between the average performance and the guaranteed worst case performance.

3.4.5 Two-sided inequality constraints

In the robust optimization paradigm, the constraints are immunized separately with respect to their specific uncertainty sets. In the case of the two-sided constraint

$$\underline{b} \le \sum_{j=1}^{n} \tilde{a}_j x_j \le \bar{b}$$

the robust formulation involves the two constraints

$$\underline{b} \le \bar{a}^T x + (P^T x)^T \xi, \ \forall \xi \in \Xi$$

and

$$\bar{a}^T x + (P^T x)^T \xi \le \bar{b}, \ \forall \xi \in \Xi.$$

Assume the polyhedral uncertainty set $\Xi = \{\xi \mid B_1(0,k_1) \cap B_\infty(0,k_\infty)\}$. In view of Theorem 2, the two robust constraints generate their own robust counterpart

$$\underline{b} \le \bar{a}^T x - k_1||P^T x - v||_\infty - k_\infty||v||_1 \tag{3.12}$$

$$\bar{a}^T x + k_1||P^T x - w||_\infty + k_\infty||w||_1 \le \bar{b}. \tag{3.13}$$

In view of Proposition 1, those two inequalities boil down into independent sets of inequalities. The point of this section is that the two sets of extra inequalities don't need to be independent. Actually the same set is valid for both, thus saving significantly on the problem size.

To see this, it suffices to notice that one need not to choose different vectors v and w in the robust counterparts (3.12) and (3.13). Indeed, for a fixed x, the constraints are equivalent to

$$k_1||P^T x - v||_\infty + k_\infty||v||_1 \le a^T x - \underline{b}$$
$$k_1||P^T x - w||_\infty + k_\infty||w||_1 \le \bar{b} - a^T x.$$

To check satisfiability of these constraints, it suffices to replace v and w by the value that minimizes the left-hand sides. This can be conveniently formalized into a theorem similar to Proposition 1.

Proposition 3. *The robust counterpart of the robust two-sided constraint*

$$\underline{b} \leq \bar{a}^T x + (P^T x)^T \xi \leq \bar{b}, \text{ for all } \xi \in \Xi = \{\xi \mid B_1(0,k_1) \cap B_\infty(0,k_\infty)\},$$

is

$$k_1 ||P^T x - w||_\infty + k_\infty ||w||_1 \leq \min\{a^T x - \underline{b}, \bar{b} - a^T x\}.$$

It can be represented by the system of linear inequalities

$$k_1 t + k_\infty e^T w \leq a^T x - \underline{b} \tag{14a}$$

$$k_1 t + k_\infty e^T w \leq \bar{b} - a^T x \tag{14b}$$

$$w + te \geq P^T x \tag{14c}$$

$$w + te \geq -P^T x \tag{14d}$$

$$w \geq 0, \quad t \geq 0.$$

The practical importance of this theorem is that the robust counterpart of a two-sided inequality is a set of inequalities with only one more inequality than in the case of a simple one-sided inequality.

A similar treatment applies to a two-sided inequality with ellipsoidal uncertainty. We then would get the robust counterpart

$$k||P^T x||_2 \leq \min\{a^T x - \underline{b}, \bar{b} - a^T x\}.$$

3.4.6 Equality constraints

The meaning of an equality constraint with uncertain coefficients is questionable. Using our standard representation of the uncertain parameters, we can formulate the robust equality constraint $\tilde{a}(\xi)^T x = b$, with $\tilde{a}(\xi) = \bar{a} + P\xi$ and $\xi \in \Xi$. The uncertainty sets that have been considered so far have a non-empty interior and $0 \in \Xi$. Consequently $\tilde{a}^T x = \bar{a}^T x + (P^T x)^T \xi = b$ for all $\xi \in \Xi$ if and only if $(P^T x)^T \xi \equiv 0$ for all $\xi \in \Xi$. This implies $P^T x = 0$. In other words, the vector x should lie in the null space of P^T. This is conveniently summarized in a proposition.

Proposition 4. *Suppose the uncertainty set Ξ has a non-empty interior. The robust equality constraint*

$$\bar{a}^T x + (P\xi)^T x = b, \ \forall \xi \in \Xi$$

is equivalent to the system of equations

$$\bar{a}^T x = b \ \text{ and } \ P^T x = 0.$$

The condition on x looks very restrictive. A possibility is that the presence of uncertain coefficients in an equality constraint signals an error in passing from a deterministic version of the problem to an uncertain one. For instance, some inequality constraints in the deterministic version are known to be necessarily tight at the optimum. This implicit knowledge makes it possible to write them as equalities in the deterministic formulation, but the true inequality formulation should be used in the uncertainty case.

A more difficult situation occurs in multistage problems in which some decisions are recourses that can be adjusted once the uncertainty is revealed. We shall see an illustration of this problem in our energy environmental planning example in a later section dealing with robust optimization in the dynamic case.

3.5 Dynamic problem with recourse

Multistage problems under uncertainty introduce a quantum of difficulties in that the successive decisions can be made on the basis of the information revealed so far. Those decisions will be named *recourses* thereafter, to emphasize the relation with the revealed information. Recourses are functions defined on richer and richer spaces. To cope with this extreme difficulty, it is reasonable to restrict the recourses to functions in a limited class, a class simple enough to allow the formulation of robust constraints, but rich enough to capture a meaningful part of the recourse possibilities. The proposed class is the set of linear functions of the revealed values of the uncertain factor ξ. Recourses in this class will be named Linear Decision Rules, LDR in short. This concept has been introduced for control problem (31) and for stochastic programming (29; 23). It is used in Robust Optimization under the name of Affinely Adjustable Robust Counterpart (AARC) (8).

3.5.1 Linear decision rules

Consider a typical constraint in a two-stage problem:

$$a_1(\xi_1)^T x_1 + a_2(\xi_1,\xi_2)^T x_2 \leq b(\xi_1,\xi_2).$$

We assume that x_1 and x_2 are variables with dimension, respectively, n and m. The coefficients a_1 and a_2 have also, respectively, dimension n and m. The dependence of coefficients a_1, a_2 and b with the uncertainty underlying (ξ_1,ξ_2) is given by the functions

$$a_1(\xi_1) = \bar{a}_1 + P_1\xi_1 \tag{15a}$$
$$a_2(\xi_1,\xi_2) = \bar{a}_2 + P_{21}\xi_1 + P_{22}\xi_2 \tag{15b}$$
$$b(\xi_1,\xi_2) = \bar{b} + b_1^T\xi_1 + b_2^T\xi_2. \tag{15c}$$

In the last equality, b_1 and b_2 have same dimension than ξ_1 and ξ_2.

To capture the adaptive property of the recourse in stage 2, the variable x_2 is replaced by a LDR, that is by a linear (more accurately, an affine) function of ξ_1, that is:

$$x_2(\xi_1) = \bar{x}_2 + D\xi_1.$$

In the above expression, the components of $\bar{x}_2$ and D are the new decision variables in the formulation of the problem with uncertain parameters. Note that those variables are to be determined prior to knowing the value of ξ_1, but the actual value of the recourse $x_2(\xi_1)$ will be determined after the value of ξ_1 has become known.

The robust equivalent of the constraint of the two-stage problem is then

$$(\bar{a}_1 + P_1\xi_1)^T x_1 + (\bar{a}_2 + P_{21}\xi_1 + P_{22}\xi_2)^T (\bar{x}_2 + D\xi_1) \\ -(\bar{b} + b_1\xi_1 + b_2\xi_2) \leq 0, \ \forall(\xi_1, \xi_2) \in \Xi.$$

To build the robust counterpart, one must solve the optimization problem

$$\max_{(\xi_1,\xi_2)\in\Xi} (P_1^T x_1 + P_{21}^T \bar{x}_2 + D^T \bar{a}_2 - b_1)\xi_1 + (P_{22}^T \bar{x}_2 - b_2)^T \xi_2 \\ +(P_{21}\xi_1 + P_{22}\xi_2)^T D\xi_1 \tag{3.16}$$

in which the variables x_1, $\bar{x}_2$ and D are parameters. We notice that this problem is quadratic with second order term

$$(P_{21}\xi_1 + P_{22}\xi_2)^T D\xi_1.$$

Remark 2. It is worth noticing that a LDR introduces uncertainty of its own. A constraint with deterministic coefficients, e.g., with $P_1 = P_{21} = 0$, $P_{22} = 0$, $b_1 = 0$ and $b_2 = 0$, becomes uncertain because of the term $(D^T \bar{a}_2 - b_1)\xi_1$.

3.5.2 Problems with uncertain recourse parameters

If the coefficient $a_2(\xi_1, \xi_2)$ of the recourse variable is not fixed, the matrices P_{21} and P_{22} are not identically zero. Since D is not restricted to be in a certain class, the quadratic term in the objective of (3.16) can be indefinite. The maximization problem in (3.16) becomes non-convex and one cannot resort to duality theory to build a simple robust counterpart, as it has be done until now. There is an exception to this bad situation. If the uncertainty set Ξ is an ellipsoid, it can be shown that the problem (3.16) is equivalent to a convex problem on the cone of the positive semi-definite matrices. It becomes possible to derive a robust counterpart. We refer the reader to (11) and (8) for more details. We just mention that there exist efficient codes, e.g., the open source Sedumi (38), that can solve very efficiently problems with constraints on the set of positive semi-definite matrices.

3.5.3 Problems with fixed recourse

To remain in the realm linear programming, we introduce the following assumption.

Assumption 2 *The $a_2(\xi)$ vector associated to the recourse is fixed ($a_2(\xi) \equiv a_2$).*

Proposition 5. *Assume the dynamic problem has fixed recourse. The robust counterpart of a dynamic constraint with LDR and a polyhedral uncertainty set is given by a set of linear constraints.*

Proof. From the fixed-recourse assumption, we have $P_{21}\xi_1 + P_{22}\xi_2 \equiv 0$. Since the quadratic term of the objective disappears, we can apply Proposition 2 and express the robust counterpart as a set of linear constraints. □

The fixed-recourse assumption confines the uncertainty to the first stage coefficients and to the right-hand sides. This is obviously limitative, but still relevant to interesting applications problems, for instance in supply chain management problem (6) or on the management of a hydraulic valley with uncertainty on the water supply (3; 2).

In some dynamic problems, we encounter constraints $a^T x = b$, with a deterministic and b uncertain. We shall get an example of it in our case problem. There, $a^T x$ is a production that is achieved by combining factors in proportion given by x and b is a demand. The only way to match an uncertain demand is to make the x vector uncertain and have it adjusted to the random behavior of b. This is achieved by a LDR incorporating the random factor in b. Two options are open. The first one is to keep working with the equality constraint and achieve global robustness as described in subsection 3.4.6. This approach has been followed in (35). An alternative consists in relaxing the equality into the inequality $a^T x \geq b$, guaranteeing that the demand will be satisfied at the possible cost of disposal if the production exceeds the demand. This approach has been followed in (6).

3.6 Case study: uncertain demand with fixed recourse

In this section, the case study deals with the model with uncertain demand and no uncertainty in the pollutant transfer coefficients. This is a multistage problem with fixed recourse for which the LDR methodology of the previous section is appropriate. To contrast Robust Optimization with more classical approaches, we have implemented a simple Stochastic Programming version. We built a discrete distribution to approximate the stochastic demand process used in the validation process. The number of branches is such that the so-called *deterministic equivalent*, i.e., the extensive formulation of the problem relatively to the event tree, has about the same size as the robust counterpart.

3.6.1 LDR and relaxed demand constraints

We reformulate Problem (3.1) to eliminate equality constraints, either by simple elimination (constraints (1e)) or by relaxation (constraints (1d)). We get the following relaxed version of (3.1)

$$\min_{x\geq 0, y\geq 0} \sum_r (O_r \sum_{p,t} x_{p,t,r} + I_r \sum_{p,t} y_{p,t,r} + M_r \sum_{p,t} (res_{p,t,r} + \sum_{\tau\leq t} y_{p,\tau,r})) \tag{17a}$$

$$x_{p,t,r} \leq res_{p,t,r} + \sum_{\tau\leq t} y_{p,\tau,r} \quad \forall t \in \mathscr{T}, p \in \mathscr{P}, r \in \mathscr{R} \tag{17b}$$

$$\sum_{t,r} x_{p,t,r} \geq d_t \quad \forall t \in \mathscr{T} \tag{17c}$$

$$\sum_{\rho\in\mathscr{R}} \sum_p (E_{p,\rho} x_{p,t,\rho}) G_{\rho,r} \leq Q \quad \forall t \in \mathscr{T}, r \in \mathscr{R} \tag{17d}$$

This more compact formulation has 72 variables and 51 constraints. We now define, for all pairs $(p \in \mathscr{P}, r \in \mathscr{R})$, the following linear decision rules for the production variables $x_{p,t,r}$ and the investment variables $y_{p,t,r}$.

$$x_{p,1,r} = \alpha^0_{p,1,r} + \alpha^1_{p,1,r}\eta_1 \tag{18a}$$

$$x_{p,2,r} = \alpha^0_{p,2,r} + \alpha^1_{p,2,r}\eta_1 + \alpha^2_{p,2,r}\eta_2 \tag{18b}$$

$$x_{p,3,r} = \alpha^0_{p,3,r} + \alpha^1_{p,3,r}\eta_1 + \alpha^2_{p,3,r}\eta_2 + \alpha^3_{p,3,r}\eta_3 \tag{18c}$$

$$y_{p,1,r} = \beta^0_{p,1,r} \tag{18d}$$

$$y_{p,2,r} = \beta^0_{p,2,r} + \beta^1_{p,2,r}\eta_1 \tag{18e}$$

$$y_{p,3,r} = \beta^0_{p,3,r} + \beta^1_{p,3,r}\eta_1 + \beta^2_{p,3,r}\eta_2. \tag{18f}$$

In that definition the random variables η_i are the ones of the demand uncertainty defined in (3.3). Thus the variables x and y are function of the demand uncertainty. Note that the investment in t is decided before the demand at t is known, while the production is set after, hence the difference in the decision rules.

If we replace the variables in the constraints of (3.17) by their linear decision rules, we obtain the LDR formulation

$$\min_{\alpha,\beta,\xi} \sum_r (O_r \sum_{p,t} \alpha^0_{p,t,r} + I_r \sum_{p,t} \beta^0_{p,t,r} + M_r \sum_{p,t} (res_{p,t,r} + (4-p)\beta^0_{p,t,r}) \tag{19a}$$

$$\alpha^1_{p,1,r}\eta_1 + \alpha^0_{p,1,r} - \beta^0_{p,1,r} \le res_{p,1,r} \quad \forall p,r \tag{19b}$$

$$\alpha^2_{p,2,r}\eta_2 + (\alpha^1_{p,2,r} - \beta^1_{p,2,r})\eta_1 + \\ +\alpha^0_{p,2,r} - \beta^0_{p,1,r} - \beta^0_{p,2,r} \le res_{p,2,r} \quad \forall p,r \tag{19c}$$

$$\alpha^3_{p,3,r}\eta_3 + (\alpha^2_{p,3,r} - \beta^2_{p,3,r})\eta_2 + (\alpha^1_{p,3,r} - \beta^1_{p,2,r} - \beta^1_{p,3,r})\eta_1 + \\ +\alpha^0_{p,3,r} - \beta^0_{p,1,r} - \beta^0_{p,2,r} - \beta^0_{p,3,r} \le res_{p,3,r} \quad \forall p,r \tag{19d}$$

$$(\sum_{p,r} \alpha^1_{p,1,r} - \hat{d}_1)\eta_1 + \sum_{p,r} \alpha^0_{p,1,r} \ge \bar{d}_1 \tag{19e}$$

$$(\sum_{p,r} \alpha^2_{p,2,r} - \hat{d}_2)\eta_2 + (\sum_{p,r} \alpha^1_{p,2,r} - \hat{d}_1)\eta_1 + \sum_{p,r} \alpha^0_{p,2,r} \ge \bar{d}_2 \tag{19f}$$

$$(\sum_{p,r} \alpha^3_{p,3,r} - \hat{d}_3)\eta_3 + (\sum_{p,r} \alpha^2_{p,3,r} - \hat{d}_2)\eta_2 + \\ +(\sum_{p,r} \alpha^1_{p,3,r} - \hat{d}_1)\eta_1 + \sum_{p,r} \alpha^0_{p,3,r} \ge \bar{d}_3 \tag{19g}$$

$$\sum_{i\in\mathscr{R}} \sum_p (E_{p,i}\alpha^1_{p,1,r})G_{i,r}\eta_1 + \sum_{i\in\mathscr{R}} \sum_p (E_{p,i}\alpha^0_{p,1,r})G_{i,r} \le Q \quad \forall r \tag{19h}$$

$$\sum_{i\in\mathscr{R}} \sum_p (E_{p,i}\alpha^2_{p,2,r})G_{i,r}\eta_2 + \sum_{i\in\mathscr{R}} \sum_p (E_{p,i}\alpha^1_{p,2,r})G_{i,r}\eta_1 + \\ + \sum_{i\in\mathscr{R}} \sum_p (E_{p,i}\alpha^0_{p,2,r})G_{i,r} \le Q \quad \forall r \tag{19i}$$

$$\sum_{i\in\mathscr{R}} \sum_p (E_{p,i}\alpha^3_{p,3,r})G_{i,r}\eta_3 + \sum_{i\in\mathscr{R}} \sum_p (E_{p,i}\alpha^2_{p,3,r})G_{i,r}\eta_2 + \\ + \sum_{i\in\mathscr{R}} \sum_p (E_{p,i}\alpha^1_{p,3,r})G_{i,r}\eta_1 + \sum_{i\in\mathscr{R}} \sum_p (E_{p,i}\alpha^0_{p,3,r})G_{i,r} \le Q \quad \forall r \tag{19j}$$

$$\alpha^0_{p,1,r} + \alpha^1_{p,1,r}\eta_1 \ge 0 \quad \forall p,r \tag{19k}$$

$$\alpha^0_{p,2,r} + \alpha^1_{p,2,r}\eta_1 + \alpha^2_{p,2,r}\eta_2 \ge 0 \quad \forall p,r \tag{19l}$$

$$\alpha^0_{p,3,r} + \alpha^1_{p,3,r}\eta_1 + \alpha^2_{p,3,r}\eta_2 + \alpha^3_{p,3,r}\eta_3 \ge 0 \quad \forall p,r \tag{19m}$$

$$\beta^0_{p,1,r} \ge 0 \quad \forall p,r \tag{19n}$$

$$\beta^0_{p,2,r} + \beta^1_{p,2,r}\eta_1 \ge 0 \quad \forall p,r \tag{19o}$$

$$\beta^0_{p,3,r} + \beta^1_{p,3,r}\eta_1 + \beta^2_{p,3,r}\eta_2 \ge 0 \quad \forall p,r. \tag{19p}$$

In that formulation all constraints depend on the random variables ξ and must be given a robust counterpart. For each constraint we choose the uncertainty set to be the intersection of the $B_1(0,k\sqrt{m})$ and $B_\infty(0,k)$ balls, where m is the number of random variables in the constraint. We also select different immunization factors k according to the type of constraints. Namely, we use k_{cap} for the capacity constraints (19b)–(19d), k_{dem} for the demand constraints (19e)–(19g) and k_{emi} for the air quality constraints (19h)–(19j). In the experiments, we set $k_{cap} = k_{emi} = 1$ and we test different values for k_{dem}. The robust equivalent of (3.19) has 419 variables and 502 constraints. It corresponds to an increase with a multiplicative factor 4 of size of the deterministic version (3.17).

We report the results in Table 3.13.

Table 3.13 Variable demand: behavior on the sample of 1000 scenarios of a robust solution with LDR.

	Deterministic	LDR with robust		
		$k_{dem} = 0.4$	$k_{dem} = 0.6$	$k_{dem} = 0.7$
Objective function				
Predicted cost performance	162.06	165.84	167.09	167.40
Observed cost performance	160.87	165.07	166.14	167.47
Constraints on the demand				
Scenarios with demand violation(s) in %	62.0	20.4	9.9	0
Conditional average relative violation in %	2.5	0.8	0.6	-
Average number of violations per scenario	2.0	1	1	-
Constraints on the air quality				
Total number of violated air quality constraints	0	53	39	82

The conclusions to be derived from Table 3.13 are twofold. If we increase the immunization level k_{dem}, the robust solution better and better tracks the (relaxed) demand constraint. On the other hand, a larger immunization factor induces more variability in the LDR and increases risks of violations in the (otherwise deterministic) air quality constraints. Note also the degradation of the cost performance when the immunization increases.

3.6.2 LDR and exact demand constraints

In the previous section, we transformed the equality demand constraint in the deterministic model (1d) into the inequality constraint (17c) to match the uncertainty in the right-hand side. We now follow the approach of Section 3.4.6 and keep the equality in the demand constraint. In view of Proposition 4, the linear decision rule coefficients must satisfy the following conditions. Namely, we replace the demand equations (19e)–(19g) by the following set of equations

$$\sum_{p,r} \alpha^0_{p,1,r} = \bar{d}_1, \quad \sum_{p,r} \alpha^1_{p,1,r} = \hat{d}_1, \tag{20a}$$

$$\sum_{p,r} \alpha^0_{p,2,r} = \bar{d}_2, \quad \sum_{p,r} \alpha^1_{p,2,r} = \hat{d}_1, \quad \sum_{p,r} \alpha^2_{p,2,r} = \hat{d}_2, \tag{20b}$$

$$\sum_{p,r} \alpha^0_{p,3,r} = \bar{d}_3, \quad \sum_{p,r} \alpha^1_{p,3,r} = \hat{d}_1, \quad \sum_{p,r} \alpha^2_{p,3,r} = \hat{d}_2, \quad \sum_{p,r} \alpha^3_{p,3,r} = \hat{d}_3. \tag{20c}$$

These conditions ensure that the demand constraints will be globally satisfied for all possible demands in the 3-dimensional space ($\mathbb{R}^3$). The model has now 412 variables and 487 constraints.

In the numerical example, we keep replacing the other constraints (the capacity constraints (19b)–(19d), the air quality constraints (19h)–(19j) and the non-negativity constraints on the production and capacities) with their equivalent robust counterpart. To make the results comparable, we set $k_{cap} = k_{emi} = 1$ as before. The last set of constraints deals with non-negativity. To enforce those constraints with maximum chances we use an immunization level k equal to 1. The results are displayed in Table 3.14. Surprisingly enough, the results are very much alike the

Table 3.14 Variable demand: behavior on the sample of 1000 scenarios of a globally robust solution with respect to the demand constraint.

Predicted cost performance	167.40
Observed cost performance	167.47
Scenarios with demand violation(s) in %	0
Total number of violated air quality constraints	82

relaxed case with $k_{dem} = 0.7$ in Table 3.13. This reflect the fact that it is relatively easy to track the demand by simple adjustment in the current period.

3.6.3 Stochastic programming

To apply stochastic programming, we model the demand process by a finite event tree. To end up with a problem with a size comparable to (3.19), we use a tree with 3 branches at each node and thus 27 scenarios in total. We assume that the probability factors η_1, η_2 and η_3 are i.i.d. with a uniform distribution. The best approximation of the uniform distribution by a finite distribution with 3 elements consists in partitioning the range space $[-1, 1]$ into 3 elements and choose the mid-point value of each subinterval to be the representative of the subinterval. The probability for each representative is $\frac{1}{3}$. The tree is represented in Figure 3.3.

If we start from the compact problem (3.17), the stochastic formulation has 624 variables and 663 constraints (419 variables and 502 constraints for the LDR formulation).

Table 3.15 Variable demand: behavior on the sample of 1000 scenarios of a stochastic programming solution built on an approximating event tree with 3 branches per node.

	Deterministic	Stochastic
Predicted cost performance	162.06	165.18
Observed cost performance	160.87	164.80
Scenarios with demand violation(s) in %	62.0	50.4
Conditional average relative violation in %	2.5	1.0
Average number of violations per scenario	2.0	1.4

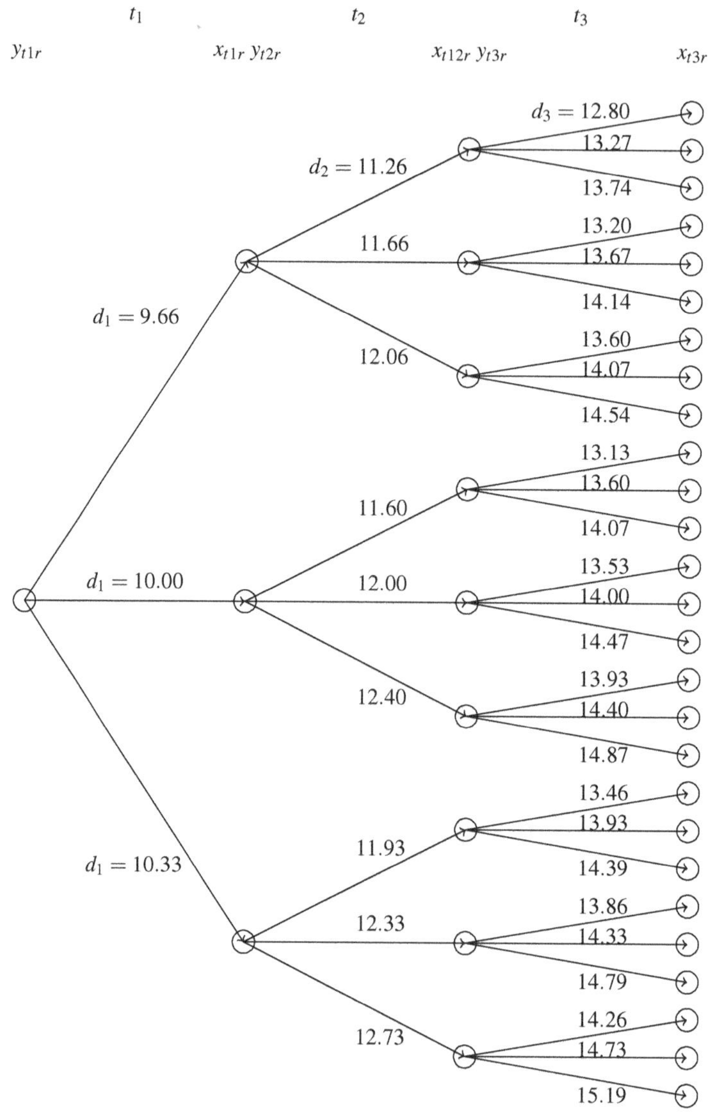

Fig. 3.3 Scenario decision tree

Table 3.15 suggests that the model cannot really cope with the uncertainty in the demand constraints. Indeed, production is determined relatively to the representative demand at the node. In the simulations, the demand is never equal to the value assigned to the node. With the symmetric distribution it lies above this value in 50% of the cases. However, we observe that the stochastic programming solution improves upon the deterministic solution, and contrary to the robust solution with LDR, it never induces air quality constraints. This is so because productions at each node are chosen to meet those constraints. In the simulations, the productions are either kept unchanged or lowered to match lower demand; so, no air quality constraint violation is to be expected.

It would be preposterous to derive definitive conclusions from this short example, but we can stress few facts. Stochastic Programming is at pain to meet constraints, while Robust Optimization controls violations pretty well. Stochastic programming treats constraints at privileged points: the nodes on the event tree, but Robust Optimization deals with the continuum of events in the event tree. To improve the control on the constraint in Robust Optimization, it suffices to increase the immunization level. A similar objective of better constraint satisfaction with Stochastic Programming calls for a larger event tree to better approximate the stochastic process, an approach that quickly leads to an numerically intractable optimization problem.

On the other side, Stochastic Programming delivers a richer output, provided one has reliable and tractable information on the probability distribution. This takes the form of the objective performance on each branch on the event tree, that gives an interesting approximation of the probability distribution of the optimal function value. It is even possible to add constraints, such as CVaR (conditional value at risk), to take into account the amount of risk.

3.7 Case study: uncertainty in demands and pollutant transfers

We consider now the more involved case where both the demands and the pollutant transfer coefficients are uncertain. As it has been pointed out earlier, LDR introduce an uncertainty of their own in the constraints in which they appear. If some of these constraints have themselves uncertain coefficients, the product of an uncertain coefficient with the uncertain decision variable (defined by the LDR) generates a bilinear form in the uncertainty. One cannot anymore use duality as in Section 3.4 to exhibit the robust counterpart.

To cope with this difficulty, we propose two empirical approaches, one that fully remains in the realm of Robust Optimization and a hybrid one that mixes Robust Optimization and Stochastic Programming.

3.7.1 A fully robust optimization approach

In the fully robust optimization approach, we eliminate the LDR in the air quality constraints by the following gimmick. We introduce an artificial variable to bound the value produced by the LDR. In our problem of interest the artificial variable represents a deterministic upper bound on the productions prescribed by the LDR. In view of the nature of air quality constraints with nonnegative coefficients only, higher productions entail higher emissions and, after pollutant transmission, deteriorate the local air quality. Therefore, it makes sense to replace the LDR value by an upper bound to comply with the worst case requirement.

This idea is implemented as follows. The air quality constraints (19h)–(19j) in the LDR formulation (3.19) are replaced by

$$\sum_{i\in\mathscr{R}}\sum_{p}(E_{p,i}x_{p,t,r})G_{i,r} \leq Q \quad \forall r,t \tag{21a}$$

$$\alpha^0_{p,1,r}+\alpha^1_{p,1,r}\eta_1 \leq x_{p,1,r} \quad \forall p,r \tag{21b}$$

$$\alpha^0_{p,2,r}+\alpha^1_{p,2,r}\eta_1+\alpha^2_{p,2,r}\eta_2 \leq x_{p,2,r} \quad \forall p,r \tag{21c}$$

$$\alpha^0_{p,3,r}+\alpha^1_{p,3,r}\eta_1+\alpha^2_{p,3,r}\eta_2+\alpha^3_{p,3,r}\eta_3 \leq x_{p,1,r} \quad \forall p,r. \tag{21d}$$

where the matrix G is uncertain. Because the upper bound x is deterministic, we now can deal with uncertain coefficients G in (21a) and write their robust counterpart. Note that constraints (21b)–(21d) together with the non-negativity constraints (19k)–(19m) form a group of two-sided constraints. We use Proposition 3 to cut down the number of constraints and variables in the robust counterpart of those two-sided constraints. Note that the production LDR variables are replaced by an upper bound in the air quality constraints (21a), but not in the capacity constraints (19b)–(19d). The formulation has 487 variables and 542 constraints

The numerical experiments aim to compare the LDR solution computed in subsection 3.6.1 and the solution of the fully robust optimization approach. In the latter, we used different immunization levels k_{emi} for the air quality constraints. In all cases we used the same immunization level $k_{dem} = 0.7$ for the demand constraints. The results are displayed in Table 3.16.

Table 3.16 Variable demand and variable transfer coefficients: behavior on the sample of 1000 scenarios of a robust solution with LDR.

	Standard LDR	LDR with robust const.		
		$k_{emi}=0$	$k_{emi}=0.5$	$k_{emi}=1$
Objective function				
Predicted cost performance	167.40	167.66	169.21	170.72
Observed cost performance	167.47	167.51	169.28	170.78
Constraints on the demand				
Scenarios with demand violation(s) in %	0	0	0	0
Conditional average relative violation in %	-	-	-	-
Average number of violations per scenario	-	-	-	-
Constraints on the air quality				
Total number of violated air quality constraints	829	802	150	11

The reader will notice that the new approach with an the immunization level in the air quality constraint set to $k_{emi} = 0$, yields a solution such that the number of violated air quality constraints is slightly less (802 instead of 829) than with the standard LDR approach, while the cost performance is slightly worse (167.66 instead of 167.40). In both cases, the model treats the air quality constraint in a deterministic way, but in the new approach the LDR production variable is replaced by an upper bound in the air quality constraint. This more conservative approach is a safer for the constraint but also more costly. However the difference almost negligible. If we increase the immunization level k_{emi}, the solution with the new approach ensures better and better protection against violations of the air quality constraints.

3.7.2 A hybrid approach: stochastic programming with robust constraints

We now assume both the coefficients of the source-receptor matrix G and the demands are uncertain. As in the previous section, the demands are defined by (3.3). Because we cannot use LDR in a constraint with other uncertain coefficients, we choose to represent the demand uncertainty via a finite event tree. We shall use stochastic programming to define the decision to be taken at each node of the tree. The new feature is that those decisions will be requested to be robust with respect to the air quality constraints. Therefore, we shall have to introduce different robust counterparts of those constraints at each node of the tree. As before, we use the event tree displayed in Figure 3.3. The hybrid formulation with a demand event tree and robust air quality constraints at the tree nodes has 819 variables and 819 constraints. The results are displayed in Table 3.17.

Table 3.17 Variable demand and variable transfer coefficients: behavior on the sample of 1000 scenarios of a solution obtained by the mixed approach.

	Deterministic	Hybrid		
		$k=0.5$	$k=1.0$	$k=1.2$
Objective function				
Predicted cost performance	165.18	166.35	167.46	167.88
Observed cost performance	164.80	165.98	167.09	167.51
Constraints on the demand				
Scenarios with demand violation(s) in %	50.4	50.4	50.4	50.4
Conditional average relative violation in %	1.0	1.0	1.0	1.0
Average number of violations per scenario	1.40	1.36	1.36	1.36
Constraints on the air quality				
% of simulations with air quality violations	54.6	27.1	1.5	0
Average number of air quality violations	1.7	1.5	1.2	0

If we compare the results of the LDR solution in Table 3.14 with those of the hybrid approach in Table 3.17, we observe that the latter can achieve much better control of the air quality constraints (even though there was no exogenous uncertainty in the first case and some in the other), while the former enables full control of the demand constraint, something that stochastic programming cannot achieve.

3.8 Probability of constraint satisfaction

Prior to applying the robust optimization to the example, it is worth relating robust optimization to chance-constrained programming. This alternative approach, that was introduced in the late fifties by Charnes and Cooper (22), replaces the deterministic constraint by a constraint in probability. Namely, it considers that a solution is admissible if the probability that the constraint be satisfied with this solution is higher than a certain threshold. Since robust optimization is not built on probability grounds, comparing the two approaches seems to be out of order. Surprisingly enough, it turns out that one can assess bounds on the probability of satisfaction of a constraint by a robust solution, at the cost of mild assumptions on the probability distributions of the uncertain parameters. The result stems from a theorem to be found in Chapter 3 of (5). We give here a slightly stronger version than the one to be found in the literature.

3.8.1 Bounds on the probability of constraint satisfaction

Theorem 3. *Let ξ_i, $i = 1,\dots,m$ be independent random variables with values in interval $[-1,1]$ and with average zero: $E(\xi_i) = 0$. If z_i, $i = 1,\dots,m$ are deterministic coefficients, we have for all $k \geq 0$*

$$Prob\left\{\xi \mid \sum_{i=1}^{m} z_i \xi_i > k\sqrt{\sum_{i=1}^{m} z_i^2}\right\} \leq \exp(-\frac{k^2}{1.5}).$$

Remark 3. The assumption in Theorem 3 deals with the support of the random variables, their expectation and their independence. No other probabilistic assumption is made, in particular, nothing concerning distributions.

The following two lemmas are used in the proof of Theorem 3.

Lemma 4 (Chebytchev inequality).

$$Prob(X \geq a) \leq e^{-a} E(e^X).$$

Proof.

$$\begin{aligned} E(e^X) &= E(e^X|X < a)\text{Prob}(X < a) + E(e^X \mid X \geq a)\text{Prob}(X \geq a) \\ &\geq E(e^X|X \geq a)\text{Prob}(X \geq a) \qquad (\text{because } e^X > 0 \Rightarrow E(e^X \mid X < a) \geq 0) \\ &\geq e^a \text{Prob}(X \geq a). \end{aligned}$$

The last inequality is from

$$E(e^X|X \geq a) \geq E(e^a|X \geq a) = e^a.$$

□

Lemma 5. *The inequality* $e^{\frac{t^2}{\alpha}} + t \geq e^t$ *with* $1 \leq \alpha \leq \sqrt{\frac{8}{3}}$ *is valid for all* $t \geq 0$.

Proof. The function $e^{t^2/\alpha} + t - e^t$ is decreasing in α for all $t > 0$. To prove the property, we consider the extreme value of α. We assume in the following $\alpha = \sqrt{\frac{8}{3}}$. The derivatives of function $g(t) = e^{t^2/\alpha} + t - e^t$ are

$$g'(t) = \frac{2}{\alpha} t e^{t^2/\alpha} + 1 - e^t$$

and

$$g''(t) = \frac{2}{\alpha}(\frac{2}{\alpha}t^2 + 1)e^{t^2/\alpha} - e^t = e^t \left(\frac{2}{\alpha}(\frac{2}{\alpha}t^2 + 1)e^{t^2/\alpha - t} - 1 \right). \tag{3.22}$$

We note that $g'(0) = 0$ and $g''(0) = 2/\alpha - 1 \geq \sqrt{3/2} - 1 > 0$. We show first that $g''(t) \geq 0$ for all $t \geq 0$. Thus we show that the term in parenthesis in the right hand side of (3.22) is positive. From the inequality $e^x \geq 1 + x$, we have

$$\frac{2}{\alpha}(\frac{2}{\alpha}t^2 + 1)e^{t^2/\alpha - t} - 1 \geq h(t) = \frac{2}{\alpha}\left(\frac{2}{\alpha}t^2 + 1\right)\left(1 + \frac{t^2}{\alpha} - t\right) - 1.$$

The right component is a fourth degree polynomial that is convex because $\alpha < 4$ and

$$h''(t) = \frac{12}{\alpha^2}\left(\frac{4}{\alpha}t^2 - 2t + 1\right) = \frac{12}{\alpha^2}\left((1-t)^2 + (\frac{4}{\alpha} - 1)t^2\right) > 0.$$

The polynomial

$$h'(t) = \frac{2}{\alpha}\left(\frac{8}{\alpha^2}t^3 - \frac{6}{\alpha}t^2 + \frac{6}{\alpha}t - 1\right) - \sqrt{\frac{3}{2}}\left(3t^3 - 3\sqrt{\frac{3}{2}}t^2 + 3\sqrt{\frac{3}{2}}t - 1\right)$$

is thus increasing; its value at 0 is $-\sqrt{3/2}$ and it tends to $+\infty$ when t tends to $+\infty$. It has a unique real root in $\bar{t} = 0.366$. The function has its minimum in $\bar{t}$ and thus $g''(t) \geq h(t) \geq h(\bar{t}) = 0.0208$.

We proved that $g''(t) \geq 0$ on the interval $[0,1]$ and thus that the function $g'(t)$ is increasing. Since $g'(0) = 0$, we conclude that $g'(t) \geq 0$ and thus that $g(t)$ is increasing. Finally since $g(0) = 0$, we have $g(t) \geq 0$ on $0 \leq t \leq 1$. □

Proof. Now we can prove Theorem 3. This proof is derived from (5). From Lemma 4 we write

$$\text{Prob}(\sum_i z_i \zeta_i > k||z||) \leq e^{-k||z||} E(e^{\sum_i z_i \zeta_i}).$$

Since the variables ζ_i are independent, then $E(e^{\sum_i z_i \zeta_i}) = E(\prod_i e^{z_i \zeta_i}) = \prod_i E(e^{z_i \zeta_i})$. We also have

$$
\begin{aligned}
E(e^{z_i\zeta_i}) &= 1+\sum_{k\geq 2} E(\frac{(z_i\zeta_i)^k}{k!}) \quad \text{(because } E(\zeta_i)=0\text{)}\\
&\leq 1+\sum_{k\geq 2}\frac{|z_i|^k}{k!} \quad \text{(because } |z_i\zeta_i|\leq |z_i|\text{)}\\
&\leq e^{|z_i|}-|z_i| \leq e^{\frac{z_i^2}{\alpha}}.
\end{aligned}
$$

The last inequality comes from Lemma 5. Thus we can write

$$
\text{Prob}(\sum_i z_i\zeta_i > k||z||) \leq e^{-k||z||}\prod_i e^{\frac{z_i^2}{\alpha}} = e^{-k||z||+\frac{||z||^2}{\alpha}}.
$$

As the relation

$$
\text{Prob}(\sum_i z_i\zeta_i > k||z||) = \text{Prob}(\gamma\sum_i z_i\zeta_i > \gamma k||z||)
$$

is true for all $\gamma > 0$, then we have

$$
\begin{aligned}
\text{Prob}(\sum_i z_i\zeta_i > k||z||) &\leq \min_{\gamma>0} e^{-\gamma k||z||+\gamma^2\frac{||z||^2}{\alpha}}\\
&\leq e^{-\frac{\alpha^2 k^2}{4}} \leq e^{-\frac{k^2}{1.5}}.
\end{aligned}
$$

Remark 4. The standard result gives the bound $\exp(-k^2/2)$. An experimental study of the function $\exp(t^2/\alpha)+t-\exp(t)$ shows that the maximum value of α, for which Lemma 5 remains true, is 1.79. The bound in the proposition is tighten to $\exp(-t^2/1.25)$.

Let us apply the result when the uncertainty set is the ball $B_2(0,k)$ and the variable support for ξ_j is the interval $[-1,1]$. We consider the robust constraint

$$
\bar{a}^T x + (P\xi)^T x \leq 0, \; \forall \xi \in B_2(0,k).
$$

From Lemma 1

$$
\max_{\xi\in\Xi}(P^T x)^T\xi \leq \max_{\xi}\{(P^T x)^T\xi \mid ||\xi||_2 \leq k\} = k||P^T x||_2.
$$

Letting $z = P^T x$, Theorem 3 yields

$$
\text{Prob}\left\{(\bar{a}+P\xi)^T x \leq 0 \mid \xi\in\Xi\right\} \geq \text{Prob}\left\{(P^T x)^T\xi \leq k||P^T x||_2\right\} \geq 1-\exp(-\frac{k^2}{1.5}).
$$

Note that for $k = 2.63$ we get the bound 0.99 on the probability.

Using Theorem 3, we can also bound the probability associated with an uncertainty set defined as the intersection of two balls in the ℓ_1 and ℓ_∞ norms respectively. This done in the next corollary.

Corollary 1. *Let ξ_i, $i = 1,\dots,m$ be independent random variables with values in the interval $[-1,1]$ and average zero: $E(\xi_i) = 0$. If z_i, $i = 1,\dots,m$ are deterministic coefficients, we have for all $k \geq 0$*

$$Prob\left\{ \xi \mid \sum_{i=1}^{m} z_i \xi_i > k \sum_{i=1}^{m} |z_i| \right\} \leq \exp(-\frac{k^2}{1.5m}).$$

Proof. The proof results from the relation between the norms ℓ_1 and ℓ_2 in $\mathbb{R}^m$ that implies

$$B_2(0, k/\sqrt{m}) \subseteq B_1(0,k).$$

Replacing k by $k/\sqrt{m}$ in Theorem 3 we get the result. □

3.8.2 Uncertainty set versus constraint satisfaction

To get a flavor of the implication of Theorem 3 and its corollary, it is worth comparing the volume of the certainty set $\Xi = B_\infty(0,1)$ with the volumes of the balls $B_1(0,k\sqrt{m})$ and $B_2(0,k)$, for some k guaranteeing a large probability of satisfying the uncertain constraint. Since the ball $B_1(0,k\sqrt{m})$ is a rotation of the ball $B_\infty(0,k)$, it has the same volume as the ball $B_\infty(0,k)$, and hence a volume larger than the ball $B_\infty(0,1)$ by a factor k^m (for $k \geq 1$). Notwithstanding, the ball $B_1(0,k\sqrt{m})$ intersects the ball $B_\infty(0,1)$ for $k \leq \sqrt{m}$. Hence the uncertainty set $\Xi = B_\infty(0,1) \cap B_1(0,k\sqrt{m})$ is smaller than the certainty set $B_\infty(0,1)$, but not in large proportion.

The situation with the ball $B_2(0,k)$ is dramatically different. This fact is discussed in (5) and is summarized by the inequality

$$\frac{\mathrm{Vol} B_2(0,k)}{\mathrm{Vol} B_\infty(0,1)} = \frac{(k\sqrt{\pi})^m}{2^m \Gamma(m/2+1)} \leq \left(k\sqrt{\frac{e\pi}{2m}}\right)^m.$$

For $k = 2.63$ and $m = 30$ the ratio on the right hand-side is strictly less than 1 and goes super-exponentially fast to zero as the dimension of the space of uncertainty factor goes to infinity. Hence in large dimension the uncertainty set may be incommensurably smaller than the certainty set, yet a solution which is robust with respect to this uncertainty set may achieve a very high probability of satisfying the constraint.

At this point, it is worth wondering whether it is appropriate to rely on an intuitive justification of the robust optimization based on the idea that it is necessary to select a large uncertainty set to achieve a high probability of constraint satisfaction. For sure, the probability of satisfying the constraint is at least as large as the probability associated with the uncertainty set, but the reverse is not true. We had a first confirmation of this fact by comparing the volume of the $B_2(0,k)$ uncertainty set with the volume of the certainty set $B_\infty(0,1)$. We now provide an example that shows that on some circumstances, it is possible to achieve very high probability while the uncertainty set has very small probability, even a zero probability!

Consider the constraint

$$\sum_{i=1}^{m} a_i x_i \leq b \tag{3.23}$$

with uncertain coefficients $a_i = \bar{a}_i + \hat{a}_i \xi_i$, where $\xi_i \in \Omega$. (Note that it is a particular case of the model $a = \bar{a} + P\xi$ where P is the diagonal matrix with main diagonal $\hat{a}$.) We define an uncertainty set $\Xi \subset \Omega$ and work with the robust constraint

$$\sum_{i=1}^{m} (\bar{a}_i + \hat{a}_i \xi_i) x_i \leq b, \ \forall \xi \in \Xi.$$

Suppose we can assess a probability distribution for ξ on Ω. We have the implication

$$\sum_{i=1}^{m} (\bar{a}_i + \hat{a}_i \xi_i) x_i \leq b, \ \forall \xi \in \Xi \text{ and } \mathrm{Prob}(\xi \in \Xi) \geq 1 - \alpha$$
$$\Downarrow$$
$$\mathrm{Prob}\left(\sum_{i=1}^{m} (\bar{a}_i + \hat{a}_i \xi_i) x_i \leq b\right) \geq 1 - \alpha.$$

Truly enough, the reverse implication does not hold. Yet, the direct implication suggests that choosing an uncertainty set with large probability might be a good strategy. The following example reveals that the strategy can be overly conservative.

In this example we assume the a_i are independent random variables

$$a_i = \begin{cases} 1 \text{ with probability } 1/2 \\ 0 \text{ with probability } 1/2. \end{cases} \tag{3.24}$$

This uncertain constraint is cast into a robust optimization framework as follows. For the time being, forget about the probability distribution and write the coefficients in the form $a_i = \frac{1}{2} + \frac{1}{2}\xi_i$ with $-1 \leq \xi_i \leq 1$. Let the uncertainty set be the ellipsoid $\Xi = \{\xi \mid ||\xi||_2 \leq k\}$. The robust counterpart of the constraint

$$\frac{1}{2}\sum_i x_i + \frac{1}{2}\sum_i x_i \xi_i, \ \forall \xi \in \Xi$$

is

$$\frac{1}{2}\sum_i x_i + \frac{k}{2}\sqrt{\sum_i x_i^2} \leq b.$$

Since the ξ_i are independent random variables with zero mean with range $[-1, 1]$, we can apply Theorem 3 and get

$$\mathrm{Prob}\left\{ \tilde{\xi} \mid \sum_{i=1}^{m} z_i \tilde{\xi}_i > k \sqrt{\sum_i^{m} z_i^2} \right\} \leq \exp(-\frac{k^2}{1.5})$$

for arbitrary z_i. For $k = 2$ we get the bound 0.07 on the probability of constraint violation. If we apply the theorem to the robust counterpart, we have that a solution to

$$\frac{1}{2}\sum_i x_i + \sqrt{\sum_i x_i^2} \le b$$

has a probability $1 - 0.07 = 0.93$ to satisfy the probabilistic constraint

$$\sum_{i=1}^{m} a_i(\xi_i)x_i \le b$$

for any probability distribution satisfying the (weak) hypothesis of the theorem. In particular, this is true for (3.24).

On the other hand, the set $\Xi = \{\xi \mid ||\xi||_2 \le 2\}$ *does not contain any single realization* of the random variable ξ as soon as ξ has dimension larger than 4. In other words $\text{Prob}(\xi \in \Xi) = 0$. So, we imposed x to be robust with respect to an uncertainty set[2] having probability 0, but we still guarantee that the robust solution satisfies the uncertain constraint with a probability at least 0.93. Clearly, Robust Optimization does much more than the intuition suggests.

Let us pursue the discussion with this example. Consider now the same version of the problem but with binary variables x_i and $b = 60$. The robust counterpart (still with $k = 2$)

$$\frac{1}{2}\sum_i x_i + \sqrt{\sum_i x_i^2} \le 60$$

is equivalent to a bound on the number variables that can set to the value 1. We easily find that this number is $N = 100$. We now ask the question: how good is the lower bound 0.93 on the true probability of constraint satisfaction? Without loss of generality, we assume that the robust solution is $x_i = 1$, $i = 1, \ldots 100$ and zero otherwise. The left-hand side in the constraint $\sum_{i=1}^{m} a_l x_l \le 60$ is a binomial random variable, with parameter 1/2. From the tables, we get that the probability of satisfaction is 0.98. It is certainly an improvement upon 0.93, but still it is quite an achievement to get 0.93 with a robust approach, in view of the weak assumption on the true distribution in Theorem 3.

3.9 Extension: Globalized robust optimization

As we have seen it again and again, robust optimization deals with worst cases with respect to uncertainty sets. Robust optimization concentrates on solutions that remain feasible for all realizations within the uncertainty sets, but is silent about realizations that lie outside. In particular it does not take into account the magnitude of

[2] Note that in this example, the ellipsoidal set has either probability 0, if k is small or probability one.

the violation. Globalized robust optimization (4) proposes an extension that admits possible constraint violations, but control their magnitude.

3.9.1 The concept of globalized robust optimization

A globalized robust solution must satisfy the following two criteria:

1. The solution is robust for all realizations in the uncertainty set.
2. If a realization falls outside the uncertainty set, a violation of the constraint is tolerated, but this violation must "remain under control".

The fuzzy concept of remaining under control must be clarified, but we already see that a globalized robust solution is robust. The concept is then more restrictive, but it covers all cases, inside and outside the uncertainty set. To make the definition operational, one needs to be more specific about the meaning of "being under control". In (4) the authors suggest to consider the distance from a current realization to the uncertainty set. If this distance is positive, i.e., if the realization is strictly exterior to the uncertainty set, a violation is acceptable, but *its magnitude should be less than a fixed multiple of this distance*. This definition applies to the realizations within the uncertainty set, because, the distance is null, and no violation is permitted.

To formalize the idea, we first define an arbitrary distance to the uncertainty set. As in the previous examples, we consider the constraint

$$(\bar{a}+P\xi)^T x = \bar{a}^T x + (P^T x)^T \xi \leq b$$

to be satisfied for any ξ in the uncertainty set. We assume that $\Omega = \mathbb{R}^m$ and that Ξ is convex. We introduce the convex distance function $d(\xi, \Xi)$ between ξ and the set Ξ. A solution is called *globally robust* if it satisfies

$$\bar{a}^T x + (P^T x)^T \xi \leq b + \alpha d(\xi, \Xi), \forall \xi \in \Omega$$

where $\alpha > 0$ is a user parameter. This parameter is chosen in function of the tolerated violation. In this formulation, ξ can be given any value[3] in $\mathbb{R}^m$.

To check whether a solution x meets the globalized robustness requirements, it suffices to replace the uncertain terms by its maximal value

$$\bar{a}^T x + \max_{\xi \in \Omega} \{(P^T x)^T \xi - \alpha d(\xi, \Xi)\} \leq b.$$

To make things more precise, let us specify that the distance function is generated by a norm. Let $\delta : \mathbb{R}^m \to \mathbb{R}_+$ be this norm, and define the distance as

$$d(\xi, \Xi) = \min_{\xi' \in \Xi} \delta(\xi - \xi').$$

[3] This requirement may be excessive and unrealistic. One could think that the set is bounded, possibly with large bounds. This possibility is shortly discussed at the end of this section.

The maximization operation in the left-hand side of the globalized robust equivalent constraint is a convex problem (maximizing a concave function). It can be written as

$$\begin{aligned}\max_{\xi\in\mathbb{R}^m}\{(P^Tx)^T\xi-\alpha\min_{\xi'\in\Xi}\delta(\xi-\xi')\}\\ &=\max_{\xi\in\mathbb{R}^m}\{(P^Tx)^T\xi+\alpha\max_{\xi'\in\Xi}\{-\delta(\xi-\xi')\}\}\\ &=\max\{(P^Tx)^T\xi-\alpha\delta(\xi-\xi')\mid\xi\in\mathbb{R}^m,\xi'\in\Xi\}.\end{aligned}$$

We have thus the globalized robust equivalent written as

$$\bar{a}^Tx+\max\{(P^Tx)^T\xi-\alpha\delta(\xi-\xi')\mid\xi\in\mathbb{R}^m,\xi'\in\Xi\}\le b. \tag{3.25}$$

3.9.2 Globalized robustness with linear programming

We shall use this expression with a choice of the norm that keeps the whole formulation in the realm of linear programming. To this end, we select a polyhedral uncertainty set and an appropriate norm that can be described by a finite number of linear inequalities, such as those generated by the balls in the ℓ_∞ and/or ℓ_1 norms. The distance to this uncertainty set will also be defined with respect to ℓ_∞ and/or ℓ_1, so that everything can be translated into linear inequalities. In principle, the combination of the ℓ_∞ and ℓ_1 norms used in defining the distance may be totally independent of the structure of the uncertainty set, but we choose to analyze a case where the uncertainty set and the distance function derive both from the same norm.

Consider the uncertainty set $\Xi=\{\xi\in\Omega\mid||\xi||_1\le k_1,||\xi||_\infty\le k_\infty\}$ defined by the intersection of the ball with radius k_∞ in the norm ℓ_∞ and the ball with radius k_1 in the norm ℓ_1. (In practice we often choose $k_1 - k_\infty\sqrt{m}$.) This polyhedron can be used to defined a norm as follows. Consider the homothetic set

$$\Xi(t)=\{\xi\in\Omega\mid||\xi||_1\le k_1t,||\xi||_\infty\le k_\infty t\}$$

with $t>0$. We can use this polyhedron to define the norm

$$\delta(\mu)=\min_{t>0}\{t\mid\mu\in\Xi(t)\}=\max\{\frac{1}{k_1}||\mu||_1,\frac{1}{k_\infty}||\mu||_\infty\}.$$

The distance from a point ξ to the set Ξ, is now defined by the optimization problem

$$d(\xi,\Xi)=\min_{v}\{\delta(\xi-v)\mid v\in\Xi\}.$$

Figure 3.4 illustrates the distance from ξ to the set Ξ with the selected norm in the 2-dimensional space with $k_1=k_\infty\sqrt{2}$. The blue dotted curves represent the sets of points that are at the same distance of Ξ. The point $v\in\Xi$ is one of the closest point from ξ. (The vector v is not unique because the distance function is not strictly convex.) The computed distance is $d(\xi,\Xi)=\delta(v,\xi)=2.53$.

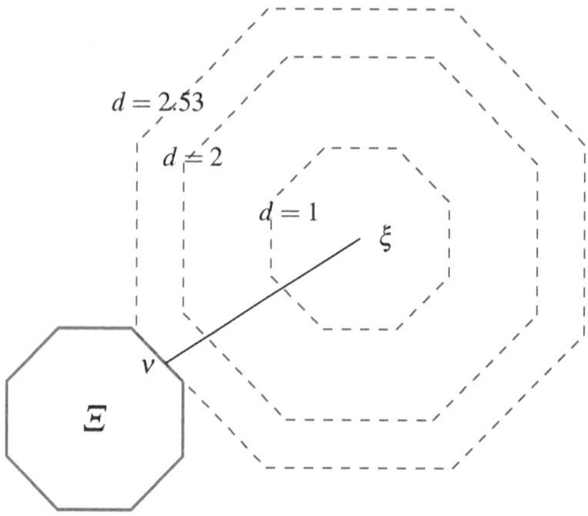

Fig. 3.4 Distance $d(\xi, \Xi)$ with the composite norm

Proposition 6. *Let $\tilde{a}^T x \leq b$ be a constraint with uncertain elements $\tilde{a} = \bar{a} + P\xi$. Let $\Xi = \{\xi \in \Omega \mid ||\xi||_1 \leq k_1, ||\xi||_\infty \leq k_\infty\}$ be the uncertainty set and let $d(\xi, \Xi) = \min_{\xi' \in \Xi} \delta(\xi - \xi')$ be the distance of a point ξ to Ξ, where $\delta(\mu) = \min_{t>0}\{t \mid \mu \in \Xi(t)\} = \max\{\frac{1}{k_1}||\mu||_1, \frac{1}{k_\infty}||\mu||_\infty\}$ is the norm induced by the polyhedron Ξ. Finally, let α be the coefficient of globalized robustness.*

The globalized robust equivalent of the globalized robust constraint

$$\bar{a}^T x + (P^T x)^T \xi \leq b + \alpha d(\xi, \Xi), \forall \xi \in \Omega$$

is given by the two constraints

$$\bar{a}^T x + k_1 ||P^T x - u||_\infty + k_\infty ||u||_1 \leq b \tag{26a}$$

$$k_1 ||P^T x - v||_\infty + k_\infty ||v||_1 \leq \alpha. \tag{26b}$$

Proof. Let us study the maximization problem yielding the worst case in the definition of the robust equivalent and show it to be a linear programming problem. First, let us introduce the auxiliary variable $\zeta \in \mathbb{R}$ and write the maximization problem yielding the globalized robust equivalent as

$$z^* = \max_{\xi, \xi', \zeta} \{(P^T x)^T \xi - \alpha \zeta \mid \delta(\xi - \xi') \leq \zeta, \ \zeta \geq 0, \ \xi' \in \Xi, \ \xi \in \Omega\}.$$

Note that we have added the redundant constraint $\zeta \geq 0$. The objective maximization ensures ζ to be as small as possible, enforcing the equality $\delta(\xi' - \xi) = \zeta$. Introducing the new variable $\eta = \xi - \xi'$ and recalling that the norm δ is the maximum of two norms, we obtain the alternative formulation

$$
\begin{aligned}
z^* = \max_{\xi',\eta,\zeta} \ & (P^T x)^T(\xi' + \eta) - \alpha\zeta \\
& ||\eta||_\infty \leq k_\infty \zeta, \ ||\eta||_1 \leq k_1 \zeta, \ \zeta \geq 0 \\
& ||\xi'||_\infty \leq k_\infty, \ ||\xi'||_1 \leq k_1.
\end{aligned}
$$

For fixed ζ, the problem is separable in ξ' and η. By Lemma 2, we have

$$
\begin{aligned}
z^* &= \max_{\zeta \geq 0} \min_{u,v} \left\{ k_1||P^T x - u||_\infty + k_\infty||u||_1 + \zeta(k_1||P^T x - v||_\infty + k_\infty||v||_1 - \alpha) \right\} \\
&= \min_{u,v} \max_{\zeta \geq 0} \left\{ k_1||P^T x - u||_\infty + k_\infty||u||_1 + \zeta(k_1||P^T x - v||_\infty + k_\infty||v||_1 - \alpha) \right\}.
\end{aligned}
$$

The interchange between the max and min operators is allowed by the strong duality theorem in linear programming. The last expression can be simplified upon noticing that the inner maximization is either unbounded if $k_1||P^T x - v||_\infty + k_\infty||v||_1 > \alpha$, or takes the value $k_1||P^T x - u||_\infty + k_\infty||u||_1$ if $k_1||P^T x - v||_\infty + k_\infty||v||_1 \leq \alpha$ (because $\zeta = 0$ in that case). Thus

$$
z^* = \min_{u,v} \{ k_1||P^T x - u||_\infty + k_\infty||u||_1 \mid k_1||P^T x - v||_\infty + k_\infty||v||_1 \leq \alpha \}.
$$

As shown in Proposition 1, the above linear programming problem can be explicitly recast in terms of a linear objective and linear inequality constraints. Moreover, if feasible, it achieves its optimal value. This makes it possible to write the deterministic equivalent as

$$
\begin{aligned}
\bar{a}^T x + k_1||P^T x - u||_\infty + k_\infty||u||_1 &\leq b \\
k_1||P^T x - v||_\infty + k_\infty||v||_1 &\leq \alpha.
\end{aligned}
$$

□

Remark 5. A theorem similar to Proposition 1 can be stated for the globalized robust equivalent (3.26) with norms l_1 and l_∞. We leave it to the reader to perform the simple transformations that lead to equivalent linear inequalities.

Remark 6. In the definition of the globalized robustness, we assumed that the set Ω of all possible realizations is $\mathbb{R}^m$. This makes the globalized robustness condition rather demanding. It may be worth considering that $\Omega \subset \mathbb{R}^m$ is constrained by box constraints, such as $\Omega = \{\xi \mid \underline{\xi} \leq \xi \leq \bar{\xi}\}$. In that case, the globalized robust equivalent calls for two additional nonnegative variables and the constraints

$$
\begin{aligned}
\bar{a}^T x + k_1||P^T x - u - w + t||_\infty + k_\infty||u||_1 + \bar{\xi}^T w - \underline{\xi}^T t &\leq b \\
k_1||P^T x - v - w + t||_\infty + k_\infty||v||_1 &\leq \alpha \\
w \geq 0, \ t &\geq 0.
\end{aligned}
$$

If we set $w = 0$ and $t = 0$, we retrieve (3.26), confirming that reducing the span of Ω enlarges the set of solutions.

Let us add few words of comments. Because of the additional constraint (26b), the set of globalized robust solutions is more restricted than the set of standard robust solutions. The globalized robustness uses the violation tolerance factor α. The smaller this parameter, the smaller must be the constraint violation. On the contrary, if α tends to infinity, constraint (26b) becomes inactive and we retrieve the plain robust solution.

In this presentation we used polyhedral uncertainty sets. Similar results can be obtained with ellipsoidal uncertainty sets. An interesting application to multi-echelon, multi-period inventory control is reported in (7).

3.9.3 Case study: Globalized robustness with uncertain demands

We now apply the concept of globalized robust optimization to the LDR formulation described in Subsection 3.6.1. We remind the reader that in Subsection 3.6.1 only the demands were uncertain and that the experiments were performed with different values of k_{dem}. Here we propose to test different values of α on the particular case $k_{dem} = 0.4$. We report the simulation results for each globalized robust solution in Table 3.18.

Table 3.18 Variable demand: behavior on the sample of 1000 scenarios of a globalized robust solution with LDR.

	LDR solutions with $k_{dem} = 0.4$		
	$\alpha = 0.3$	$\alpha = 0.2$	$\alpha = 0.1$
Predicted cost performance	165.84	166.26	166.82
Observed cost performance	165.07	165.71	166.62
Scenarios with demand violation(s) in %	20.4	16.0	1.1
Conditional average relative violation in %	0.8	0.6	0.4
Average number of violations per scenario	1	1	1
Total number of violated air quality constraints	69	69	69

We notice that for $\alpha = 0.3$ we retrieve the standard robust solution. Lower values of α permit a better control of the demand violations.

The model formulation has 433 constraints and 509 variables (419 variables and 502 constraints for the standard LDR formulation).

3.10 Conclusion

The main goal of this chapter was to present an alternative approach to the dealing with uncertainties in environmental and energy planning. To this end, we introduced basic concepts in robust optimization and we applied them to an illustrative example.

For the sake of the exposition, we chose an overly simplified example of modest size with three periods only. In practice, the models that are commonly used in the area of environmental and energy planning are much larger and much more complex, with thousands of variables and constraints. One should expect that the robust optimization methodology generates robust counterparts of much larger size, possibly to the point that they become impractical. This view must be qualified, by differentiating the type of uncertainties that are taken into consideration. The main difficulty stems from the multi-period feature of the models with uncertainties described as stochastic process. At this point, there is a sharp distinction to be made, depending on whether the realizations of the stochastic process are progressively revealed to the decision-maker, or only at once at the end of the terminal stage. If the exact value of the process remains unknown till the horizon of the model, the decision process is essentially static: all present and future decisions are to be taken at once, but the eventual effect of uncertainty is revealed in a second phase with no possibility of recourse. We treated an example of this sort in Section 3.3. The fact that the size of robust counterpart increases linearly with the dimension of the uncertainty factors suggests that, in this situation, even large models could be handled by robust optimization.

The situation is much more dramatic if the uncertainty is progressively revealed in time. Then, the decision process must be adapted to the revealed information, a fact that introduces a major, fundamental difficulty for all known methods, except in some special cases[4]. Linear decision rules for problems with fixed recourses may provide an acceptable approximation of an adaptive behavior and yield interesting hindsight, as those obtained in Section 3.6. We even suggested that robust optimization could be used in connection with traditional approaches to handle more complex cases. Other possibilities are to be considered. LDR is a crude approximation of the anticipation mechanism in the decision process. But is is already a highly complex one in regard of the actual decision process in the environmental and energy planning problems (and in many other problems!). Moreover, in an implementation phase of the model solution, the optimal LDR will almost surely never be implemented as such. Rather, at each time stage, a fresh version of the problem will be considered to account for the new situation. This leads to view the LDR not as a practical "open loop control", but like a tool yielding a plausible anticipation of future evolution and providing valuable information in the design of the first stage decisions.

An interesting alternative, though probably a computationally costly one, would be to work in a folding horizon framework and scenario simulations. Suppose a scenario is selected, which prescribes up to the horizon the realization of the various uncertainties. The decision-maker only knows the probabilistic way the scenario is built, but not the scenario itself. The planning model may have a rather rich description of uncertainty in its first stage, in a robust optimization framework, and a cruder description for the later stages based on a LDR. The first stage optimal solution of this model is implemented in connection with the realization of the stochastic pro-

[4] For instance, Markovian decision processes with reduced state space cardinality can be treated very efficiently by dynamic programming

cess in the first stage. This determines the state of the system at the beginning of the second stage, where a problem of the same nature as before, and a shorter horizon, is considered anew. This approach has been implemented in (2; 3; 6) and could prove interesting in more general models.

The secondary goal of this chapter was to give a simple enough introduction to the burgeoning field of robust optimization. We deliberately confined our presentation to the linear programming context, except for a few mention of ellipsoidal uncertainty sets. In the latter case, the robust counterparts fall in the realm of second order cone programming (SCOP) for which highly efficient solution methods exist. The striking fact is that in the case of ellipsoidal uncertainty sets the robust counterpart is obtained via similar duality-based arguments. Further generalizations can be made that take advantage of the power of conic programming with linear constraints and self-dual cones. Most recent contributions in robust optimization are concerned with these extensions and/or use them intensively. They are beyond the scope of this introduction, but they cannot be ignored. The hope is that this introduction will make it easier for the reader to get around with the growing literature.

Finally, we would like to mention that the probabilistic results in Section 3.8 have been considerably strengthened in some recent contributions. The crude hypotheses of Theorem 3 of random factors with zero mean and a symmetric range around the mean can be replaced by a range on the mean and a bound of the variance. Many other results of comparable nature are reported in (5). They give more evidences of the strong link between chance constrained programming and robust optimization.

Acknowledgements This work was partially supported by the European research project FP7-PLANETS

References

[1] E. Adida and G. Perakis. A robust optimization approach to dynamic pricing and inventory control with no backorders. *Mathematical Programming*, 107(1-2):97 – 129, 2006.

[2] R. Apparigliato. *Règles de décision pour la gestion du risque : Application à la gestion hebdomadaire de la production électrique*. PhD thesis, Ecole Polytechnique, Paris, 2008.

[3] R. Apparigliato, J.-P. Vial, and R. Zorgati. Optimisation robuste linéaire : application à la gestion court terme d'une vallée hydraulique. Working paper H-R32-2007-00658-FR, EDF R&D, Département Osiris, 1, avenue du Général de Gaulle, F-92141 Clamart, 2007.

[4] A. Ben-Tal, S. Boyd, and A. Nemirovski. Extending the scope of robust optimization: Comprehensive robust counterparts of uncertain problems. *Mathematical Programming Series B*, 107(1-2):63–89, 2006.

[5] A. Ben-Tal, L. El Ghaoui, and A. Nemirovski. *Robust Optimization*. Princeton University Press (Book under review), 2009.

[6] A. Ben-Tal, B. Golany, A. Nemirovski, and J.-Ph. Vial. Retailer-supplier flexible commitments contracts: a robust optimization approach. *Manufacturing and Service Operations Management*, 7:248–271, 2005.
[7] A. Ben-Tal, B. Golany, and S. Shtern. Robust multi-echelon, multi-period inventory control. Technical report, 2008. To appear in *European Journal of Operational Research*.
[8] A. Ben-Tal, A. Goryashko, E. Guslitzer, and A. Nemirovski. Adjustable robust solutions of uncertain linear programs. *Mathematical Programming*, 99(2):351–376, 2004.
[9] A. Ben-Tal and A. Nemirovski. Robust convex optimization. *Mathematics of Operations Research*, 23:769 – 805, 1998.
[10] A. Ben-Tal and A. Nemirovski. Robust solutions of linear programming problems contaminated with uncertain data. *Mathematical Programming Series A*, 88:411 – 424, 2000.
[11] A. Ben-Tal and A. Nemirovski. *Lectures on Modern Convex Optimization: Analysis, Algorithms and Engineering Applications*. MPS-SIAM series on Optimization. Society for Industrial and Applied Mathematics and Mathematical Programming Society, 2001.
[12] A. Ben-Tal and A. Nemirovski. Selected topics in robust convex optimization. Technical report, Faculty of Industrial Engineering and Management, Technion, Technion city, Haifa 32000, Israel, 2006. (To appear in *Mathematical Programming*).
[13] D.P. Bertsekas. *Dynamic Programming and Optimal Control*. Athena Scientific, Belmont, Mass., 1995.
[14] D. Bertsimas, D. Pachamanova, and M. Sim. Robust linear optimization under general norms. *Operations Research Letters*, 32:510–516, 2004.
[15] D. Bertsimas and M. Sim. Price of robustness. *Operations Research*, 52:35–53., 2004.
[16] D. Bertsimas and A. Thiele. A robust optimization approach to inventory theory. *Operations Research*, 54(1):150–168, 2006.
[17] D. Bienstock. Experiments in robust portfolio optimization. Technical report, Center for Financial Engineering, Columbia University, January 2007.
[18] J. R. Birge and F. Louveaux. *Introduction do stochastic programming*. Springer-Verlag New York, Inc., 1997.
[19] G. C. Calafiore. Ambiguous risk measures and optimal robust portfolios. Technical report (to appear in *SIAM Journal on Optimization* 2007), Dipartmento di Automatica e Informatica, Politecnico di Torino, Italy, 2007.
[20] G. C. Calafiore and L. El-Gahoui. On distributionally robust chance-constrained linear programs. *Journal of Optimization Theory and Applications*, 130:1–22, 2006.
[21] S. Ceria and R. Stubbs. Incorporating estimation errors into portfolio selection: Robust portfolio construction. *Journal of Asset Management*, 7(2):109 – 127, 2006.
[22] A. Charnes and W.W. Cooper. Chance constrained programming. *Management Science*, 6:73–89, 1959.

[23] A. Charnes, W.W. Cooper, and G.H. Symonds. Cost horizons and certainty equivalents: an approach to stochastic programming of heating oil. *Management Science*, 4:235–263, 1958.

[24] X. Chen, M. Sim, and P. Sun. A robust optimization perspective of stochastic programming. Working paper, NUS Business school, 2005.

[25] G. B. Dantzig and A. Madansky. On the solution of two-stage linear programs under uncertainty. In Jerzy Neyman, editor, *Fourth Berkeley Symposium on Mathematical Statistics and Probability (June 20-July 30, 1960)*, volume 1, pages 165–176. University of California Press, 1961.

[26] G.B. Dantzig. Linear programming under uncertainty. *Management Science*, 1(3-4):197–206, 1956.

[27] Fragnière E. and Haurie A. Markal-geneva: a model to assess energy-environment choices for a swiss canton. In *Carraro C, Haurie A (eds) Operations research and environmental management*. Kluwer Academic Books, Dordrecht, 1996.

[28] L. El Ghaoui and H. Lebret. Robust solutions to least-square problems to uncertain data matrices. *SIAM Journal of Matrix Analysis and Applications*, 18:1035–1064, 1997.

[29] S. J. Garstka and R. J.-B. Wets. On decision rules in stochastic programming. *Mathematical Programming*, 7:117–143, 1974.

[30] D. Goldfarb and G. Iyengar. Robust portfolio selection problems. *Mathematics of Operations Research*, 28:1 – 38, 2003.

[31] C.C. Holt, F. Modigliani, and H.A. Simon. A linear decision rule for production and employment scheduling. *Management Science*, 2(1):1–30, 1955.

[32] G. Iyengar. Robust dynamic programming. *Mathematics of Operations Research*, 30:257 – 280, 2005.

[33] P. Kall and S. W. Wallace. *Stochastic Programming*, volume 2nd edition. John Wiley & Sons, Chichester, 1994.

[34] L.B. Miller and H. Wagner. Chance-constrained programming with joint constraints. *Operations Research*, 13, 1965.

[35] A. Ouorou and J.-P. Vial. A model for robust capacity planning for telecommunications networks under demand uncertainty. In *Proceedings of the 6th International Workshop on Design and Reliable Communication Networks, DRCN 2007*, 2007.

[36] A. Prékopa. On probabilistic constrained programming. In *Proceedings of the Princeton Symposium on Mathematical Programming*, pages 113–138. Princeton University Press, Princeton, 1970.

[37] A. L. Soyster. Convex programming with set-inclusive constraints and applications to inexact linear programming. *Operations Research*, 21:1154–1157, 1973.

[38] J.F. Sturm. Using SeDuMi 1.02, a Matlab Toolbox for Optimization over Symmetric Cones (updated for version 1.05). Technical report, 2001.

Chapter 4
Coping with Climate Risks in Indonesian Rice Agriculture: A Policy Perspective

Rosamond L. Naylor and Michael D. Mastrandrea

Abstract Policymakers responsible for agriculture, rural economic growth, and food security worldwide are confronted by a myriad of climate-related risks. This chapter provides a framework for using climate information in the design of policy to manage such risks, highlighting several tools of analysis that can be applied in the context of both climate variability and global climate change. The concepts are developed through a case study of the rice sector in Indonesia, a country directly affected by climate variability related to El Niño-Southern Oscillation events. The risk assessment model is based on the probability of climate events, critical thresholds of damage related to those events, and the role of policy in reducing climate-related impacts on agricultural systems. Because risk assessment involves estimation of both the probability of climate events and the expected consequences of those climate events, Bayesian analysis is applied to show how climate information can be used to update subjective probabilities over short and long time scales. Bayesian updating can help reduce the chances that policymakers will make the wrong policy decision given all of the available information. Even with these tools, however, the chapter shows that there is always the chance for Type I and Type II errors in policymaking, because the climate system is impossible to predict with certainty.

4.1 Introduction

Farmers throughout the world operate under high-risk conditions. Each season, they grapple with a variety of unanswered questions: When will the rains come? Will

Rosamond L. Naylor and Michael D. Mastrandrea
William Wrigley Senior Fellow and Director of the Program on Food Security and the Environment, and Research Associate at the Woods Institute for the Environment, Stanford University.
Their work draws on substantial research collaborations over time with Walter Falcon, David Battisti, Dan Vimont, and Marshall Burke. The authors thank these scholars and Kate Niemer Johnson for their helpful comments and assistance on this chapter. The work was funded by the NSF Human and Social Dynamics Program, award 0433679.

J.A. Filar and A. Haurie (eds.), *Uncertainty and Environmental Decision Making*, International Series in Operations Research & Management Science 138,
DOI 10.1007/978-1-4419-1129-2_4,

rainfall be sufficient for crop growth? Will there be heat waves or frosts during the growing cycle? What will be the market price for their crops at the time of harvest? Will a crop disease or pest infestation lower yields? Risk plays a dominant role in agricultural decision-making, but its role in regional food security—defined here as access to affordable food for all people at all times—is even more pronounced. Policymakers face major challenges, particularly in less developed countries, in guaranteeing widespread access to food at reasonable prices given the inherent uncertainties in weather and the changing nature of global food markets resulting from policies and production in other countries. The world food crisis that arose in the spring of 2008—which was characterized by unusually high and unstable prices for key staple commodities and which led to food riots in several countries—is but one reminder of the food security challenges faced by governments (Naylor and Falcon, 2008).

In this chapter, we focus specifically on agricultural risks and uncertainties related to climate variability and global climate change from a policy viewpoint. Policymakers have little control over the weather, which is driven by very short-run (hourly to daily) patterns in atmosphere and ocean circulation. With good scientific information, however, policymakers in many regions can anticipate longer-run (monthly, yearly, decadal) climate variability and climate change reflected in patterns of temperature and precipitation. Such climate fluctuations involve structural dynamics in the physical system that can be modeled and projected with varying degrees of certainty over different spatial and temporal scales. To the extent that climate variability and change in the mean state can be projected, governments can then facilitate adaptation; that is, they can augment markets by implementing policies to promote domestic food security via trade (e.g., arrange for food imports when crop production is expected to decline domestically), investments (e.g., fund crop research or improvements in irrigation infrastructure), and early-warning systems or safety-net programs for vulnerable populations within their countries.

There are large literatures in both the climate projection and food policy fields,[1] but the integration of climate information into food policy analysis has not been well developed. Our goal in this chapter is to provide a framework for using climate information in the design of food policy, highlighting several tools of analysis that can be applied in the context of both climate variability and global climate change. Despite widespread attention on the potential impacts of climate change following the most recent assessment by the Intergovernmental Panel on Climate Change (IPCC, 2007), the easiest entry point for evaluating climate-related risks in agriculture and food security is arguably in the climate variability domain. Patterns of climate variability, particularly *El Niño-Southern Oscillation (ENSO)* dynamics, have been measured for decades; observational data are easily accessible on the internet and have been used widely in food and agriculture impact assessments (see for example, Cane et al., 1994; Nicholls, 1986; Mjede et al, 1998; Hansen et al., 1998; Podesta et al, 1999; Pfaff et al., 1999; Broad and Agrawala, 2001; Naylor et al. 2001, 2007; Falcon et al., 2004). Furthermore, the short-term nature of ENSO

[1] The literature in each of these fields is too extensive to list here; see, for example, papers cited in IPCC (2007) and The World Bank (2007).

dynamics relative to the much longer time-scale of global climate change provides cyclical data from which to estimate risks and devise frameworks for prediction. We thus begin with climate variability and discuss a set of analytical issues that are key to applying climate information in policy formulation. We then move to the topic of global climate change from a food policy perspective.

The chapter starts with a general discussion of climate risks and uncertainties related to agricultural production, food policy, and food security. The principles of risk analysis and the use of climate information in food policy analysis are then developed using Indonesian rice agriculture as an illustrative example. Indonesia provides an excellent template for the methodology because the country is located on the western edge of the Pacific Ocean and is affected directly by ENSO dynamics. Rice, the main staple crop for the country's ~220 million citizens, depends importantly on the timing and amount of rainfall throughout the year, and thus the food security implications of climate variability are significant. Indonesia is also the world's fourth most populous country and one of the true success stories in terms of poverty reduction during the past 40 years. However, with half of its population still living under $2 per day, food security remains a key policy concern. The chapter then moves briefly into the global climate change arena and concludes with some comments on how the principles developed for the Indonesian rice example might be generalized to other regions and food systems.

4.2 Climate Risks in Agriculture

Risk is broadly defined here as the probability (P) of an event occurring times the consequence (C) of that event on human or natural systems. For the purpose of this chapter, risk is defined as the vulnerability of agricultural systems (crop production and food security) to changes in precipitation and temperature patterns caused by climate variability and climate change.

4.2.1 Climate Variability and ENSO

Climate variability encompasses inter-annual or decadal fluctuations in precipitation and temperature patterns driven by coupled ocean-atmosphere circulation dynamics. Such dynamics can affect planting dates, crop yields and rotations, farm income, and food prices at local to global scales.

ENSO is a dominant and recurring pattern of climate variability in the eastern equatorial Pacific that is characterized by anomalies in sea-surface temperature (referred to as *El Niño* and *La Niña* for warming and cooling periods, respectively), and by seasonal fluctuations in sea level pressure between Tahiti and Darwin, Australia (*Southern Oscillation*). Changes in sea surface temperature and sea level pressure in the Pacific are related to wind intensity along the equator and concurrent shifts in the depth of warm water (measured by the warm water volume above the *thermocline* — the transition point between the upper well-mixed ocean layer and the deep ocean) between east and west. Neutral ENSO conditions are characterized by moderately strong easterly winds and a deep thermocline in the western Pacific, causing convection over the western portion of the Pacific and the onset of normal monsoon rains over the Indonesian archipelago (Figure 1). During El Niño (warm mode) periods of the cycle, the easterly winds slacken and sometimes even reverse direction, shifting the concentration of warm water from the western to the central Pacific Ocean. As a result, convection occurs over the central Pacific, and dry conditions tend to prevail in the western Pacific. In extreme El Niño years, warm water is concentrated in the eastern Pacific where there is also a deeper thermocline and reduced upwelling near the surface—and thus lower productivity of pelagic fisheries such as Peruvian anchoveta (Pauly and Tukayama, 1987).

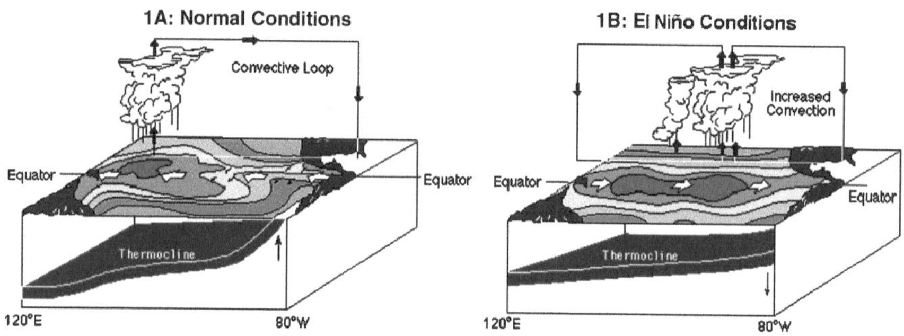

Fig. 4.1 The coupled ocean-atmosphere system determining ENSO conditions. Fig. 1A shows neutral conditions. Fig. 1B shows (warm mode) El Niño conditions.

There are three key points to note concerning the ENSO pattern of climate variability and its impact on agriculture. First, the cycle has seasonal regularity, as depicted in Figure 4.1A for 1997 (strong El Niño conditions) and 1998 (transition to La Niña conditions). ENSO conditions can be detected starting in March or April, and they typically build through the remainder of the calendar year, peaking around December. Second, there is a recurrent pattern to ENSO in general, with El Niño years followed by La Niña years (interspersed with neutral years) oscillating on 3-7 year time scales (Figure 4.1B)—much like a bathtub of warm water sloshing back and forth. And third, ENSO patterns have strong climate teleconnections throughout the world with related impacts on agricultural systems in countries as diverse as Zimbabwe, Australia, the U.S., Argentina, and Indonesia (Cane et al., 1994; Nicholls, 1986; Hansen et al., 1998; Podesta et al, 1999; Naylor et al., 2001). The strongest and most direct ENSO-agriculture connections are found in countries along the equator bordering the Pacific Ocean.

The figures 11.2 and 11.3 below show Niño 3.4 sea surface temperature anomalies (SSTAs) in degrees C on the y-axes, recorded by buoys in the Pacific region

(5 ° N-5 ° S)(170-120 ° W) and measured against a long-term sea surface temperature base (1971-2000). [2]

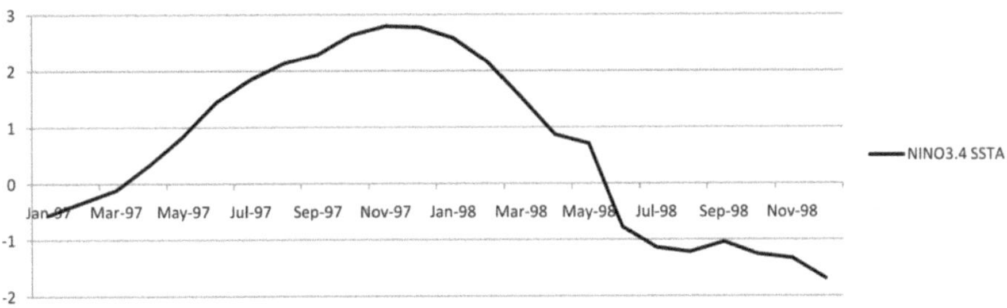

Fig. 4.2 Seasonality of El Niño/La Niña events, 1997-98.

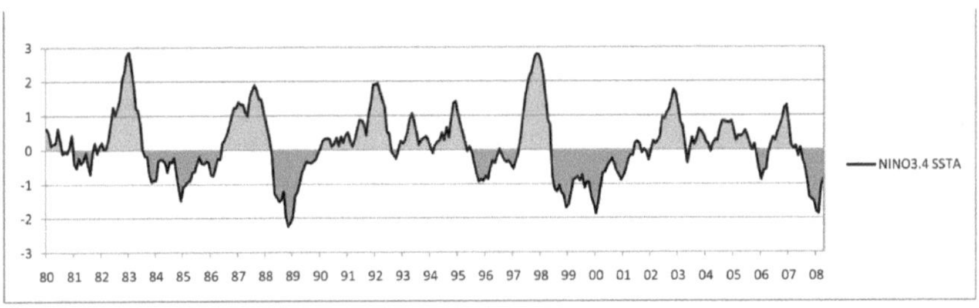

Fig. 4.3 Oscillation of ENSO events over years, January 1980 to August 2008.

4.2.2 Global Climate Change

Climate risks in agriculture stem from longer-term changes in the mean climate state as well as variability[3]. Anthropogenic climate change—often referred to as global

[2] El Niño conditions are characterized by SSTAs in the range of 1.5 to 3.0 ° C, while La Niña conditions are in the -1.0 to -2.0 ° C range. ENSO conditions can also be measured by sea level pressure represented by the Southern Oscillation Index. Source: National Weather Service Climate Prediction Center: `http://www.cpc.ncep.noaa.gov/data/indices/` (accessed January 12, 2009)

[3] Climate variability will continue to exist on top of a change in the mean climate state in the longer-run future. At present, the best assumption is that the current patterns of climate variability will prevail in future decades with climate change, although this is a topic of ongoing research within the climate science community (David Battisti, personal communication). Only one of the global climate models (the Hadley Center Model) contributing to the IPCC 4th Report in 2007 contains explicit ENSO dynamics.

warming—is defined by significant shifts in average temperature and precipitation levels regionally and worldwide (IPCC 2007). The agricultural sector is likely to be affected more directly than any other sector by rising temperatures, sea level rise, and changing precipitation patterns throughout the world. Agriculture also contributes to global climate change through carbon release from land clearing, nitrous oxide emissions from fertilizer use (denitrification), and methane release from rice fields and ruminant livestock (e.g., cattle and sheep) (Steinfeld et al. 2007).

Predicting climate conditions decades in advance involves uncertainty about future emissions of greenhouse gases at the global scale; the sensitivity of atmosphere, ocean, and terrestrial dynamics to greenhouse gas concentrations; and biophysical and behavioral feedbacks over time. Nonetheless, there is broad consensus among the some 20 global climate models (GCMs, also known as general circulation models) considered by the IPCC on three points (IPCC 2007). First, all regions will become warmer. The marginal change in temperature will be greater at higher latitudes, although tropical ecosystems are likely to be more sensitive to projected temperature changes due to the evolution of species under more limited seasonal temperature variation (Deutsch et al. 2007). Average seasonal temperatures for the main cropping periods in the tropics and sub-tropics are very likely to exceed the hottest years on record by the end of the century (Battisti and Naylor, 2009). Second, soil moisture is expected to decline with rising temperatures and increased evapotranspiration in many areas of the sub-tropics, leading to sustained drought conditions in some areas and flooding in other areas where rainfall intensity increases but soil moisture decreases. Third, sea level will rise globally with thermal expansion of the oceans and ice melt. Projections of regional and seasonal precipitation patterns several decades out reflect substantial variation (Allen and Ingram, 2002; Chou and Neelin, 2004; IPCC 2007). Most models project increased rainfall with warming in the deep tropics and at high latitudes, and decreased rainfall over large land areas of the subtropics—that is, regions that are currently wet are expected to become wetter, and dry regions are expected to become drier. However, the annual cycles of rainfall are unclear from the GCMs.

4.3 Risk Assessment

In order to evaluate the risks of climate variability and changes in the mean state on agricultural systems, the probability of any climate event occurring must be multiplied by the expected consequence of such an event on crop production, farm incomes, or food security. Figure 4.4 illustrates four categories of risk-related outcomes: those with low probability and low consequence (low risk); those with low probability and high consequence (medium risk); those with high probability and low consequence (medium risk); and those with high probability and high consequence (high risk). Policymakers are most concerned with the high-risk category, but attention is also warranted for medium-risk events, particularly when the projected consequences are large (e.g., the impact of Hurricane Katrina on New Orleans in 2005 due to weak levies and flooding).

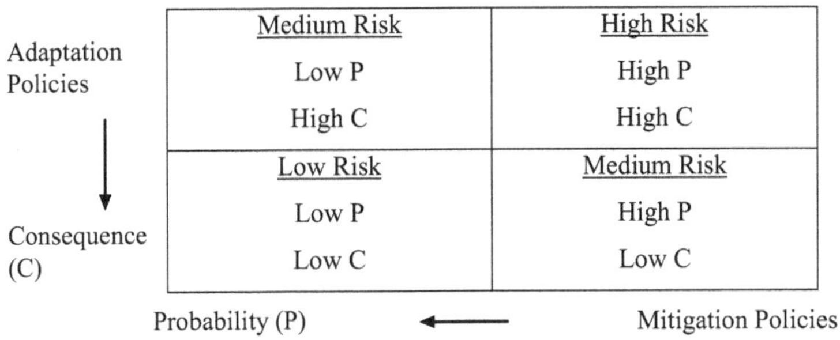

Fig. 4.4 Assessing risks: [probability of climate change] × [consequence for agriculture]

There are two ways to reduce risk according to the framework shown in Figure 4.4 (Pittock and Jones, 2001, IPCC 2007). The first is through adaptation to climate variability or climate change, which reduces the consequences. Adaptation can be planned (e.g., action to reduce consequences as a result of anticipated change, such as investing in drought- or heat-tolerant crops) or autonomous (e.g., measures taken to reduce the consequences irrespective of climate projections, such as changing planting dates according to the actual timing of rainfall). The other way of reducing risk is through mitigation, which lowers the probability of climate changes occurring in the first place. In this chapter, we focus exclusively on planned climate adaptation related to food and agricultural policy, as opposed to autonomous adaptation or mitigation. It is important to note, however, that mitigation and adaptation must be complementary strategies, as it is very likely that increasingly severe consequences from climate change will eventually exceed limits to adaptation.

Because risk assessment involves estimation of both the probability of climate events and the expected consequences of those climate events, it involves a process of knowledge updating, particularly with respect to the likelihood of climate events and how the impact of those climate events might propagate through the agricultural system (Jones, 2003). Bayesian analysis can be used over long time scales (e.g., how average temperatures or precipitation patterns are changing regionally over time). It can thus serve as a tool for quantifying the degree of climate uncertainty in risk assessment over time.

Bayesian analysis can be used to update climate information on short time scales (e.g., how an El Niño event is developing throughout the year) and over long time scales (e.g., how average temperatures or precipitation patterns are changing regionally over time) and can quantify the degree of climate uncertainty in risk assessment over time.

In addition to climate updating, the consequences of climate changes on agriculture and the role of policy in altering the magnitude of those consequences through planned adaptation should also be included in the risk assessment framework (Carter et al., 1994; Jones, 2001; Anand, 2002). Projecting future consequences thus involves additional uncertainties with respect to biophysical crop responses, farmers' behavior, the introduction of new agricultural technologies and infrastructure, changes in markets, and policy feedbacks in the system.

Determining which climate outcomes have high or low consequences is often a subjective issue depending on political and/or societal priorities. The likelihood of exceeding thresholds that are critical for crop production, agricultural income generation, or food prices is one robust way, within a risk framework, of assessing potential damages (Carter et al., 1994; Parry et al., 1996; Kenny et al, 2000; Jones, 2001, 2003). A wide variety of threshold concepts—biophysical, behavioral, and user-designed thresholds—are used in climate impact studies (Jones, 2000). Each concept indicates a point beyond which the biophysical, socioeconomic, or institutional system in question is significantly affected by, or fundamentally changes in response to, patterns of climate variability or change. Figure 4.5 illustrates the threshold concept using climate variability as the mechanism of change; the threshold chosen for this illustration is successive months of drought in the main cropping season.

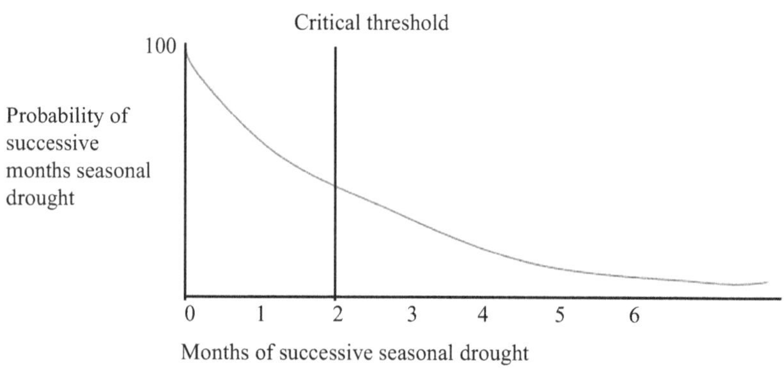

Fig. 4.5 Illustration of critical threshold of exceedance

Assessing food security risks according to the probability of climate events, critical thresholds of damage related to those events, and the role of policy in reducing climate-related impacts on agricultural systems provides a discrete mechanism for analysts to use new information and identify important targets for policy intervention. This approach differs from optimization, which typically entails a sense of "command-and-control" over the variables being optimized. When high levels of uncertainty exist, it is always possible to learn more about the dynamics of a system, making the search for an optimal solution somewhat unrealistic (Ludwig et al., 2001). The approach here also avoids the "black box" problem of obscuring important variables in the analysis.

Although a framework can be developed that uses updated information to aid in policy decisions, policymakers do not have perfect foresight, and there remains a risk that they will opt for the "wrong" strategy. What are the consequences of going down the wrong path? The answer can be captured by the concept of Type I and Type II errors, which occur when decision makers wrongly accept or reject a particular null hypothesis. Assume, for example, a null hypothesis that there will be no change in climate from the average state (e.g., neutral conditions are assumed for the ENSO cycle in the short run, and no change in the mean climate state or other sources of climate variability is assumed in the long run). A Type I error would occur if this hypothesis was rejected when it should have been accepted (false positive); that is, policymakers would assume a high likelihood of climate change and allocate funds to adapt to such expected change even though no change occurs in the end. The cost in this case would be largely financial (e.g., in the form of investments for imports, R&D, and infrastructure), and farmers might suffer over time from acquiring inappropriate technology or by losing market shares to imported commodities. A Type II error would occur if the hypothesis was accepted when it should have been rejected (false negative). Policymakers would assume a minimal likelihood of change in climate, and hence would not allocate resources to climate adaptation, when in fact such change occurs. The cost in this case is potentially more serious. The country might not have the capability to produce sufficient food, arrange for timely food imports, distribute staple foods domestically, or generate adequate incomes to purchase food, thus leaving its population, particularly lower income groups, at risk of malnutrition over short- to long-run periods.

Food and agricultural policy in the context of climate variability and climate change is indeed a risky process, and good policymakers worry about both Type I and Type II errors. In the sections that follow, we develop an example of Indonesian rice agriculture to explore how the risk principles outlined above can be put into operation in a policy context.

4.4 Indonesian Rice Agriculture

The Indonesian archipelago has about 17,500 islands straddling the equator, 9000 of which are inhabited. The country lies at the edges of the Pacific, Eurasian, and Australian tectonic plates and is thus an active site for volcanic eruptions, earthquakes, and tsunamis. It has distinct ecological zones, characterized by Wallace's Line that runs north-south along the Sunda Shelf with predominantly Asian species to the west and Australian species to the east. Indonesia's tropical climate, geography, large forested areas, and extensive coastline (~80,000 km) support the world's second highest level of biodiversity after Brazil. Moreover, with 400 language groups, more Muslims than any other country, and a rich diversity of Catholics, Protestants, Hindus, Buddhists, and animists, Indonesia is one of the most culturally diverse countries in the world. It is thus a fascinating country to study in its own right, and also an excellent model for assessing the risks of climate variability on agriculture given its location on the western edge of the Pacific.

Indonesia's cropping systems vary from Bali's manicured irrigated rice systems, to NTT's rainfed corn and sorghum based systems, to the Maluka's dryland cassava systems, to Borneo's oil palm plantations, to Papua's sago palm and sweet potato systems. Rice is the main staple food in most, but not all, parts of the country and the rice sector is the largest employer in the agricultural economy overall. The islands of Java and Bali, which contain over half of Indonesia's population and account for roughly 55 percent of the nation's rice production, are dominated by rice agriculture, particularly in the monsoon season. Because of its importance in national food security, rice is the focus of our risk framework on climate and agriculture.

Rice production in Indonesia is strongly influenced by annual and interannual variations in precipitation caused by the Austral-Asia monsoon and El Niño-Southern Oscillation (ENSO) dynamics. Indonesia consistently experiences dry climatic conditions and droughts during the warm phase of the ENSO cycle (El Niño), with significant consequences for rice output, rural incomes, and staple food prices.[4] Over the longer run, rising concentrations of greenhouse gases will likely create additional climate impacts on Indonesian rice agriculture. The combined forces of climate variability and climate change could have dramatic effects on agricultural production in Indonesia (Naylor et al., 2007). As a result, policymakers are beginning to focus more seriously on the vulnerability of rural populations to climate events (UNDP, 2007).

Understanding the seasonality of rainfall and rice production is key to assessing climate-related risks, because rice plantings follow the rains in the "run-of-the-river" irrigated systems and rainfed systems that are typical on Java, Bali, and other rice-producing provinces. Figure 4.6 illustrates the regional pattern in the monsoon movement, with onset beginning in the northwest, progressing toward the southeast, and terminating initially in the east. The peak of the wet season typically occurs in December-January when the northwest monsoon (which extends from the western Indian Ocean to the date line) sweeps across Java and Bali toward Australia (Hamada et al., 2002; Hendon, 2003). The largest absolute variation in rainfall occurs at the end of the dry season during monsoon onset (September-November) (Hendon, 2003).

In neutral ENSO years, the main wet-season rice crop in Java is planted between late October and early December when there is sufficient moisture to prepare the land for cultivation and to facilitate the early rooting of transplanted seedlings. The main planting season occurs before the peak of the winter monsoon (December-January), because excessive water at the vegetative growth stage hampers rooting and decreases tiller production (De Datta, 1981). During the 90-120 day grow-

[4] Studies of ENSO impacts on crop production and famines have been published for various regions of Indonesia. See for example Kirono and Khakhim (1998), Kirono and Tapper (1999), Fox (2000), Naylor et al. (2001), and Falcon et al. (2004).

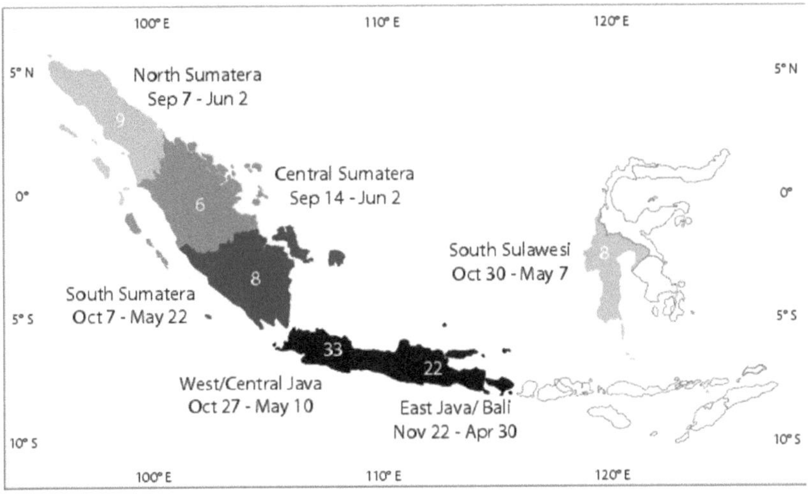

Fig. 4.6 Indonesia's rice producing provinces, with corresponding monsoon onset and termination dates, and corresponding percentage of total national rice production (2004) represented by each province in white numbers. Onset dates are days past Aug 1 when accumulated rainfall equals 20 cm, averaged over reporting rainfall stations in the region for the years 1979-2004; termination date refers to day on which 90% of that year's rainfall has accumulated. (Taken from Naylor et al., 2007).

out period from transplanting to harvest, at least 20 cm of cumulative rainfall is needed to moisten the ground sufficiently for planting, and about 100 cm of rainfall is needed throughout the season for cultivation (IRRI, 2004). A smaller dry season planting takes place in April-May after the wet season crop is harvested.

The potential consequences of El Niño events on rice production and food security in Indonesia relate mainly to timing, as illustrated in Figure 4.7 for West Java, one of the country's major rice bowls. Strong El Niño events can cause a delay in monsoon onset by as much as two months, thus delaying rice plantings or causing farmers to switch to crops that require less moisture early in the production cycle, such as maize (Naylor et al., 2001; Falcon et al., 2002). Since the rains eventually do come, there is little impact of ENSO on rice yields. A delay in planting in turn extends the "hungry season" (*paceklik*, the season of scarcity) before the main rice harvest. In addition, delayed planting of the main wet season crop may not be compensated by increased planting later in the crop year, leaving Indonesia with reduced rice area and a larger than normal annual rice deficit. In the absence of policy intervention, these two effects can drive prices up in domestic and international markets—with disproportionate impacts on poor net consumers of rice (Falcon et al., 2004). During La Niña years, rains typically arrive earlier than normal, allowing for early planting and harvesting of rice.

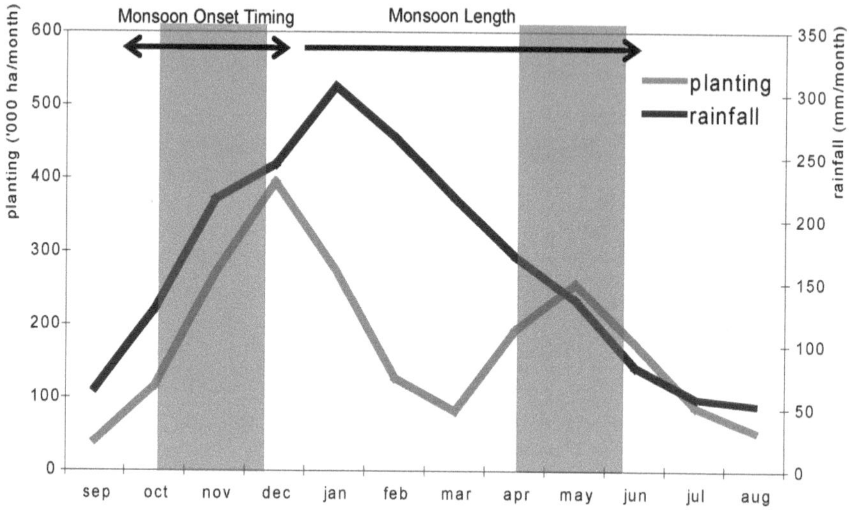

Fig. 4.7 Annual cycle of rainfall (top line and right-hand axis) and rice planting (bottom line and left-hand axis) in West Java based on observational data from 1971-2005 (Indonesian Central Bureau of Statistics database). There are two periods of rice planting that correspond to rainfall and "run-of-the-river" irrigation. The sensitive periods for risk assessment are the timing of monsoon onset and the monsoon length (and termination), shown by the vertical shaded bars.

Given the observed consequences of a monsoon delay for the Indonesian rice economy in El Niño years, related research defined a 30-day delay in monsoon onset as a critical threshold for risk assessment (Naylor et al., 2007).[5] The likelihood of exceeding this threshold is high in strong El Niño years when the SSTA is in the 1.5 to 3.0 °C range. Based on historical data, a 30-day delay in the monsoon onset causes rice production on Java and Bali to fall by 1.12 million tons on average for the January-April harvest season alone. Production declines of this magnitude when scaled up to the national level are a serious matter for Indonesian policymakers; in strong El Niño years, it is not uncommon for the government to import 10-20% of total world rice trade (Falcon et al., 2004; Naylor et al., 2007).

The ability to use climate predictions in agricultural policy decisions is thus critical. If prior information on major El Niño events can be applied wisely, policymakers can prevent periods of rice shortages, elevated prices, malnutrition among the poor, and political unrest by arranging food imports, distributing rice appropriately

[5] Onset is defined as the number of days after August 1 when cumulative rainfall reaches 20 cm, and delay is defined as the number of days past the mean onset date (averaged over the period 1979-2004). Statistical analysis of the observational record shows a correlation between onset delay and total rainfall in September-December (when the main rice crop is planted) of -0.94 for West/Central Java and -0.95 for East Java/Bali, indicating that delayed monsoon onset is associated strongly with decreased total rainfall in this period (Naylor et al., 2007).

within the country, and facilitating the consumption of alternative staples such as cassava or maize.

With this background in mind, we move to the quantification of climate risk. The first step is to develop an empirical model of El Niño impacts on Indonesian rice production, and to show how this model can be used in forecasts for the purposes of food policy. The next step is to develop a method for estimating the probability of El Niño events with Bayesian updating. Within this risk assessment framework, it becomes clear that climate-related uncertainties can lead to large social costs depending on the policy response to projected El Niño events.

4.5 Projecting the Impact of El Niño Events on Rice Production

In creating an empirical model of El Niño effects on rice production in Indonesia, a few key questions come immediately to mind: How should the model be constructed to capture the seasonality of rice production and ENSO events? And at what spatial scale should the model be designed in order to address the food security concerns of Indonesian policymakers? It becomes clear, based on Figure 4.8, that using calendar year data to analyze El Niño impacts would not be successful. Instead, using crop year data that extend from September (beginning of the main rice planting season) to August (end of the second dry season), and which match El Niño cycles that influence the timing and amount of rainfall with accumulating force from September to December, would lead to more significant results. Fortunately, Indonesian agricultural statistics are available by trimesters (January-April, May-August, September-December) as illustrated in Figure 4.8, and these trimesters correspond closely with rice production and ENSO cycles. SSTA data during the May-August trimester provides some clues as to whether or not an El Niño event is emerging, as we discuss in the following section. Earlier work shows that the August Niño3.4 SSTA is a good predictor of crop-year rice production in Indonesia, accounting for over 60% of the year-to-year variability (Naylor et al., 2001; Falcon et al., 2004).

The spatial scale of analysis is also important and depends fundamentally on the question being asked. If the question is how individual farmers and communities can prepare for and adapt to El Niño events, then creating a model at the sub-province district (*kabupaten*) or local (*kecamatan*) level makes sense[6].However, if the question focuses on national food security, rice prices, and rice supplies across different provinces of Indonesia, then the scale of analysis should be national or provincial. Analysis at the larger scale is particularly appropriate if El Niño predictions are being used to assist policymakers in planning decisions about rice imports, rice dis-

---

[6] The International Research Institute for Climate and Society (IRI) at Columbia University is pursuing this decentralized approach in Indonesia. See `http://portal.iri.columbia.edu/portal/server.pt/gateway/PTARGS{\_}0{\_}0{\_}4688{\_}430{\_}0{\_}43 http{\%}3B/portal.iri.columbia.edu{\%}3B9086/irips/projectview.jsp?id=69` (accessed on September 24, 2008)

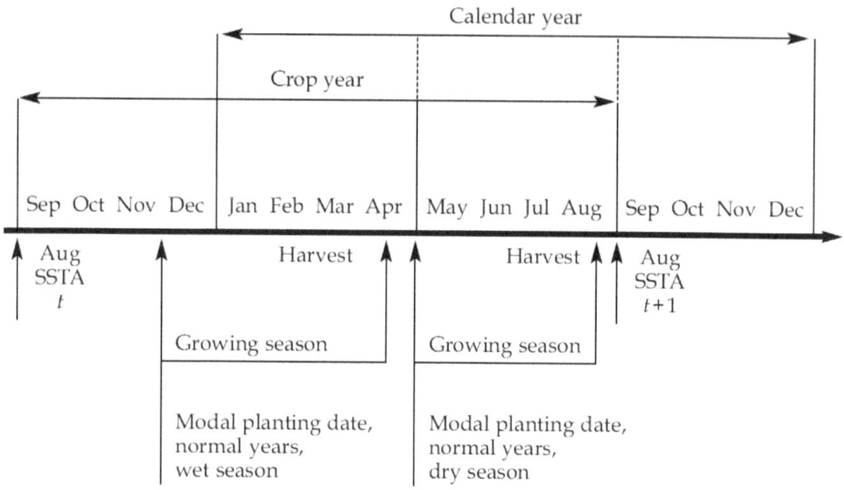

Fig. 4.8 Crop calendar for Indonesian rice production (taken from Falcon et al, 2004)

tribution among provinces, and national agricultural policy initiatives. With only 6 to 7 percent of global rice production traded in international markets each year (the vast majority is produced and consumed within individual Asian countries), shortfalls in domestic rice production in any given year can be difficult to cover on a timely basis through trade without prior planning. And as Asian governments can attest, shortfalls and sudden price spikes for rice can generate food riots and threaten the political stability of governments, as seen in 1973-4 and again in 2008 (Naylor and Falcon, 2008).

Using national or provincial-scale data for the past 25 years, a simple model can be created to estimate the level of paddy (unhusked rice) production and the contribution of ENSO to variation in paddy production over time.[7] Paddy production in Indonesia increased from 38.4 mmt in 1983/4 to 54.3 mmt in 2006/7. Because continual growth in production—driven primarily by technological change—overshadows year-to-year variation in production, the time trend must be removed. The model captures cropped area, yields and total production of paddy and can be disaggregated on a seasonal and provincial basis. Equation (10.1) shows the model of national paddy production as a function of SSTAs and time/technology with data covering the 1983/4 to 2006/7 crop years.[8]

$$\text{PaddyProduction}_{\text{Sep(t)}-\text{Aug}(t+1)} = \alpha - \beta(\text{Nino}\,3.4\text{SSTA}_{\text{Aug}(t)}) + \delta(time) - \phi(time^2) \quad (4.1)$$

[7] Paddy production data for Indonesia are available from 1983 forward from the Central Bureau of Statistics (Biro Pusat Statistik (BPS), various years). Paddy is the product from the farm; milled rice is what consumers purchase and what is traded in international markets. Milled rice is equivalent to about two-thirds the volume of unhusked paddy.

[8] This model is more formally developed in Falcon, Naylor et al (2004). A quadratic function for the independent variable on time is used because the rate of technological change decelerates over time.

with

Adj-R^2 = 0.94, DW = 2.34

and

Independent variable	Coefficient ('000)	t-stat
Constant (α)	36,600	43.55
Aug3.4 SSTA (β)	1,117	-3.58
Time (δ)	1,237	7.99
Time*2 (ϕ)	-20	-3.38

The model shows a significant causal influence of ENSO on production variability in Indonesia; a 1 °C increase in the SSTA in the Central Pacific leads to a 1.12 million ton decline in paddy production. Over the past 25 years the August SSTA has varied by as much as 3.4 °C from year to year, and thus a 1.1 million ton effect per degree can have a severe impact on total paddy production. When disaggregating the above model by season and province, the most significant results hold for the main monsoon-growing season in the most predominant paddy regions of Java and Bali (Falcon et al., 2004).

This model can also be used for forecasting. To predict seasonal or crop-year paddy production from equation (1), the August SSTA is plugged into the right hand side of the equation once the coefficients have been established from the time-series regression analysis (Falcon et al., 2004). For example, the model can be used to predict paddy production for the full crop year from September (t) through August ($t+1$) using the August Niño 3.4 SSTA in period t. The use of this model in predictions is straightforward. The August Niño 3.4 SSTA is compiled by NOAA and can be accessed online on a weekly basis.[9] The weekly averaged SSTA for the month of August (t) can then be applied to equation (1) to predict the direction and magnitude of change in paddy production by the end of the crop year ($t+1$) relative to the previous year (t) (Falcon et al., 2004).[10] Although it must be emphasized that the forecasts are probabilistic in nature, the relationship between SSTAs and paddy production in Indonesia is strong enough to warrant using this information for strategic purposes of contracting imports or planning for domestic paddy distribution on a regional basis, particularly when the August SSTA is $> |1|$. It is important, however, to update the model with the most current production numbers in order to capture the technological trend. Figure 11.7 shows the difference in production projections with two years of additional data. The updated data indicate that technological change in rice production remains strong, and that a deceleration in production should not be assumed.

The use of the observed August SSTA to predict paddy production for the upcoming crop year (September (t) to August ($t+1$)) provides significant lead-time

[9] See National Weather Service Climate Prediction Center:
`http://www.cpc.ncep.noaa.gov/data/indices/` (accessed January 12, 2009).

[10] Seasonal predictions can be estimated using the appropriate disaggregated equations by trimester as shown in Falcon et al (2004).

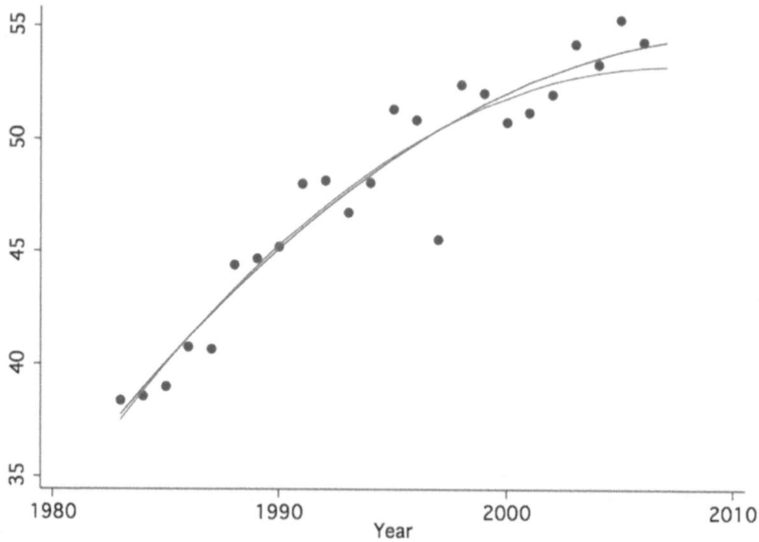

Fig. 4.9 Actual and predicted paddy production (mmt) for Indonesia (vertical axis) over time. The gray line shows model predicted values with 1983-2005 data, the lighter line shows model predicted values using 1983-2007 data, and the darker dots show actual production values. The difference between the two lines is attributed to a more robust technology trend than was earlier predicted. Source: BPS (various years); modeled by M. Burke, Stanford University.

for policymakers to plan for potential deficits or surpluses. It is also possible, as described below, to predict ENSO events with even greater lead-time using Bayesian updating to take advantage of all new information on the emergence of El Niño or La Niña conditions during the months leading up to the monsoon season.

4.6 Assessing the Probability of an El Niño Event

Given the potential impacts of El Niño events on seasonal and annual rice production, policymakers in Indonesia are anxious to know each year: Is an El Niño coming? There are two general approaches to ENSO forecasting that are currently employed: physically derived climate dynamic models (dynamic forecasting) and statistical relationships derived from time series and cross section data (empirical forecasting).[11] Dynamic forecasting generally relies on coupled atmosphere-ocean models that simulate circulation patterns controlling interannual variations in sea surface temperature and sea level pressure. Empirical relationships can range in complexity from simple regression relationships to much more complex models

[11] Mason and Mimmack (2002) provide a review of forecasting methods developed prior to 2002.

incorporating many physical variables such as sea surface temperature anomalies across time and space. ENSO forecasts generally have a time horizon of a few months to two years. Uncertainty and the spread among model forecasts increase as the forecast lead-time lengthens.

ENSO forecasts are generally presented in deterministic terms, without clear quantification of the uncertainty associated with a given forecast. One method to capture—and even reduce—uncertainty is through probabilistic forecasting using Bayes' Theorem (Bayes, 1764)[12]. Bayesian analysis provides a means for quantifying the subjective probability of a hypothesis being true—that is, what one believes to be the chance of an event occurring based on evidence, theory, and intuition—and updating that subjective probability as new information becomes available (a process commonly referred to as Bayesian updating). As shown in Appendix A, Bayes' Theorem focuses on the probability of a hypothesis (H) being true, given a set of evidence (E).

The Bayes' Theorem equation can be used to calculate the subjective probability that a certain year will be an El Niño year (our hypothesis, H), and to update that probability as evidence of El Niño conditions (E)—in the form of sea surface temperature anomalies (SSTAs)—accumulates over the course of the year. A relatively straightforward example can be used to demonstrate this process, using observations of monthly Niño3.4 SSTAs available on line for the period 1950-2007. First we use observed monthly SSTAs to calculate the probability of a given year being an El Niño year. The prior probability of a given year being an El Niño year, P(H), is calculated as the fraction of years from 1950-2007 that were El Niño years (34.5%). (El Niño years are defined here as those with at least five consecutive months of Niño3.4 SSTAs greater than 0.5 °C.) This probability is then updated month-by-month using Bayes' Theorem, as each month's SSTA becomes available.[13] In this calculation we use SSTAs for May-September of each year, the period during which El Niño events are physically gaining strength in the Pacific region prior to the main rice-planting season in Indonesia. Appendix A shows the formal definition of Bayes' Theorem and provides the details of this calculation with a numerical example, as well as a complete table of calculated forecasts for 1950-2007.

Figure 4.10 presents Bayesian forecasts of the probability of an El Niño event occurring based on May-September SSTA data for the years 1980-1986; this period was chosen because of its strong El Niño and non-El Niño (neutral and La Nina) cycles. The figure also shows monthly SSTA data for the entire annual cycle to illustrate the accuracy of the forecast method. In general, Bayesian updating provides accurate forecasts of El Niño conditions by June/July. For example, 1981, 1984, and 1985 exhibited consistently negative SSTAs throughout the summer, and the forecast of an El Niño event for each year quickly dropped close to zero. Likewise, 1982 exhibited positive and rapidly increasing SSTAs throughout the summer, and the forecast quickly rose toward 100%.

---

[12] There are alternative approaches to probabilistic forecasting; see for example the International Research Institute (`http://iri.columbia.edu/climate/ENSO/currentinfo/figure3.html`, accessed on September 12 2008).

[13] For the SSTA data source, see note 9 above.

However, as illustrated by 1980 and 1983, the forecast involves greater uncertainty for years in which SSTAs are positive in late spring but fall during the course of the summer. Forecasts for these particular years showed a rising probability of an El Niño event initially, but then the probability fell towards zero. Conversely, the forecast for 1986—a year in which SSTAs were negative at the start of the summer but quickly rose into El Niño territory—took longer to reach a high probability. Uncertainty over El Niño forecasts always decreases as the monsoon season begins in September/October.

In years when the May to September SSTAs change direction, policymakers would be wise to take advantage of the Bayesian technique by remaining cautious in their response to ENSO conditions. Rather than jumping on the bandwagon that "an El Niño is coming" too soon, policymakers should wait until late August or September to confirm the climate trend. If they wrongly assume that El Niño conditions will prevail when in fact neutral ENSO conditions result in the end (Type II error), they are likely to spend unnecessary funds to import rice for the main harvest season—a move that could lead to excess rice in the market at a time when domestic farmers rely on relatively high prices for their incomes. Under such circumstances, the government might need to store excess rice in order to keep producer prices stable—another costly budget decision—or they might simply let prices drop which would benefit consumers but hurt producers. Conversely, if policymakers remain complacent in their forecast that no El Niño will develop when in fact one does (Type I error), the consequences could be equally if not more severe. Without arranging for rice imports or domestic reallocation within the country, policymakers could face rice shortages, higher food prices, food riots, and rising malnutrition among the poor.

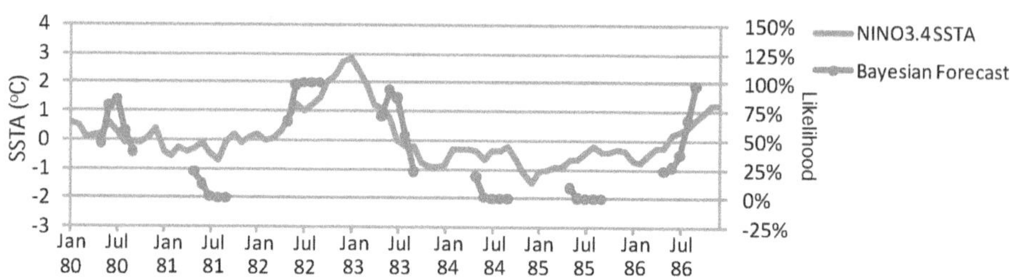

Fig. 4.10 Niño3.4 SSTAs (darker line and left axis), and yearly Bayesian forecasts of the probability of a given year being an El Niño year (lighter line and right axis). Forecasts are based on May-September SSTAs.

The forecast method demonstrated above is very simple, and there are a variety of ways in which it could be improved to reduce the potential for Type I and II errors. For example, information from year ($t-1$) can be used to inform the initial forecast for year (t) because, for example, there is a low probability that one strong El Niño year will be followed immediately by a second strong El Niño year. Additional information could be used in the updating process, such as monthly SOI data or SSTA data from a broader geographic band (e.g., 30 ° N to 30 ° S), to inform the forecast.

Finally, a more complex approach to Bayesian updating can be applied to ENSO forecasting by combining forecasts generated from a set of dynamic and empirical approaches. For example, Coelho et al. (2004) present a Bayesian approach to forecasting December Niño3.4 SSTA with a 5-month lead time that relies on the integration of historical information and dynamic model forecasts. For a given year, the prior distribution is defined using a simple empirical model expressing December Niño3.4 SSTA as a function of July Niño 3.4 SSTA ($R^2 = 0.76$). This prior distribution is updated using the combined (ensemble) mean forecast of December SSTA from a set of coupled models, as well as the relationship between these ensemble mean forecasts and observed SSTAs. This combination produces an updated posterior distribution with greater forecast skill than either of the forecast methods taken individually.

Although the latter approach may produce higher forecast skill, it is important to keep in mind the level of scientific understanding of the policymakers in question. Most policymakers do not have the climate science sophistication to integrate dynamic and empirical models, and may not view such results as credible. In Indonesia, policymakers tend to favor the simple relationship: "A 1 ° C increase in the SSTA in the Central Pacific leads to a 1 million ton decline in paddy production". They are also adept at following the progression of SSTAs over the course of the summer leading up to the monsoon season. But understanding which dynamic model to apply, and how to interpret the uncertainties of a combined empirical-dynamic approach becomes more difficult for the policy community. Even when applying simple empirical relationships, policymakers sometimes wrongly interpret "reasonable probabilities" as "fact". Updating climate information as it becomes available helps to quantify the increasing probability of a climate event occurring.

4.7 Implications for Long-Run Climate Forecasting

Given the usefulness of Bayesian updating for short-run ENSO forecasting, can similar methods be applied for forecasting and adapting to changes in the mean climate state? Unfortunately the process is not as straightforward for long-term forecasts because predictable cycles of temperature and precipitation do not exist. There is substantial variation over time. The most realistic climate forecasts for the purposes of this chapter rely on GCM output from the IPCC (2007). GCM model results of future climate variables can be combined in an ensemble approach to depict the full range of uncertainty in long-term forecasts (e.g. see Tebaldi, 2006; Naylor et al., 2007; Battisti and Naylor, 2009). The distributions of future climate variables, which can be calculated at different spatial and temporal scales, can then be used as an indicator of the type of changes that the policy community should anticipate, monitor, and potentially adapt to in order to minimize damage.

There is a fundamental disconnect between the time horizon of a policymaker (typically in office for a few years or possibly a few decades) and the time horizon of global climate change (decades, centuries). Policymakers can anticipate change in the mean state, but it is more difficult to know if one is experiencing such change while in office. How does one know when the mean climate state (temperature or precipitation) exceeds the historical bounds of variability and that adaptation investments are needed? This is not a simple task in real time. As one observer from South Africa noted, "We'll know when we no longer have cattle and there are only camels on the land."

There are three main difficulties in saying with confidence that a new era of climate has begun. First, a historical baseline of climate that includes natural variability must be established; as climate data are updated under changing climate conditions, the baseline will shift. Second, selecting the recent period for measuring a changed mean state—one that includes variability, either the same pattern or different from the historical baseline—is controversial. Is a period of 10 years sufficient to measure changes in average climate from the observational record? 20 years? 50 years? There are regions of the world that have undergone more than a decade of intense drought, such as the African Sahel from the 1960s to the 1990s, an area that has since become wetter (Kandji et al., 2006). And finally, measuring changes in the average annual cycle of temperature or precipitation can lead to false conclusions; it is critical to measure seasonal changes in the annual cycle as well. For example, climate projections indicate a high probability that the annual average level of precipitation will remain constant for Indonesia by the middle of the 21*st* century, but the annual cycle of rainfall is expected to shift such that more rain will fall in a short wet season and will be followed by a much longer dry season (Naylor et al., 2007)[14].

Updating climate information over long time periods for policy purposes thus requires direct involvement by both the policy and climate science communities to identify metrics of importance to society and thresholds beyond which real damage is likely to occur. In the Indonesian example, a 30-day delay in the monsoon onset was identified as a critical threshold because of its large impact on domestic rice supplies and prices and the absence of government intervention (*ibid.*). Once the relevant climate variable(s) are identified, climate scientists can work to provide projections, and if such projections are not possible, they can work with policymakers to refine the target. Given the projections of a significant seasonal change in the annual cycle of precipitation in Indonesia, policymakers there can monitor such change and create adaptation strategies as time progresses. For example, they might

---

[14] In the case of projecting changes in the annual cycle of precipitation over the Indonesian archipelago, climate science collaborators David Battisti and Dan Vimont used empirical downscaling techniques. For more information see Vimont, Battisti and Naylor (2008).

want to invest in water storage facilities (reservoirs and linked irrigation systems) to take advantage of periods of more intense rainfall and cover longer dry periods. They also might want to invest in drought tolerant crops, or provide incentives for alterations in cropping systems that match both climate conditions and market demand.

In this particular case, the Bayesian null hypothesis would be a change in the annual cycle of precipitation that affects crop production, food availabilities, and incomes throughout the year. The prior would be established on the basis of the observed annual cycle going back in time for decades, and this prior would be updated with new information as the years progress. The likelihood of the null hypothesis being true could thus increase over time as more information becomes available on the pattern of rainfall over the course of the year. This analysis is very different from the Bayesian analysis of El Niño events described above because long-run change in the climate's mean state has not yet been established.

4.8 Conclusions and Implications for Other Regions and Systems

Policymakers responsible for agricultural production and trade, rural economic growth, and food security throughout the world are confronted by a myriad of climate-related risks. Here we have developed a case study of the Indonesian rice sector to illustrate how the policy community might cope with such risks. Our example highlights several conditions for assessing risks in agricultural policy, including the need to understand the seasonality of crop production, the relevant patterns of precipitation and rainfall that affect crop area and yields, and the roles of domestic production, distribution, and trade for domestic food security. Understanding the physical, human, and policy environment from first-hand experience—as opposed to modeling expected risk outcomes from afar—is more likely to lead to sensible policy formulation and implementation in the end.

Although the Indonesian example serves as an excellent illustration of the type of risk assessment that can benefit policymakers in their efforts to manage the potential consequences of climate variation and long-term change, it cannot be generalized to all countries. Climate systems in many regions are extremely noisy (without predictable cycles like ENSO controlling seasonal precipitation), and many areas have limited data with which to conduct sound empirical analysis. Nonetheless, the conceptual framework for risk assessment and management used in the Indonesian case study can be applied generally to most situations.

A key lesson from the case is that the integration of sound science and policy greatly benefits the risk management process in the case of climate variability and climate change. This integration entails the policy community learning enough about climate science to ask the right questions and interpret the answers wisely. Conversely, it entails the climate science community understanding enough about the interests of agricultural policymakers to supply climate information that is relevant to the system in question.

Finally, in the process of assessing climate risks, it is important that policymakers appreciate the probabilistic nature of the exercise. Even when there is over a 90% chance that a certain climate outcome such as a major El Niño event will occur, there is still a 10% chance that such an outcome will not occur. Policy makers typically have a strong preference for certainty, and whether or not they actually take action based on climate predictions depends in large part on their risk aversion and on the relative consequences of being wrong in both directions (action and no action). The relative consequences often highlight policy tradeoffs—for example budgetary costs versus increased malnutrition among the poor, or lower incomes in one region of the country versus another. Unfortunately, climate predictions are not hard facts, particularly with longer forecast horizons. Just as human behavior is impossible to predict with absolute certainty, so too is the climate system.

Appendix : Using the Bayesian Theorem to Forecast El Niño Events, 1950-2007

Bayes' Theorem provides a quantitative approach to measuring belief in a hypothesis (H) being true given the available evidence on hand (E), or $\mathrm{P}(H|E)$. For example, one could test the hypothesis that there will be an El Niño (null hypothesis) based on accumulating evidence (E) of SSTAs over the course of the ENSO cycle. According to the formal definition of the theorem, $\mathrm{P}(H|E)$ can be calculated as the probability of E if H were true, $\mathrm{P}(E|H)$, multiplied by the ratio of the unconditional probability of H, $\mathrm{P}(H)$, divided by the marginal probability of E. The latter is calculated as the sum of the probability of E given each possible state (null or alternative hypothesis, i.e., El Niño or no El Niño), $\sum \mathrm{P}(E|H_i)$, multiplied by the probability of each state, $\mathrm{P}(H_i)$.

$\mathrm{P}(H)$ is often referred to as the prior probability, which is then updated to produce the posterior probability, $P(H|E)$. In other words, how likely is H to be true, given E.

$$\mathrm{P}(H|E) = \frac{\mathrm{P}(E|H)\mathrm{P}(H)}{\sum \mathrm{P}(E|H_i)P(H_i)}$$

Applying the theorem to a more intuitive illustration, the Bayesian updating equation shown above can be used to reduce uncertainty in climate forecasts as El Niño conditions develop prior to the monsoon season. Here we provide an example of how the Bayesian equation can be used in the calculation of El Niño forecasts for a major El Niño year, 1997-98. We also provide forecasts for the years 1950-2007 and highlight El Niño years.

During the strong El Niño year of 1997 (shown in Figure 11.2), the May Niño 3.4 SSTA was 0.84 °C. Given this evidence, E, we update the prior probability, $P(H)$, as follows. As stated above in the main text, we assume P(H) initially equals 34.5%, the fraction of years from 1950-2007 designated as El Niño years. The remaining term of the numerator of Bayes' Theorem, $\mathrm{P}(E|H)$, is calculated as the probability of observing an SSTA of 0.84 °C in an El Niño year. We calculate this probability by assuming the distribution of the SSTA in May during El Niño years is normal and estimating the mean and standard deviation of this distribution based on the observed May SSTA in El Niño years occurring between 1950 and 2007. With this distribution, the probability of a May SSTA of 0.84 °C during an El Niño, $\mathrm{P}(E|H)$, is 44.2%. To calculate the denominator, we calculate $\sum \mathrm{P}(H_i)\mathrm{P}(E|H_i)$, which is equal to

$$\mathrm{P}(H)P(E|H) + \mathrm{P}(\mathrm{not}H)\mathrm{P}(E|\mathrm{not}H).$$

We already know $\mathrm{P}(H)$ and $\mathrm{P}(E|H)$, as both appeared in the numerator. $\mathrm{P}(\mathrm{not}H) = 1 - \mathrm{P}(H)$, or 65.5%, and $\mathrm{P}(E|\mathrm{not}H)$ is calculated in the same way as $\mathrm{P}(E|H)$, except that in this case we must calculate the probability of observing an SSTA of 0.84 °C in a non-El Niño year.

Again, we estimate the mean and standard deviation of a normal distribution for the SSTA in May during non-El Niño years using observed data from 1950-2007.

Table 4.1 1950-2007 El Niño Forecasts:

Year	Start	May	June	July	August	September
1950	34.48%	0.40%	0.02%	0.00%	0.00%	0.00%
1951*	34.48%	26.08%	19.87%	41.35%	91.70%	97.55%
1952	34.48%	28.06%	4.66%	0.70%	0.51%	0.16%
1953*	34.48%	44.21%	62.94%	91.42%	88.11%	99.56%
1954	34.48%	22.26%	3.46%	0.06%	0.00%	0.00%
1955	34.48%	1.87%	0.06%	0.00%	0.00%	0.00%
1956	34.48%	30.59%	7.24%	0.14%	0.00%	0.00%
1957*	34.48%	54.30%	73.64%	97.52%	99.99%	100.00%
1958*	34.48%	53.51%	9.52%	13.09%	42.43%	4.74%
1959	34.48%	45.17%	24.13%	1.93%	0.16%	0.01%
1960	34.48%	31.10%	14.39%	6.98%	6.74%	4.06%
1961	34.48%	41.28%	71.61%	33.92%	6.36%	0.20%
1962	34.48%	13.24%	6.37%	2.59%	1.66%	0.10%
1963*	34.48%	26.57%	24.41%	91.54%	99.78%	99.99%
1964	34.48%	3.55%	0.06%	0.00%	0.00%	0.00%
1965*	34.48%	52.71%	88.48%	99.87%	100.00%	100.00%
1966*	34.48%	32.12%	61.54%	79.09%	68.64%	42.85%
1967	34.48%	24.14%	30.40%	9.19%	1.56%	0.07%
1968*	34.48%	11.57%	10.14%	23.45%	49.94%	43.39%
1969*	34.48%	66.89%	89.35%	90.15%	98.76%	99.90%
1970	34.48%	33.66%	6.67%	0.09%	0.00%	0.00%
1971	34.48%	10.63%	0.88%	0.06%	0.00%	0.00%
1972*	34.48%	62.01%	93.14%	99.90%	100.00%	100.00%
1973	34.48%	13.95%	0.45%	0.00%	0.00%	0.00%
1974	34.48%	3.55%	0.72%	0.07%	0.01%	0.00%
1975	34.48%	4.23%	0.03%	0.00%	0.00%	0.00%
1976*	34.48%	12.55%	5.81%	4.68%	10.14%	49.43%
1977*	34.48%	50.18%	66.97%	77.67%	72.62%	87.76%
1978	34.48%	19.12%	3.47%	0.29%	0.01%	0.00%
1979	34.48%	39.78%	49.97%	13.57%	5.48%	69.58%
1980	34.48%	47.04%	79.63%	84.24%	57.09%	39.41%
1981	34.48%	22.26%	11.68%	0.92%	0.02%	0.01%
1982*	34.48%	66.11%	97.76%	99.96%	100.00%	100.00%
1983*	34.48%	70.86%	93.76%	86.52%	54.24%	22.99%
1984	34.48%	19.12%	1.26%	0.14%	0.01%	0.00%
1985	34.48%	9.20%	0.57%	0.05%	0.01%	0.00%
1986*	34.48%	24.14%	27.05%	37.81%	67.56%	97.18%
1987*	34.48%	69.60%	98.94%	100.00%	100.00%	100.00%
1988	34.48%	0.93%	0.00%	0.00%	0.00%	0.00%
1989	34.48%	8.67%	0.86%	0.09%	0.01%	0.00%
1990*	34.48%	51.46%	46.93%	46.83%	64.70%	64.96%
1991*	34.48%	61.73%	93.40%	99.69%	99.98%	100.00%
1992*	34.48%	71.33%	89.09%	96.54%	89.63%	72.18%
1993*	34.48%	70.32%	94.38%	98.51%	98.05%	99.12%
1994*	34.48%	50.61%	75.46%	82.81%	97.90%	99.35%
1995	34.48%	35.20%	31.17%	12.95%	1.20%	0.04%
1996	34.48%	18.27%	6.94%	3.37%	0.76%	0.12%
1997*	34.48%	67.44%	98.62%	100.00%	100.00%	100.00%
1998*	34.48%	64.81%	5.27%	0.02%	0.00%	0.00%
1999	34.48%	5.94%	0.11%	0.00%	0.00%	0.00%
2000	34.48%	9.48%	1.15%	0.11%	0.02%	0.00%
2001	34.48%	29.07%	32.32%	38.42%	35.36%	18.29%
2002*	34.48%	55.07%	92.81%	99.73%	100.00%	100.00%
2003	34.48%	18.27%	11.67%	22.17%	18.54%	38.08%
2004*	34.48%	51.04%	61.19%	91.93%	99.66%	99.99%
2005	34.48%	56.91%	83.00%	92.68%	91.73%	86.66%
2006*	34.48%	44.69%	61.19%	70.24%	91.65%	99.48%
2007	34.48%	27.56%	22.27%	3.51%	0.19%	0.00%

Based on this calculation, $P(E|\text{not}H)$ equals 11.2%. Comparing $P(E|H)$, 44.2%, and $P(E|\text{not}H)$, 11.2%, indicates that an SSTA of $0.84\,^\circ$C is much more likely to be observed in an El Niño year than in a non-El Niño year, which suggests we should strengthen our belief that the year in question is an El Niño year. This bears out numerically using Bayes' Theorem. Putting these terms together:

$$P(H|E) = \frac{0.442 \times 0.345}{0.442 \times 0.345 + 0.112 \times 0.655}$$

The table above provides the full set of these calculations for 1950-2007, with El Niño years starred.

References

[1] Allen M.R., Ingram W.J. 2002. Constraints on future changes in climate and the hydrologic cycle. *Nature* **419**, 224-232.

[2] Anand, P. 2002: "Decision-making when science is ambiguous," *Science* **295**: 1839.

[3] Battisti and Naylor, 2009, "Historical Warnings of Future Food Insecurity with Unprecedented Seasonal Heat", *Science* **323**, 240-244.

[4] Bayes, T. 1764: "An Essay Toward Solving a Problem in the Doctrine of Chances," *Philosophical Transactions of the Royal Society of London* **53**, 370-418.

[5] Broad, K. and Agrawala, S., 2000: "The Ethiopia food crisis – uses and limits of climate Forecasts," *Science* **289**: 1693-1694.

[6] Cane, M.A., Eshel, G., and Buckland, R.W., 1994: "Forecasting Zimbabwean maize yield using eastern equatorial Pacific sea surface temperature," *Nature* **370**, 204-205.

[7] Carter, T. R., Parry, M. L., Harasawa, H., and S. Nishioka, 1994: *IPCC Technical Guidelines for Assessing Climate Change Impacts and Adaptations.* University College,

[8] London and Center for Global Environmental Research, Japan. 59 pp.

[9] Chou C., Neelin, J.D. 2004. Mechanisms of global warming impacts on regional tropical precipitation. *Journal of Climate* **17**, 2688-2701.

[10] Coelho, C.A.S., Pezzulli, S, Balmaseda, M., Doblas-Reyes, F.J., and Stephenson, D.B., 2004: "Forecast calibration and combination: A simple Bayesian approach for ENSO," *Journal of. Climate,* **17,** 1504-1516.

[11] De Datta, S. K., 1981: *Principles and Practices of Rice Production.* John Wiley and Sons, New York.

[12] Deutsch, C. A., J. J. Tewksbury, R. B. Huey, K. S. Sheldon, C. K. Ghalambor, D. C. Haak, and P. R. Martin. 2008. "Impacts of Climate Warming on Terrestrial Ectotherms Across Latitude", *PNAS* **105**, 6669-6672.

[13] Naylor, R.L., Falcon, W.P, Wada, N., and D. Rochberg, 2002. "Using El Niño Southern Oscillation Climate Data to Improve Food Policy Planning in Indonesia", *Bulletin of Indonesian Economic Studies* **38** (**1**), 75-91.
[14] Falcon W.P., Naylor R.L., Smith W.L., Burke M.B., McCullough E.B., 2004. "Using Climate Models to Improve Indonesian Food Security, *Bulletin of Indonesian Economic Studies* **40** (**3**), 357-79.
[15] Fraedrich, K., and N. R. Smith, 1989: "Combining predictive schemes in longrange forecasting," *Journal of Climate,* **2,** 291–294.
[16] Hamada J, Yamanaka MD, Matsumoto J, Fukao S, Winarso PA, Sribimawati T, 2002: "Spatial and Temporal Variations of the Rainy Season Over Indonesia and Their Link to ENSO," *Journal of the Meteorological Society of Japan,* 80: 285-310.
[17] Hansen, J.W., Hodges, A.W., and Jones, J.W., 1998: "ENSO influences on agriculture in the southeastern United States," *Journal of Climate,* **11**: 404-411.
[18] Hendon, Harry H., 2003: "Indonesian Rainfall Variability: Impacts of ENSO and Local Air–Sea Interaction," *Journal of Climate*, **16**, 1775–1790.
[19] Intergovernmental Panel on Climate Change, 2007, *Fourth Assessment Report: Climate Change 2007,* Cambridge University Press, available from `http://www.ipcc.`
[20] IRRI (International Rice Research Institute), 2004. *Water Requirements.* Accessed at: `http://www.knowledgebank.irri.org/troprice/Water{\_}Requirements.htm` (June 7, 2006).
[21] Jones, R. N., 2000: "Managing uncertainty in climate change projections: issues for impact assessment," *Climatic Change,* **45**: 403-419
[22] Jones, R. N., 2001: "An environmental risk assessment/management framework for climate change impact assessments," *Natural Hazards,* **23**: 197-230
[23] Jones, R.N., 2003: *Managing climate change risks*. OECD working paper from a workshop series on "Benefits of Climate Policy: Improving Information for Policy Makers". Paris: OECD, 17 pp.
[24] Kenny, G. J., Warrick, R. A., Campbell, B. D., Sims, G. C., Camilleri, M., Jamieson, P.
[25] D., Mitchell, N. D., McPherson, H. G., and M. J. Salinger., 2000: "Investigating climate change impacts and thresholds: an application of the CLIMPACTS integrated assessment model for New Zealand agriculture," *Climatic Change,* **46**: 91-113.
[26] Ludwig, D., Mangel, M., and B. Haddad., 2001: "Ecology, conservation and public Policy," *Annual Review of Ecology and Systematics*, **32**: 481-517.
[27] Mason, S. J., and G. M. Mimmack, 2002: "Comparison of some statistical methods of probabilistic forecasting of ENSO," *Journal of Climate*, **15**, 8–29.
[28] Mjelde, J.W., Hill, H.S.J., and Griffiths, J.F., 1998: "A review of current evidence on climate forecasts and their economic effects in agriculture," *American Journal of Agricultural Economics,* **80**: 1089-1095.
[29] Metzger, S., M. Latif, and K. Fraedrich, 2004: "Combining ENSO forecasts: A feasibility study," *Monthly Weather Review.,* **132,** 456–472.

[30] Naylor, R. N., Battisti, D. S., Vimont, D. J., Falcon, W. P., and M. B. Burke, 2007. "Assessing risks of climate variability and climate change for Indonesian rice agriculture", *PNAS* **104** (19), 7752-7757.
[31] Naylor, R. and W. Falcon, 2008. "Our Daily Bread: A Review of the Current World Food Crisis", *Boston Review*, September/October, 13-18.
[32] Naylor R L, Falcon W P, Rochberg D, Wada N, 2001: *Climatic Change* **50**:255-265.
[33] Nicholls, N. 1986. "Use of the Southern Oscillation to forecast Texas winter wheat and sorghum crop yields," *Agricultural and Forest Meteorology,* **38**: 9-15.
[34] NOAA. 2008. `http://www.cpc.noaa.gov/data/indices/`
[35] Parry, M. L., Carter, T. R., and M. Hulme, 1996: What is a dangerous climate change? *Global Environmental Change,* **6**: 1-6.
[36] Pauly, D. and I. Tukayama (Eds.) 1987: The Peruvian Anchoveta and Its Upwelling System: Three Decades of Change. International Center for Living Resources Management (ICLARM), Review 15, Manila, The Philippines: 351 pp.
[37] Pfaff, A., Broad, K., and Glantz, M., 1999: "Who benefits from seasonal-to-interannual climate forecasts?" *Nature,* **397**: 645-646.
[38] Pittock, A. B. and R. N. Jones., 2001: "Adaptation to what and why?" *Environmental Monitoring and Research* **61**: 9-35.
[39] Podesta, G.P., Messina, C.D., Grondona, M.O., and Magrin, G.O., 1999: "Associations between grain crop yields in Central-Eastern Argentina and El Niño-Southern Oscillation," *Journal of Applied Meteorology,* **38**: 1488-1498.
[40] Tebaldi, C and Knutti, R, 2006, "The use of the multi-model ensemble in probabilistic climate projections," *Philosophical Transactions of the Royal Society,* 365, 2053-2075.
[41] Steinfeld, H., Gerber, P., Wassenaar, T., and V. Castel, 2007. *Livestock's Long Shadow: Environmental Issues and Options.* Part IV: Livestock's Role in Climate Change and Air Pollution. Rome: The United Nations Food and Agricultural Organization.
[42] UNDP (United Nations Development Program), 2007. The Other Half of Climate Change: Why Indonesia Must Adapt to Protect Its Poorest People. Jakarta, Indonesia: UNDP, 20 pp. (available at: `http://www.undp.or.id/pubs/docs/UNDP{\%}20-{\%}20The{\%}20Other{\%}20Half{\%}20of{\%}20Climate\ \{\%}20Change{\%}20EN.pdf`, accessed on January 28, 2009).
[43] Vimont, D., Battisti, D. and R. Naylor, 2008. "Downscaling Indonesia precipitation using large-scale meteorological fields", University of Wisconsin (submitted to *International Journal of Climatology*), November.

Chapter 5
A Primer on Weather Derivatives

Pauline Barrieu and Olivier Scaillet

Abstract We give a short introduction to weather derivatives. We discuss what they are made for. We describe the markets on which they are exchanged, and how they are used to promote agricultural risk transfer in developing countries via the World Bank program. We also treat some specific issues such as basis risk, pricing and design.

JEL codes: G12, G13, Q54.

Keywords: weather derivatives, World Bank program.

5.1 Introduction and definition

5.1.1 What is weather risk?

Weather is not only an environmental issue but also a key economic factor, as recognized by the former US Commerce Secretary, William Daley, in 1998, when he stated that at least $1 trillion of the world economy is weather sensitive. Moreover, it is estimated that 20% of the world economy is directly affected by weather. In a business survey, 160 out of 200 utility companies cited weather as a key determinant of their operating revenues, and half of them as a key determinant under performance below normal. More importantly, the risk exposure is not homogeneous across the globe and some countries, usually those heavily dependent on agriculture, are more sensitive than others. It also includes a large range of phenomenons such as modifications in temperature, wind, rainfall or snowfall.

Pauline Barrieu
London School of Economics

Olivier Scaillet
HEC Genève and Swiss Finance Institute

J.A. Filar and A. Haurie (eds.), *Uncertainty and Environmental Decision Making*, International Series in Operations Research & Management Science 138, DOI 10.1007/978-1-4419-1129-2_5,

Weather risk has some specificities compared to other sources of economic risk: in particular, it is a local geographical risk, which cannot be controlled. The impact of weather is also very predictable: the same causes will always lead to the same effects. Moreover, weather risk is often referred to as a volumetric risk, its potential impacts being mainly on the volume and not (at least directly) on the price. This explains why hedging of weather risk via the trading of commodities futures is difficult and imperfect. For example oil futures price does not depend solely on demand (cold winter) and can be high even if demand is low in case of a war for instance. Volumetric risk is imprecisely compensated by the price variation in the futures position.

Usually, when subject to some risk, it is possible to hedge against it by contracting some insurance policies. But, this is not really a possibility for weather risk for two main reasons: first it is more a high frequency - low severity risk but also the same weather event can generate economic losses for some agents and some gains for others. As an example, we can think of a touristic place. If it rains one day during the summer, this is a bad day for providers of outside activities but this will not lead to tremendous losses. On the contrary, if the whole season is cold and wet, this could lead to the bankruptcy for some local businesses. Some rain in this case could be interesting for coffee shops and indoor businesses. Therefore, weather risk is a risk that is part of everyday life, having limited economic consequences on an everyday basis but with some huge potential consequences in its accumulation or repetition. Insurance is not a well-suited solution, as it could be for dramatic weather events such as hail, storm, or drought. To deal with this risk, some financial contracts depending on weather conditions (temperature, rainfall, snowfall...) were created and introduced on the financial market 10 years ago. They are called *weather derivatives.*

Note that weather derivatives differ from Cat (catastrophe) derivative which target rare events of a catastrophic type (high severity with low probability). A natural catastrophe is usually defined as an event yielding a loss of at least five million dollars and affecting a significant number of insured people. Cat derivatives, and in particular cat-bonds, are closely related to insurance contracts and, as such, suffer from the same drawbacks like exposition to moral hazard and adverse selection. In some cases they are considered as true insurance products by tax regulators.

5.1.2 Weather derivatives

Weather derivatives are financial instruments whose value and/or cashflows depend on the occurence of some meteorological events, which are easily measurable, independently authentifiable, and sufficiently transparent to act as triggering underlyings for financial contracts. Typically location is clearly identified and measurement is provided by independent and reliable sources. The underlying meteorological events can be considered as noncatastrophic. Use of weather derivatives generalizes standard financial risk management practice based on hedging foreign exchange and interest rate risks via financial derivatives.

According to the Weather Risk Management Association (WRMA), the financial market related to weather has two main facets: the management of the financial consequences of adverse weather for those with natural exposure to weather, and the commercial trading of weather risk, both in its own right and in conjunction with a variety of commodities. The first weather transactions took place in 1997 between Enron and Koch Industries. These transactions were the result of a long thinking process by Koch, Willis and Enron aimed at finding a means of transferring the risk of adverse weather. These deals followed the deregulation of the energy market in the US (transition from an oligopolistic position to a status of mere participant to a competitive market) and was based upon some temperature indices to compensate the energy producer in case of a mild winter. The first European deals took place in 1998 between Enron and Scottisch Hydro Electric on a similar basis.

WRMA conducts every year a survey of the weather market. The results, even if they reflect only a part of the whole market, are interesting and can give an idea of the market trend. The last available survey (2006) shows that even if weather risk can affect the whole economy, the energy sector is still the main part of the market, with companies such as British Gas, Hess Energy or Aquila Energy. Financial institutions and hedge funds, such as ABN Amro or Merrill Lynch, insurance and reinsurance companies, such as AXA Re or Swiss Re are also among the large players of this market, mainly because of their important activity of the energy market (gas and oil) and of the cross-hedging opportunities weather derivatives offer as we will see later. Beside large players, there also exist lots of small couterparties such as municipalities, ski resorts, golf clubs, beverage producers and leasure parks. Success in those markets often requires writing a large number of contracts (economies of scale) and exploiting a large set of sharp skills at the trader and analyst levels. The size of the market was estimated by WRMA to be over \$45 billions in 2006, compared with \$9.7 billions in 2005. About 75% of the transactions are based upon temperatures, and 10% on rainfalls.

The most common underlying is related to the notion of Degree Day which is expressed as the difference between a reference level temperature ($65°F$ or $18°C$) and the average daily temperature T. The average is computed between the maximum and minimum recorded temperature over a particular day. A Heating Degree Day (HDD) is defined as follows:

$$HDD = (65°F - T)^+, \tag{5.1}$$

where $(\cdot)^+ = \max(\cdot, 0)$. The bigger the HDD is, the colder the temperature is, and as a consequence the larger the demand for heating will be. For example if the today average is $60°F$, we needs 5 degrees of heating ($HDD = 5$). On the contary if the today average is $70°F$, we do not need any heating, and the HDD is zero. Similarly a Cooling Degree Day (CDD) is defined as follows:

$$HDD = (T - 65°F)^+. \tag{5.2}$$

The bigger the *CDD* is, the hotter the temperature is, and the larger the demand for cooling should be. The definition of a temperature index in those terms reflects the close relationship between the energy sector and the weather derivative market. Daily results are often cumulated to give a total on a given period, such as a week (Xmas-NewYearEve, sport event), a month (sales period, harvest period), or a quarter (summer holidays, opening season of a resort).

Weather derivatives differ in their characteristics, with different contract types such as Swap, Call and Put. The contract period can be the HDD season (November to March) or the CDD season (May to September). The measurement authority, the reference index and payment conditions can vary as well. Contract schedule is generally made of three dates: transaction date (from two days to several months before the observation period), payment date (five days after the end of the period to get the data used in computing the payoff), adjustment date (three months after the payment date when final data are published and possible recording errors removed).

HDD swaps are contracts where both counterparties agree to exchange their climatic risk. It allows to smooth volatile cash flows. One counterparty accepts to pay the other when the index goes above a given level, and vice-versa if it goes below. The level is most of the time fixed so that expected positive and negative cashflows compensate exactly, and the current value of the swap is nil. For example, a beverage producer can enter a swap to get protection when summer time is too cold, and forego part of its benefits when it is too hot.

HDD calls are contracts where the buyer gets compensated when the climatic risk is above a predetermined level. The protection implies an upfront payment of a premium to the seller. For example an airline company can buy a call on the number of days where the average windspeed exceeds a given level. This protects against loss of revenues in case of delays or cancelations for a cost equal to the paid premium.

HDD puts are contracts where the buyer gets compensated when the climatic risk is below a predetermined level. Again we have an upfront payment of a premium to the seller. For example a ski resort can buy a put on the number of days where the average snowfall is lower than a given level. Loss of revenues because of a reduced number of skiers is diminished for a cost equal to the paid premium. Most of the time call and puts contracts are assorted with caps which bound the compensation payments but reduce the upfront payment.

5.2 The weather derivatives market

Most of the transactions are tailor-made and part of the OTC (Over The Counter) market. Usually OTC transactions are realized within the ISDA standards (Master Agreement standards of the International Swap and Derivatives Association) which provide standardised contracts aimed at easing OTC transaction processes. Some go through specialized brokers but most of them are done without any intermediaries. Taylor-making is not surprising as these structures suit better the management of weather risk and the needs of the various players on this market. The organized markets are however rather successful, mainly because of the transparency, liquidity and security they offer. Among them, the most predominant one is the Chicago Mercantile Exchange (CME). The CME reported 1,041,439 trades in summer 2005 and winter 2005/6. Combined 2005/6 CME trades were up over 300% compared to 223,139 trades in 2004/5 period.

5.2.1 *Organized market: the Chicago Mercantile Exchange (CME)*

CME weather derivatives are part of the CME Alternative Investment Products, with derivatives based on economic indices, housing price index or ethanol. Two major types of contracts are traded on the CME: futures and options on futures, both based upon a temperature-based index computed by Earthsat using weather data from the National Climate Data Center.

The first contracts appeared in 1999, first focusing on American cities. Now the CME offers contracts for 18 American cities (Las Vegas, Atlanta, Chicago, Cincinnati, New York, Dallas, Philadelphia, Portland, Tucson, Des Moines, Boston, Houston, Kansas City, Minneapolis, Sacramento, Salt Lake City, Detroit, Baltimore), together with 6 Canadian cities (Calgary, Edmonton, Montreal, Toronto, Vancouver, Winnipeg), 9 European cities (London, Paris, Amsterdam, Berlin, Essen, Stockholm, Rome, Madrid, Barcelona) and 2 Japanese cities (Tokyo, Osaka). Recently, the CME has experienced a fantastic growth, as shown in the figure below:

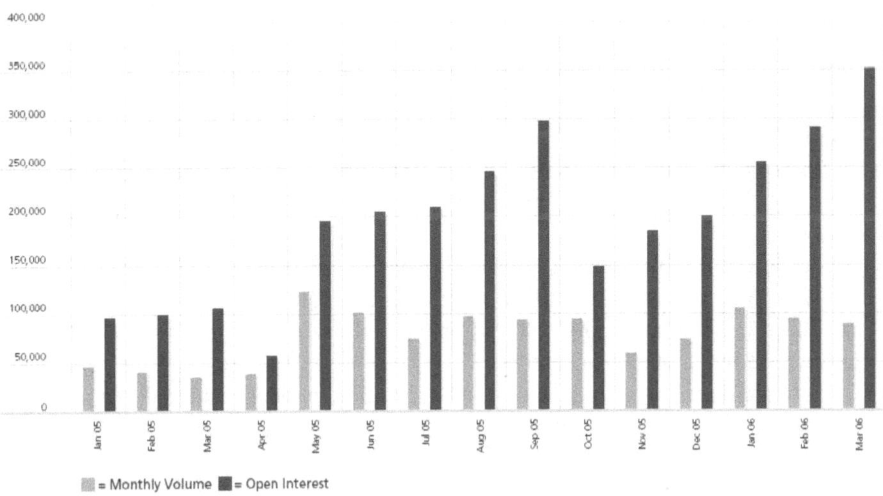

Fig. 5.1 CME - Weather volume and open interest (From the CME Brochure "CME Weather Futures and Options")

the history of the CME weather derivatives can be divided into two main parts: before 2003 and after. In 2003, everybody in the market thought the CME should close its weather activities because of an obvious lack of interest. But, surprisingly, the CME launched new contracts and modified the structure of the existing contracts, making the whole market more attractive.

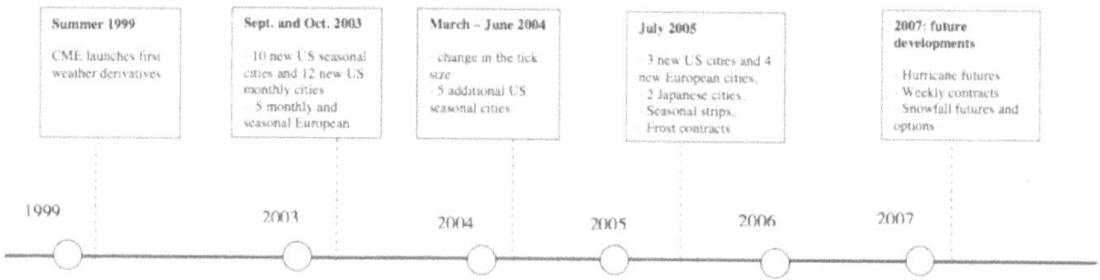

Fig. 5.2 CME - Key historical facts)

The value of a CME future contract is simply expressed as

$$\text{Value of the index } \times \$20 \tag{5.3}$$

The tick size, or monetary value of one index point, was first fixed at $100 but then lowered down in 2004, as part of the desire to increase the overall flexibility of the market and its attractiveness. For Europe, the tick size is £20 and for Japan, ¥2500.

Today, there are three main types of contracts traded on the CME, depending on the considered period:

- the monthly contracts;

- the seasonal contracts, based upon a whole season, winter covering months between November and April, and summer covering the period between May and October;

- the seasonal strip contracts, where the market participants can choose a period from 2 to 6 consecutive months in a given season. This additional flexibility was introduced in September 2005 to focus on the real risk exposure, since the volatility can vary from one month to another.

There are three main types of indices, depending on the considered period:

- For the winter months, the monthly and seasonal contracts are based upon Cumulative Heating Degree Days, computed as the aggregated value of Daily Heating Degree Days over the considered period, $(65°F - T)^+$ in the US and $(18°C - T)^+$ elsewhere, where T is the average daily temperature defined as the average between the minimum observed temperature and the maximum observed temperature during the day.

- For the summer months, the monthly and seasonal contracts are based upon Cumulative Cooling Degree Days, computed as the aggregated value of Daily Cooling Degree Days over the considered period, $(T - 65^{\circ}F)^{+}$ in the US. Elsewhere, contracts are based upon Cumulative Average Temperature.

- For Amsterdam, some specific contracts are available in winter (between November and March). They depend on the number of frost days, i.e., weekdays when frost is recorded.

The future developments of the CME are numerous. Since last March 2007, the CME has added hurricane futures and options on five U.S. defined areas - Gulf Coast, Florida, the Southern Atlantic Coast, the Northern Atlantic Coast, and the Eastern U.S, for the hurricane season beginning June 1. The underlying indexes are calculated by Carvill, a leading independent reinsurance intermediary in specialty reinsurance that tracks and calculates hurricane activity. These contracts provide an additional way to help addressing the needs of the insurance industry and other markets. In particular, insurers and others will be able to transfer their risk to the capital markets and thereby increase their capital in order to insure customers.

The CME is planning to offer weekly contracts, for the American cities only. These products enable market participants to hedge short-term intra-monthly weather risk. They will list two weekly contracts (Monday through Friday) and will be settled the following business day. The weekly contracts are different from the monthly and seasonal strip contracts essentially in the way that the weekly contract is an index of the week average temperature (not degree days).

Snowfall futures and options will be soon added to the existing products mainly based upon temperature. CME Snowfall futures will be geared to a CME Snowfall Index and will be offered initially on two U.S. cities - Boston and New York. These contracts will trade on a monthly basis from October through April.

5.2.2 European situation

Some attempts were made in Europe to launch an independent organized market for weather derivatives. In particular, the LIFFE launched three temperature-based European weather indices (Berlin, Paris, London) in July 2001, but this activity stopped later after the acquisition by Euronext.

In November 2005, Powernext, which is a European energy exchange based in France, and Meteo France launched the quotation of national temperature indices for 9 European countries (France, UK, Italy, Belgium, The Netherlands, Portugal, Spain and Switzerland). Three types of indices are provided: historical indices, observed indices and forecast (for Day D, $D+1$, $D+2$ and $D+3$) for average, minimum and maximum temperatures. The specificity of their indices relies upon the formula used to obtain a national aggregate. Various temperatures are weighted by population as to provide an (accurate) measure of risks related to the economic activity. Therefore:

$$\text{National index} = \frac{\sum n_i T_i}{\sum n_i}, \tag{5.4}$$

where n_i is the population of region i and T_i is the relevant temperature in region i. The computed index can be seen as a gross estimate of the national risk. The objective is global and probably focused on energy producers. No contract is offered yet but these indices appear at the moment as decision tools and potential underlying for OTC weather transactions.

This initiative is further developed with the launch in June 2007 of Metnext, specialized in indices for weather risk management. Météo-France and Euronext, a subsidiary of NYSE Euronext, have teamed up to launch a joint venture named Metnext specializing in innovative solutions for index-based management of weather risk. Metnext, in which Météo-France holds a 65% interest and Euronext the remaining 35%, will initially offer two types of service for businesses, insurance companies, reinsurance companies, and banks. Indices tailored to the specific needs of individual firms in industry or other areas of business will enable them to anticipate, and thus manage, their exposure to weather risk more effectively. They will be based on historic data concerning the impact of variables such as temperature, rainfall, and the direction and wind speed on management parameters. Historical and forecasted indices will be available to cover forecast periods ranging from a few days to several weeks or several months, depending on the specific needs of the firm.

5.2.3 Electronic trading platforms

Since the beginning of the weather market in the late 90*s*, electronic trading platforms have always played an important role in the development of the market, especially Enron's platform in the early days. A lot of big market players have such a platform, and in particular Spectron is rather important in Europe.

These platforms typically act as service providers for the energy trading industry. They act as a neutral third party to trades and have no ownership function. Their role can be seen as a compromise between pure OTC transactions and organized markets: by putting together various participants on the weather risk marker, these platforms increase their opportunity to sign a deal that is still rather standardized but closer to their real exposure. Moreover, new structures of deals are typically agreed upon in these platforms: the first weekly swap or the first precipitation deal were signed on Spectron.

5.3 The World Bank program based upon weather derivatives

Another dimension of weather derivatives lies in the specificities of this widespread risk, with among the most exposed those relying on agriculture. Therefore, it seems natural to see the World Bank as one of the actors on this market. Indeed, according to the 2005 (9) report , "the World Bank can play a role in assisting countries in taking actions that effectively use limited government resources to facilitate market-based agricultural risk transfer".

5.3.1 General presentation

The role of the World Bank is crucial in developing countries. However, it is currently exposed to natural disasters without the benefit of ex-ante structures to finance losses. The main challenge seems to be the innovation in risk transfers, as to lead ways to solutions to many economic and social problems.

Among the various structures of the World Bank, the Commodity Risk Management Group (CRMG) has for main objective to deal with the agricultural risk in developing countries, where agricultural risk is defined as "negative outcomes stemming from imperfectly predictable biological, climatic and price variables". The economic and social consequences of this risk being so huge in some countries, it seems logical to seek for some form of protection against this risk. However, even if insurance companies were to write insurance policies against this risk, they would be typically too expensive for small farmers, and the compensation would take too long to be effective, partly because of the claim checking procedure. As far as the local government is concerned, its role is more to protect against catastrophic risk (low frequency - high severity).

The CMRG has developed several projects throughout the globe in order to deal with agricultural risk transfer using weather derivatives. Among these projects, the pilot program in India has been particularly successful.

5.3.2 From theory to practice: a pilot project for agricultural risk transfer in India

The Indian exposure towards agricultural risk was assessed in a national survey carried out in 1991. One sixth rural households had loans from rural financial institutions as to finance their agricultural activities, but only 35% of the actual credit needs were met through these channels. The World Bank estimated then that moneylenders, the reduction in farming inputs, overcapitalization and overdiversification of the activities have all led to a suboptimal asset allocation for Indian farmers. The challenge was therefore to innovate and find low-cost ways to reach farmers and help them to better manage their risk.

The first project of a total duration of 3 years was initiated in 2003 by the World Bank, based upon the idea of a large scaling-up and mainstreaming of weather insurance for small farmers. This project was run jointly by the CRMG, BASIX, an Indian microfinance institution created in 1996, and ICICI Lombard, a Mumbai based

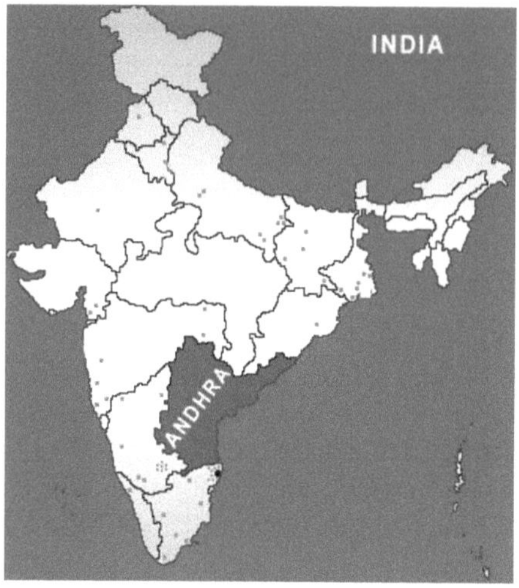

Fig. 5.3 India map

insurance company. Three different phases have to be distinguished in the launching and running process of the first weather insurance initiative in India.

The first year, 2003, was a pilot year, and the program was only targeting groundnut and castor farmers in the Andhra-Pradesh district of Mahbubnagar as indicated in the map below

The design of the insurance scheme was made by ICICI Lombard, with the technical support of the CRMG and the consultation of BASIX, the objective being to protect the farmers from drought during the groundnut and castor growing season, which corresponds to Khariff (the monsoon season, between June and September). The entire portfolio sold by BASIX was insured by ICICI Lombard with a reinsurance contract from Swiss Re. The marketing and sales were made in 4 villages selected by a local bank, according to their current involvement in microfinance. Workshops and presentation meetings were arranged by BASIX. 230 farmers (154 producers of groundnut and 76 of castor) bought insurance against drought during Khariff, and most of them were considered as small farmers, with less than 2.5 acres of land. The insurance contract bought by the farmers had only the name of insurance. It was indeed a financial derivative based upon a reference weather index. The index based upon rainfall has been carefully established by farmers and biologists as to represent the real impact of rain on the growth of either the groundnut or the castor (two indices were considered depending on the crop). The different stages of the growth were also taken into account by weighting differently the different subperiods of the considered season. If the rainfall was not sufficient, then the contracts would be automatically triggered, and the farmers would automatically received their compensation. Because of this triggering mechansim, the costs of the overall scheme are lower than those of a traditional insurance programme, and the compensation recieved promptly by the farmers, helping them to survive until the next season.

The payout structure of the groundnut weather insurance policy was as follows:

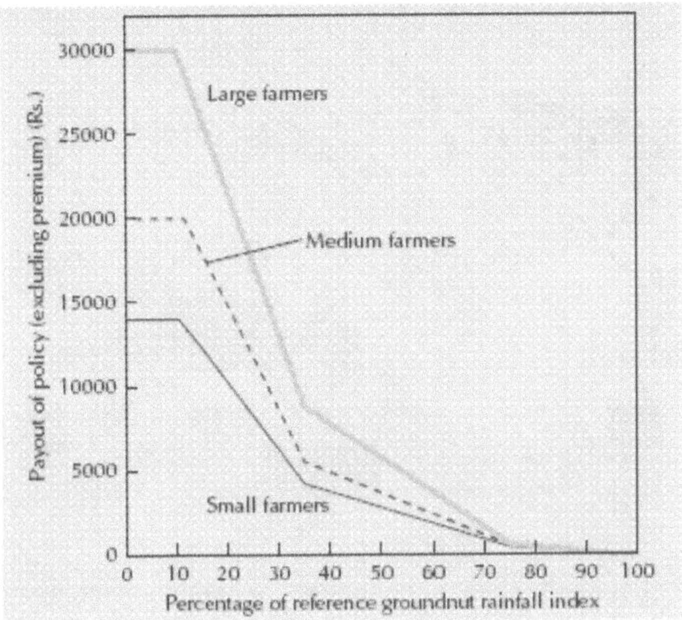

Fig. 5.4 Payout structure of the groundnut weather insurance policy

Different payoffs were considered depending on the size of the land, as to take into account scale effect.

The year 2004 witnessed an extension of the first experience to 4 new weather stations in Khamman and Antapur, in Andhra-Pradesh. Existing contracts were sligthly modified, following the feedback of farmers who bought them in 2003. In particular, more weight was put on the initial sowing period of groundnut in the computation of the rainfall index. New contracts were created for cotton farmers, targeting excess rainfall this time. 400 farmers bought these contracts from BASIX and 320 bought them directly from the insurance company ICICI Lombard, which did not buy reinsurance this time.

Further improvements were made in the third phase of the project in 2005: a total of 7685 contracts in 36 local in India were sold by BASIX, through the microfinance channel, but more generally more than 250000 farmers bought weather insurance in the country, directly from ICICI Lombard.

The overall project has been a major sucess, improving the financing condition of many small farmers in India. The scheme still exists. A long term question is certainly the study of the impact of these policies on the farmers behaviour.

5.3.3 Current and future projects

The CRMG has many other projects throughout the globe as shown in the following figure:

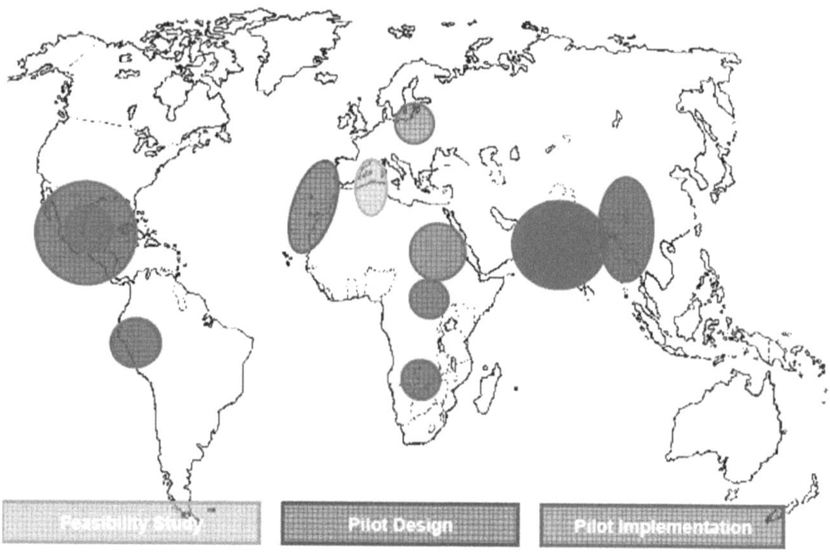

Fig. 5.5 CRMG projects throughout the globe

All these projects are based upon the same idea: the underwriting of insurance contracts dependent upon an index, i.e., weather derivatives, as to improve the overall efficiency of the risk management: cheaper product, less delay in receiving the compensation since the contracts are standardized and no claim check is required. The characterization of these indices is therefore essential for the success of these products. More precisely, these indices should be related to the agricultural risk, as this was the case in India. They are typically determined with the help of the local farmers and biologists to characterize the critical events. They involve typically:

- Rainfall measures, as to identify drought or excess rainfalls which can damage the culture, as in India for groundnut and castor;

- Some events: for instance, the so called "winterkill" index in Ukraine, counting the number of days when the temparture falls below $-16°C$. Temperatures below this level are killing the seeds in the soil and therefore can be seen as responsible for a poor harvest the consecutive season.

- Growing Degree Day, as to measure the growth and development of plants and insects during the "growing season":

$$\text{Daily Growing Degree Day} = (T - L)^+ \tag{5.5}$$

The threshold L can vary during the life of the contract. The accumulated value is a good proxy to establish the development stages of a given grop or of an insect. As a consequence, a proper scheduling of pesticide or herbicide can be made.

5.4 Specific issues related to weather derivatives

Due to the specificities of the underlying risk, weather derivatives have also some specific issues. Some of them are common to all securities written on a non-financial underlying risk, as this is the case for insurance risk securities based on a catastrophic risk or on a mortality index.

5.4.1 Basis risk

When securitizing insurance risks, some basis risk is usually introduced as the new products usually depend on an index rather than on a single firm exposure. Basis risk simply translates the risk related to the spread that can exist between the agent exposure and the hedge proposed. Indeed, before a non-financial risk can be securitized, i.e., transfered to the capital markets, it must be in general standardized. This risk can theoretically create some problems and prevent the derivative market from developing.

As a consequence, the characterization and the computation of an appropriate index is an essential step. The reference index should be clearly identified, non manipulable, consensual, and widely available, with a long history. Usually this is not an issue with weather data that have been collected for many years by public and independent institutions. Still, the construction of an index from existing data requires transparency: in particular, the temperature used for the CME contracts is not a real temperature but a daily average of two extreme (real) temperatures. The computation rule should be explicit and made by an independent authority, here Earthsat. For the Powernext index, the computation rule is more complex but available for the subscribers; a database has been reconstructed for the index so that the different market participants should be able to relate their own risk exposure to the index. In general basis risk is more pronounced in the US than in Europe where distances are smaller, but local climate heterogeneity plays a key role in both geographic areas.

Note also that basis risk might not be such an issue for large energy producers, which are globally exposed towards weather conditions, with some concentration in large cities, densely populated. In their case, hedging with large cities contracts is a good approximation. The initiative of Metnext seems to fill in this gap in the European market.

The basis risk, even if clearly identified, can be a huge limit to the development of the market by small companies. How should a company based in Glasgow hedge its weather risk? Is the London contract representative of its risk? How is the risk "difference between London and Glasgow" understood? All these common sense remarks could explain partially the increased development of the CME market after the introduction of additional locations in 2003 but also its (still) relatively limited expansion. OTC transactions seem to be better suited for the management of this highly local risk.

Management of book of OTC weather derivatives exploits partially this imperfect link between measurement locations to achieve diversification in their positions. This geographical diversification supplements directional diversification to reduce risks. This requires a monitoring of the correlation between locations. Besides there often exist limits in terms of number of transactions and amount per transaction as well as types of trading and hedging strategies in order to keep everything under control.

5.4.2 Pricing issues

The literature on the pricing of weather derivatives is really important, especially on temperature derivatives. Two aspects are tackled: the understanding and modeling of the underlying distribution and the characterization of the pricing rule. The choice of the model and of the pricing rule is heavily dependent on the accounting classification of the weather derivative, either as an insurance product or as a financial derivative.

5.4.2.1 Underlying risk modelling

The statistical analysis of meteorological data is complex because of missing data, seasonality, nonstationarity (long term trend with wild short term blips), multidimension (different locations of the measurement units), and multiple interactions. Long term weather forecasts are notably imprecise and not always suitable for valuation and risk management.

US data are easy to get freely whereas European data are relatively expensive. US historical data are available through the National Climate Data Center (NCDC) which is part of the National Oceanic and Atmospheric Administration (NOAA). This constraints the choice of the measurement authorities. The optimal length of the database depends on the regularity of the data (trend, regular seasonality). Between 10 and 30 years is considered as the norm. The problem of the choice of the optimal sample size is well known to climatologits. It concerns the determination of the so-called Optimal Climate Normal (OCN) or optimal mean window of former years used to predict expected value for next year. The National Center for Environmental Prediction (NCEP) uses a methodology which measures deviations between the chosen OCN and a forecast made from a fixed period of 10 years. Most of the time specialised data vendors provides further servicing and control such as correcting human recording errors, harmonizing data in case of a change in the measurement device, desaisonalizing, and trend modelling (global warming, surface warming of the Pacific ocean known as El Nino phenomenon).

Statistical modelling of index dynamics, often after appropriate prewhitening (desaisonalising and detrending) can be broadly divided into parametric approaches and nonparametric approaches. Parametric approaches rely on standard time series tools such as Autoregressive-Moving Average (ARMA) models, Generalized Autoregressive Heteroscedastic (GARCH) models, and Markov switching models. These models allows introducing mean reversion phenomenon in the conditional mean or variance of the temperature data ((8), (7)) Parametric approaches may also rely on continuous time models (instead of a discrete time models) such as jump-diffusion models. The workhorse in that type of modeling is the Ornstein-Uhlenbeck process ((12), (13), (14), (4)) featuring mean reversion and stationary behaviour. Extensions based on fractional Brownian motion have also been considered to include long memory effects (6)). Nonparametric or semiparametric approaches rely on resampling methods by blocks of data (block bootstrap, subsampling) or by blocks of residuals ((20)). Other approaches try to incorporate weather forecasts to improve on the modelling. Jewson & Caballero (18) use medium-range forecasts (up to 11 days), while Yoo (21) uses seasonal forecasts based on ranking temperature being above, near or below normal.

Multivariate factor analysis also helps to identify and quantify factors relevant to explain differences between North versus South location measures as well as East versus West location measures. They can often be related to the presence of lakes, oceans (Pacific versus Atlantic), deserts or dominant wind directions (North).

5.4.2.2 Characterization of a pricing rule

Given the uncertainty and the flexibility in their accounting classification, but also their relative illiquidity, several pricing methods have been suggested for weather derivatives. They can be classified into three main categories: actuarial, financial and economic. In the following, we will briefly present these various approaches, focusing on the (forward) pricing rule of a weather derivative with a payoff F at a future time. Considering forward price allows us to simplify the problem in terms of interest rates and to focus on the pricing rule itself.

Actuarial method

The first method based upon actuarial arguments uses the fair value, corrected by some margin as pricing rule. More precisely, denoting by $\mathbf{P}$ the statistical probability measure used as prior probability measure, the (forward) price of the derivative can be obtained as

$$\pi(F) = \mathbf{E}_{\mathbf{P}}(F) + \lambda \sigma_{\mathbf{P}}(F) \tag{5.6}$$

Different authors have studied the impact of the choice of the probability measure on the pricing. For instance, (17) compared two methods to compute $\mathbf{E_P}(F)$: burn analysis, where the expected value is obtained as the average over historical data and index modeling, where a theoretical model for the relevant index is calibrated over past data, before simulations are conducted to find the expected value. For the price, both methods tend to be equivalent; for the sensitivities, the index modeling is better. (20) looked at the impact of a calibration error with index modeling: **P** is in fact $\mathbf{P}(\theta)$, with θ the model parameters. The impact is measured as $\pi_{\mathbf{P}(\widehat{\theta})}(F) - \pi_{\mathbf{P}(\theta)}(F)$.

Financial method

The second method is more financial. Some papers assume the weather derivative market to be complete and therefore use the risk-neutral pricing rule:

$$\pi(F) = \mathbf{E_Q}(F), \tag{5.7}$$

where **Q** is the unique risk-neutral probability measure. The underlying assumption is really strong and cannot be justified. A milder argument consists in assuming absence of arbitrage opportunities only. From the fundamental theorem of asset pricing, this is equivalent to the existence of at least one equivalent martingale measure $\mathbf{P}^*$. Attainable contingent claims can be priced using one of these probability measures. However, writing

$$\pi(F) = \mathbf{E_{P^*}}(F) \tag{5.8}$$

is equivalent to consider weather derivatives as attainable. Obviously, this is a strong underlying assumption, and the weather market is still very incomplete by essence.

In an incomplete market framework, there exist however many different methods to price a contingent claim, without creating any arbitrage opportunity. A rather standard approach involves utility maximization. To do so, we consider the preferences of each individual agent. This implies of course to make a distinction between *individual prices* and *equilibrium prices* that can be observed on the whole market.

Any individual wants to maximize the expected utility of her terminal wealth in this framework. The maximum price she is ready to pay for the weather derivative is therefore the price such that she is indifferent, from her utility point of view, between buying it or not buying it. For this reason, the price obtained by utility maximization techniques is called *indifference price* (many references exist on this subject, the seminal paper being that of Hodges & Neuberger (1989) (15)). Denoting by u the utility function of the agent we consider, and assuming that there is no interest rates (for the sake of simplicity), the indifference buyer price of F, $\pi^b(F)$, is determined as:

$$\mathbf{E_P}\left(u\left(W_0 + F - \pi^b(F)\right)\right) = \mathbf{E_P}(u(W_0)), \tag{5.9}$$

where W_0 is the initial wealth (which may be random). This price $\pi^b(F)$, which theoretically depends on the initial wealth and on the utility function, is not (necessarily) the price at which the transaction will take place. This gives an upper bound to the price the agent is ready to pay for the contract F. The agent will accept to buy the contract at any price below $\pi^b(F)$.

In the characterization of the indifference price, there is no question on the volume of the transaction. The potential buyer has two options: either buying 1 contract or not buying it. There is another possible approach, which consists of determining the price of the contract such that agreeing a little into the contract has a neutral effect on the expected utility of the agent. This notion of *fair price* was first introduced by (10) and (11) and corresponds to the zero marginal rate of substitution price. More precisely, the fair price p is determined such that:

$$\frac{\partial \mathbf{E}_{\mathbf{P}}\left(u\left(W_0+\theta F-p\right)\right)}{\partial \theta}\Big|_{\theta=0}=0. \tag{5.10}$$

Economic approach

The *transaction price* is an equilibrium price, either between the seller and the buyer only, or between the different players in the market. Note that a transaction will take place only if the indifference buyer price is higher than the indifference seller's price, which gives a lower bound to the price the seller is ready to accept for the contract. This is a necessary condition for a transaction, and more generally for a market equilibrium, which characterizes the situation where all agents in the market maximize their expected utility at the same time by exchanging their risks (such an equilibrium is also called Pareto-optimal).

Such an equilibrium approach has been adopted by different authors, such as (7) or (16). While the first authors look at how weather forecasts can influence the demand for weather derivatives, and hence their price, the latter consider the problem of pricing in an incomplete market, with a finite number of agents willing to exchange their weather risk exposure. The price of the contract is obtained as that of the Pareto-optimal equilibrium.

5.4.3 Design issues

As shown by various failed attempts of weather risk securitization, in particular the failed issued of a weather bond by Enron in 1999 (see for instance (3) for a detailed study), the design of the new securities appears as an extremely important feature in the transaction. It may be the difference between success and failure. More precisely, as previously mentioned, the high level of illiquidity, deriving partly from the fact that the underlying asset is not traded on financial markets, makes these new products difficult to evaluate and to use. The characterization of their price is very interesting as it questions the logic of these contracts itself. Moreover, the determination of the contract structure is a problem in itself: on the one hand, the underlying market related to these risks is extremely illiquid, but on the other hand, the logic of these products itself is closer to that of an insurance policy. More precisely, even though weather derivatives have all the features of standard financial contracts, they are very different from the classical structures, as their underlying risk is related to a non-financial risk. Consequently the question of the product design, unusual in finance, is raised. Different papers have been looking at this question (see for instance (3)). We briefly present here a simple approach to optimally design the cash flow structure of a weather derivative.

5.4.3.1 Simple framework

Let us consider two economic agents, henceforth called A and B, evolving in an uncertain universe modelled by a probability space $(\Omega, \mathcal{F}, \mathbf{P})$. At a fixed future date T, agent A is exposed to a weather risk for a random amount X. In order to reduce her exposure, A wants to issue a financial product, with a payoff F, and sell it to agent B against a forward price π at time T. Both agents are supposed to be risk-averse. For the purpose of this simple study, we assume that they are working with the same kind of choice criterion, an increasing exponential utility function u defined as:

$$u(x) = -\gamma \exp\left(-\frac{1}{\gamma}x\right) \tag{5.11}$$

with risk tolerance coefficient γ_A and γ_B, respectively.
Agent A objective is to choose the optimal structure for the weather derivative (F, π) as to maximize the expected utility of her final wealth, i.e., seeking:

$$\arg\max \mathbf{E_P}\left[u_A\left(X - (F - \pi)\right)\right]. \tag{5.12}$$

Her constraint is that agent B should have an interest to enter into this transaction. At least, the F-structure should not worsen agent B expected utility. To decide whether or not she should enter this weather transaction, agent B compares two expected utility levels, the first one corresponding to the case where she simply invests her initial wealth in a bank account and the second one to the situation where she enters the F-transaction.
Thus, agent A is working under the constraint:

$$\mathbf{E_P}\left[u_B\left((F - \pi) + W_0\right)\right] \geq \mathbf{E_P}\left[u_B\left(W_0\right)\right], \tag{5.13}$$

where W_0 is the (forward) wealth of agent B before the F-transaction.

With the exponential utility functions, the problem to solve is:

$$\begin{aligned} &\min_{F,\pi} \gamma_A \mathbf{E_P}\left[\exp\left(-\frac{1}{\gamma_A}(X-(F-\pi))\right)\right] \\ \text{subject to } &\mathbf{E_P}\left[\exp\left(-\frac{1}{\gamma_B}((F-\pi)+W_0)\right)\right] \le \mathbf{E_P}\left[\exp\left(-\frac{1}{\gamma_B}W_0\right)\right]. \end{aligned} \tag{5.14}$$

Given the convexity of the program, the constraint is bounded at the optimum and the optimal pricing rule $\pi^*(F)$ of the financial product F is entirely determined by the buyer (agent B) as

$$\pi^*(F) = \gamma_B \ln \mathbf{E_P}\left[\exp\left(-\frac{1}{\gamma_B}W_0\right)\right] - \gamma_B \ln \mathbf{E_P}\left[\exp\left(-\frac{1}{\gamma_B}(F+W_0)\right)\right] \tag{5.15}$$

Agent B determines the minimal pricing rule, ensuring the existence of the transaction. The price $\pi^*(F)$ corresponds to the maximal amount agent B is ready to pay to enter the F-transaction and bear the associated risk, given her initial wealth W_0. In other words, $\pi^*(F)$ corresponds to the certainty equivalent of F for the utility function of agent B, or to the buyer indifference pricing rule.

5.4.3.2 Optimal structure

In the present simple framework, the optimal structure is given by the so-called Borch Theorem, presented below. In a quite general utility framework, (5) obtained optimal exchange of risk, leading in many cases to familiar linear quota-sharing of total pooled losses.

Proposition 1. *The optimal structure of the weather derivative, given as solution of the optimization program* (1) *is given as a proportion of the initial exposure X, depending only on the risk tolerance coefficients of both agents:*

$$F^* = \frac{\gamma_B}{\gamma_A+\gamma_B}X \qquad \mathbf{P}\ a.s. \qquad \textit{(up to a constant)}. \tag{5.16}$$

For the proof of this result, we refer the reader to (2) for instance.

Note that agent A is transferring a part of her initial risk in this contract according to her relative tolerance. Moreover, if the issuer, agent A, has no exposure, no transaction will occur between both agents. In this sense, weather derivatives have a non-speculative underlying logic, and are there to help with the true risk transfer between the agents in the market.

5.4.3.3 Extensions

There are various ways in which this modelling approach can be extended. In particular, the buyer may also be exposed towards a weather risk. The potential diversification impact for both agents is taken into account since the optimal structure in this case becomes a transfer of the risk of each agent in proportion to their relative risk tolerance. Using obvious notation, the optimal structure of the contract becomes in this case:

$$F^* = \frac{\gamma_B}{\gamma_A + \gamma_B} X_A - \frac{\gamma_A}{\gamma_A + \gamma_B} X_B \qquad \mathbf{P}\ a.s. \tag{5.17}$$

Note that even if agent A, the seller, is not initially exposed towards a weather risk, there will still be a transaction between both agents due to the exposure of the buyer.

The diversification impact using the financial markets can also be easily taken into account in this framework. Assuming that both agents can choose optimally their financial investment in the market, as to maximize their expected utility, the optimization program (1) is transformed by referring to another probability measure $\widehat{\mathbf{Q}}$, instead of the prior probablity measure $\mathbf{P}$. $\widehat{\mathbf{Q}}$ is in fact the minimal entropy probability measure and enables to account for the diversification impact offered by optimal financial investment (for more details, please refer to (2)).

Finally, more general decision criteria can be considered instead of the exponential utility. In particular, convex risk measures can be used as to assess the capital requirement needed by financial institutions of insurance companies when undertaking risk. When both agents have risk measures of the same family, then the proportional structure remains optimal as shown in (3), otherwise some additional features maybe added to the structure (see for instance (19)).

References

[1] P. Barrieu and N. El Karoui, Reinsuring climatic risk using optimally designed weather bonds, *Geneva Papers Risk and Insurance Theory*. Vol. 27, 87-113 (2002).

[2] P. Barrieu and N. El Karoui, Inf-convolution of risk measures and optimal risk transfer, *Finance and Stochastics* 9, 269-298 (2005).

[3] P. Barrieu and N. El Karoui, Pricing, hedging and optimally designing derivatives via minimization of risk measures, in *Volume on Indifference Pricing* (ed: Rene Carmona), Princeton University Press (2007).

[4] F.E. Benth and J. Saltyte-Benth, The volatility of temperature and pricing of weather derivatives. To appear in Quantitative Finance (2007).

[5] K. Borch, Equilibrium in a reinsurance market, *Econometrica* 30 (1962), 424-444.

[6] D.C. Brody, J. Syroka and M. Zervos, Dynamical pricing of weather derivatives, *Quantitative Finance*, Vol.2, p.189-198 (2002).

[7] S. Campbell and F.X. Diebold, Weather forecasting for weather derivatives, *Journal of the American Statistical Association*, Vol. 100, No. 469 p.6 – 16 (2005).

[8] M. Cao and J. Wei, Weather derivatives valuation and market price of risk, *The Journal of Futures Markets*, Vol. 24, No. 11, p. 1065-1089 (2004).

[9] Commodity Risk Management Group, Progress report on weather and price risk management work, Agriculture & Rural Development, The World Bank (http://www.itf-commrisk.org/) (2005).
[10] M. Davis, Option pricing in incomplete markets, in *Mathematics of Derivative Securities*, eds M.A.H. Dempster and S.R. Pliska, Cambridge University Press (1998).
[11] M. Davis, Pricing weather derivatives by marginal value, *Quantitative Finance*, Vol. 1, No 3, p. 305-308 (2001).
[12] R. Dischel, Black-Scholes won't do, *Risk Magazine*, E.P.R.M., October, p. 8-13 (1998).
[13] R. Dischel, Shaping history, *Risk Magazine*, E.P.R.M., October, p. 13-15 (1999).
[14] F. Dornie and M. Queruel, Caution to the wind, *Risk Magazine*, E.P.R.M., July, p. 30-32 (2000).
[15] S.D. Hodges and A. Neuberger, Optimal replication of contingent claims under transaction costs. *Review of Futures Markets*, vol. 8, 222-239 (1989).
[16] U. Horst and M. Mueller, On the spanning property of risk bonds priced by equilibrium. To appear in *Mathematics of Operations Research* (2007).
[17] S. Jewson and A. Brix, *Weather derivative valuation: the meteorological, statistical, financial and mathematical foundations*, Cambridge University Press (2005).
[18] S. Jewson and R. Caballero, The use of weather forecasts in the pricing of weather derivatives. *Meteorol. App.*, 10, 367-376 (2003).
[19] E. Jouini, W. Schachermayer and N. Touzi, Optimal risk sharing for law invariant monetary utility functions. *Mathematical Finance*, to appear (2008)
[20] O. Roustant, J.-P. Laurent, X. Bay, L. Carraro, A bootstrap approach to the pricing of weather derivatives, *Bulletin Français d'Actuariat*, (2005).
[21] S. Yoo, Weather derivatives and seasonal forecast, Working Paper, Cornell University (2003).

Chapter 6
Optimal Control Models and Elicitation of Attitudes towards Climate Damages

Philippe Ambrosi, Jean-Charles Hourcade, Stéphane Hallegatte, Franck Lecocq, Patrice Dumas, Minh Ha Duong

Abstract This paper examines the consequences of various attitudes towards climate damages through a family of stochastic optimal control models (*RESPONSE*): cost-effectiveness for a given temperature ceiling; cost-benefit analysis with a "pure preference for current climate regime" and full cost-benefit analysis. The choice of a given proxy of climate change risks is actually more than a technical option. It is essentially motivated by the degree of distrust regarding the legitimacy of an assessment of climate damages and the possibility of providing in due time reliable and non controversial estimates. Our results demonstrate that (a) for early decades abatement, the difference between various decision-making frameworks appears to matter less than the difference between stochastic and non stochastic approach given the cascade of uncertainty from emissions to damages; (b) in a stochastic approach, the possibility of non-catastrophic singularities in the damage function is sufficient to significantly increase earlier optimal abatements; (c) a window of opportunity for action exists up to 2040: abatements further delayed may induce significant regret in case of bad news about climate response or singularities in damages.

keywords: cost-effectiveness, cost-benefit, climate sensitivity, climate change damages, uncertainty, optimal climate policy, decision making frameworks.

Philippe Ambrosi
The World Bank, Climate Change Team, 1818 H St. NW, Washington D.C., USA.

Jean-Charles Hourcade, Patrice Dumas and Minh Ha Duong
CIRED, Jardin Tropical, 45bis avenue de la Belle Gabrielle, F-94 736 Nogent-sur-Marne, France.

Franck Lecocq
1) INRA, UMR 356 Économie Forestière, F-54000 Nancy, France. 2) AgroParisTech, Engref, Laboratoire d'économie forestière, F-54000 Nancy, France.

Stéphane Hallegatte
1) CIRED, Jardin Tropical, 45bis avenue de la Belle Gabrielle, F-94736 Nogent-sur-Marne, France.
2)Ecole Nationale de la Météorologie, Météo-France, Toulouse, France.

J.A. Filar and A. Haurie (eds.), *Uncertainty and Environmental Decision Making*, International Series in Operations Research & Management Science 138,
DOI 10.1007/978-1-4419-1129-2_6,

6.1 Foreword

This chapter is a reprint of a paper, with the same title that appeared in 2003 in the journal *Environmental Modeling and Assessment*[1]. The present article compared optimal greenhouse gas emissions abatement schedules over the coming decades, derived from a family of stochastic optimal control models, each reflecting a particular attitude as to the choice of a metrics for assessing climate policy benefits in the context of significant and durable uncertainties on climate risks: cost-effectiveness for a given temperature ceiling; cost-benefit analysis with a pure preference for current climate regime and full cost-benefit analysis. Our results demonstrated the crucial role of uncertainty, and when it is resolved, on short-term optimal abatement level; discounting, in this regard, was shown to have much less influence. In addition, given the cascade of uncertainties, convergent recommendations were obtained for the three decision-making frameworks considered, highlighting the rationale for early action as abatement further delayed might induce significant regret in case of bad news regarding climate response or singularities in damages.

Over the past few years, a growing consensus has built up on the necessity to address climate change, with a rising awareness about the risks of unmanaged climate change. A landmark contribution in this respect, the Stern Review[2], called for strong and early action, whose benefits far outweigh the economic costs of not acting. IPCC was awarded the Nobel Prize, and in its Fourth Assessment Report[3] also underscored the medium- to long-term climate risks of delayed action and outlined the existence of a window of opportunity over the next two to three decades to preserve the opportunities to achieve lower stabilization levels.

Since the publication of this article in 2003, the question of the optimal timing of action in the context of large uncertainties has continued to attract scholars' interest. It is not the purpose of this foreword to comprehensively review these developments. Rather, we propose to focus on a couple of contributions that resonate with the approach followed in the article, exploring some questions further or contributing to a deeper understanding of the timing issue.

Learning plays a central role in this matter. Given the long timeframes and inertias (both natural and socioeconomic) at stake, given daunting uncertainties, addressing climate change requires a sequential decision making process, one that can incorporate new information as it becomes available. Today's (rational) decision should seek to preserve the greatest possible flexibility in future choices, choices that may ultimately be made on the basis of reduced uncertainties. As IPCC put it in its Second Assessment Report, summarizing tensions between environmental and economic regrets, today's decision-maker has to balance *the economic risks of rapid abatement now (that premature capital stock retirement will later be proven*

[1] We thank the editor and publishers of this journal for authorizing this reprint.

[2] Nicholas Stern (2007). Stern Review: The Economics of Climate Change.

[3] IPCC, 2007: Climate Change 2007: Synthesis Report. Contribution of Working Groups I, II and III to the Fourth Assessment Report of the Intergovernmental Panel on Climate Change [Core Writing Team, Pachauri, R.K and Reisinger, A. (eds.)]. IPCC, Geneva, Switzerland, 104 pp.

unnecessary), against the corresponding risks of delay (that more rapid reduction will then be required, necessitating premature retirement of future capital)[4]. For long there has been no solid consensus on which of these regrets would dominate and what the resulting impact of learning on short-term decision would be (act or wait)? Accounting for latest work on learning and climate change (including recent analyses of the learning process itself and its time span), ONeil et al. come to the conclusion that *a clear consensus on a central point is that the prospect of learning does not support the postponement of emissions reductions today*[5]. This does not imply that the chapter on learning and climate is completed. In particular, as the eventuality of learning often brings in some flexibility to todays decision (even more so in a cost effectiveness framework), it is critical to get a sense of how long it could take to resolve uncertainties (in particular to manage overly optimistic expectations and their implications for action). Our results outlined the large benefits from knowing the value of climate sensitivity before 2040 in view to mitigate to some extent the significant economic regrets that are concomitant with a precautionary policy. A growing body of evidence suggests that such information might not be soon available, forcing one to follow relatively stringent emissions pathways. Among recent articles for instance, Webster et al. estimate that a substantial reduction in uncertainty in climate sensitivity, of up to 20-40%, could be possible within the next two to five decades as a result of additional climate observations, implying improved information only starting in 2030[6]. Further, Leach concludes learning on natural dynamics may be in the order of centuries[7].

We also considered a pure preference for the current climate regime (PPCR), in particular to get round the stringency of the cost-effectiveness framework (where constraints must be met whatever the cost) without resorting to a fully itemized cost-benefit analysis. This PPCR accounted for precautionary positions leading to favor the current climate regime over unknown alternatives. Our results showed that while the decision-maker might opt for less ambitious climate policies than in a cost-effectiveness framework (higher warming on long-term), short-term efforts are not much relaxed, in particular in the presence of uncertainty. One advanced explanation is that future (and richer) generations place a higher value on the stability of the current climate regime as it is specified as a superior good and as it is more and more seriously threatened by the atmospheric build up of greenhouse gas concentrations. Considering a PPCR counterbalanced to some extent the impact of discounting on

[4] Bruce JP, Lee H and Haites EF (Eds) (1996). Climate change 1995: Economic and social dimensions of climate change. Contribution of Working Group III to the Second Assessment Report of the Intergovernmental Panel on Climate Change, Cambridge University Press.

[5] O'Neill BC, Crutzen P., Grbler A., Ha Duong M., Keller K., Kolstad C., Koomey J., Lange A., Obersteiner M., Oppenheimer M., Pepper W., Sanderson W., Schlesinger M., Treich N., Ulph A., Webster M. and Wilson C (2006). Learning and climate change, Climate Policy 6:585-589.

[6] Webster M, Jakobovits L and Norton J (2008). Learning about climate change and implications for near-term policy, Climatic Change 89: 67-85.

[7] Leach A (2007). The climate change learning curve, Journal of Economic Dynamics and Control 31:1728-1752.

short-term abatement level. This has been further examined by Sterner and Persson[8], one amongst the many contributions on discounting stimulated by the Stern review. They report that future scarcities that will be induced by the changing economy and climate should lead to rising relative prices for certain goods and services, raising the estimated damage of climate change and counteracting the effect of discounting. Their numerical analysis shows that taking relative prices into account can have as large an effect on economically warranted abatement levels as can a low discount rate.

We last examined key timing issues under a cost-benefit framework and emphasized the critical importance of the shape of the damage curve (prior to the scale of damages). We found that the eventuality of an abrupt acceleration in damages (as opposed to gradually increasing damages, as is the case with most integrated assessment models) is consistent with early abatement efforts, even without assuming large catastrophes[9]. This is even more so the case in presence of uncertainty on climate sensitivity. Thresholds and non-linearities have been receiving growing attention (and certainly Weitzman's recent contribution made an even stronger case[10]) but we may still be short of a deep understanding of their origins. There are long lists of examples of non-linearities in the response of the climate system to anthropogenic forcing (e.g., slow down of the thermohaline circulation or methane discharge by thawing permafrost) or of ecosystems to climate change (e.g., coral reefs), with potentially significant, durable and irreversible impacts. More research is needed however to understand the very translation of impacts into social costs and potential shocks to fragile economies: even without assuming large disruptions of the environment, slowly deploying and hardly perceptible impacts can still have large implications for vulnerable populations, with no or few alternatives, low resources and weak institutional capacity to anticipate or react and plan adaptation measures. Growing attention to the determinants of adaptation as well as to the link between development stage and vulnerability may bring promising results.

6.2 Introduction

Little progress has been made since 1992 on what constitutes a "*dangerous anthropogenic interference with the climate system*" (1). The Third Assessment Report (TAR) of the Intergovernmental Panel on Climate Change (IPCC) delivered new

---

[8] Sterner T and Persson M (2008). An Even Sterner Review: Introducing Relative Prices into the Discounting Debate, Review of Environmental Economics and Policy 2(1):61-76; doi:10.1093/reep/rem024

[9] Also established by Dumas P and Ha-Duong M (2005). An abrupt stochastic damage function to analyze climate policy benefits. In: Haurie A and Viguier L (Eds), The Coupling of Climate and Economic Dynamics: Essays on Integrated Assessment. Kluwer Academic, Dordrecht, The Netherlands, pp. 97112.

[10] Weitzman M (2007). The Role of Uncertainty in the Economics of Catastrophic Climate Change, AEI-Brookings Joint Center Working Paper No. 07-11.

material on climate change impacts but did not venture to deliver any conclusive judgment [2]. In this context, which decision-making framework should be used to design climate policies remains an open question (see (3) (chap. 1, 2) and (4) (chap. 10) for a survey). Bounded cost; minimax regret; maximin gain; minimax loss; cost-effectiveness, tolerable windows and safe landing approaches; cost-benefit analysis, each with its own merits and limitations, are representative of the diversity of decision-making attitudes in a *sea of uncertainty* (5).

The objective of this paper is to assess how each of these attitudes translates in terms of timing of emissions abatement. It is achieved through the use of optimal control models, which can put some rationale into pending controversies and thus facilitate the emergence of compromises because they are apt to disentangle the sources of misunderstandings from the real division lines. Indeed they force the analyst to (a) identify the pathways through which climate change may impact on global welfare; (b) clarify the proxies that are used to capture the benefits of climate action, and against which the costs of this action are to be weighted; and (c) make explicit the level of confidence about scientific information and the ethical choices which underpin the selection of a given framework. Hence, after having discussed (section 7.2) how various attitudes towards climate change lead to various metrics to capture the benefits of climate policies, we successively analyze the optimal abatement pathway derived from (section 7.3) a cost-effectiveness analysis of temperature ceiling objectives, (section 7.4) a cost-benefit analysis using a pure preference for current climate regime and (section 7.5) a cost-benefit approach using a monetized quantification of damages.

6.3 Metrics for assessing benefits of climate policies

IPCC TAR indicates that global mean temperature is projected to increase by 1.4 to 5.8 °C over the period 1990 to 2100 as a consequence of greenhouse gases anthropogenic emissions.

Faced with such a large uncertainty range, debates about the application of the Precautionary Principle come to select specific metrics of the benefits of climate policies. The selection of such metrics is certainly motivated by value judgments, but it is not independent from the degree of distrust regarding the possibility of timely providing reliable and non controversial estimates. Attitudes can be grouped into three broad categories:

1. A first one considers that the uncertainty about climate impacts and damages is so high that they cannot be confidently assigned any numerical value; environmental benefits are thus set in the form of arbitrary ceilings on either greenhouse gases (GHGs) concentration, temperature, or any other multidimensional indicator. Approaches such as a safe corridor, a safe landing or a viability path also belong to this cost-efficiency framework; their outcome depends obviously on whether the constraints are set by a *convinced ecologist* or by a *skeptical ecologist* (à la Lomborg). Sharing the same distrust about predictions of climate im-

pacts, the convinced and the skeptical ecologist may search for a *reasoned compromise*, and agree on a sequential decision-making process in which an initial trajectory can be adapted in the light of new information. This common will to consider several conceivable futures and to keep open alternative options leads to substitute a stochastic to a deterministic cost-efficiency model.
2. Another attitude refuses the arbitrary setting of absolute targets and demands a cost-benefit analysis; however, being skeptic about explicit prediction and assessment of damages, it does not use an itemized assessment of these damages but a willingness to pay for avoiding various levels of climate change. In modeling terms, this is translated through the inclusion of climate change indicators (temperature or rainfall patterns, extreme events) in the utility function to express a *pure preference for current climate regime* (PPCR). This PPCR conveys precautionary ethics leading to favoring the current climate regime over unknown alternatives; it incorporates psychological motivations about endangered habitats, the amenity or bequest value of landscapes, all values considered a part of climate policy benefits in the absence of definition of climate feedbacks on economic productivity. Depending on the specification of the utility function, the environment appears (or does not appear) as a superior good (a good to which agents dedicate a growing share of their income as they become richer).
3. The last attitude leads also to the carrying out of a cost-benefit analysis but requires an itemized monetary assessment of impacts. This assessment confronts uncertainty of impact predictions and raises controversies[11] about monetary valuation (such as placing a monetary value on human life in different countries or aggregating regional estimates assuming a compensation hypothesis). Many perform such an assessment though, for lack of anything better, to place some rationale into policy debates about long term targets, be it to convince public opinion to accept subsequent unpopular measures or to resist disproportionate demands from environmentalists.

Hence, the choice of a metrics of the benefits from mitigation policies is actually much more than a mere technical option. It reflects a judgment on the quality of the available information and on its ability to serve as a common basis in the negotiation process. In other words, it implies a trade-off between accuracy and relevance [8]: accuracy because the further *down* we move along the causal chain linking GHGs emissions to climate change damages, the less confidence we place in our ability to predict the outcome of the cascade of uncertainties we are faced with; relevance because the further *up* we proceed from damages functions to GHGs concentrations ceilings, the further we get from a precise description of climate change consequences, in particular with regard to welfare and distributive aspects.

To compare the policy implications of these attitudes, we have performed a set of harmonized numerical experiments based on the *RESPONSE* model family. *RE-*

[11] For instance, the forum on valuation of ecosystem services (Ecological Economics 25(10.1) (1998)) which has been emulated by Costanza on the basis of [6]. Many respondents have indeed pointed out the risk of underestimating the environment as Toman [7] elegantly puts it: "a serious underestimate of infinity".

SPONSE is an aggregate optimal control integrated assessment model which includes reduced forms of carbon cycle and climate dynamics. Using a sequential decision-making framework, we focus on the sensitivity of first period decisions to the combination of uncertainties about climate change consequences, and to the choice of one of the three metrics described above.

These experiments have been conducted for a single baseline growth scenario (A1B, SRES) and set of abatement cost curves. The specification of the abatement cost function is meant to capture the role of socio-economic inertia as a cost multiplier; it incorporates an autonomous technological change factor (see appendix).

The objective functions only differ among variants of the model:

- *RESPONSE_T* (T for temperature) explores the first attitude and minimizes the discounted sum of abatement costs with respect to an environmental constraint on global mean temperature rise;
- *RESPONSE_P* (P for preference for current climate regime) explores the second attitude and maximizes the intertemporal welfare derived from both consumption and the amenity value of climate;
- *RESPONSE_D* (D for damages) explores the third attitude and maximizes the intertemporal welfare derived from consumption minus abatement expenditures and resulting damages.

General specifications of *RESPONSE* will be introduced in the following section with the description of *RESPONSE_T* (both in deterministic and stochastic versions); for the remaining sections, we will solely describe the specific differences (with respect to the generic model) of the version we refer to.

6.4 Lessons from a stochastic cost-effectiveness analysis: *RESPONSE_T*

Up to to the late 90's, efforts to clarify controversies about the timing of GHGs abatements have been carried out through a cost-effectiveness analysis within a sequential decision framework enabling to adjust an initial response in the light of new information (9). Whereas a delay in the bulk of abatement efforts is justified if a given GHGs concentration target is known in advance (10), an earlier and higher departure from baseline emissions trends is required if the same, say 550 ppm target, is considered as the mean of three yet unknown values (say 450 ppm, 550 ppm and 650 ppm) and if information about the real value is expected to be disclosed some decades in the future (9). Two main results emerge from this analysis which remain relevant whatever the attitude towards climate risks:

1. the role of the interplay between uncertainty about the ultimate target and the inertia of technical and environmental systems: without inertia, transition costs of switching from one emission path to another would be null, and uncertainty would not matter; in fact, inertia raises both the costs of premature abatement and that of accelerated abatement if stronger action is called for later;

2. the value of the discount rate matters less than (10.1) the set of probabilities placed on the targets, and, more specifically the weight given to the tightest one and (10.2) the date of resolution of uncertainty: the later this uncertainty is to be resolved, the earlier the abatement efforts have to be scheduled.

Whereas a cost-effectiveness analysis of concentration ceilings is policy relevant because it follows the very language of the UNFCCC, it is a poor proxy of the benefits of climate action. More specifically it does not allow for considering the uncertainty regarding climate sensitivity. This parameter is defined as the global mean surface temperature increase at the equilibrium, when CO_2 concentration is kept constant at twice the pre-industrial level. Literature sets this parameter between +1.5 °C and +4.5 °C [11] (chap. IX).

This is why it is attractive to carry out a cost-effectiveness analysis of temperature ceilings. *RESPONSE_T* performs such an analysis through an objective function (10.1) minimizing the discounted sum of abatement costs (a surrogate of a utility-maximization model where consumption is lowered by mitigation measures), with as environmental constraint (10.2) a ceiling on the global mean temperature rise relative to its 1990 value.

$$\underset{Ab_t}{Min} \sum_{t=1990}^{2300} \frac{f(Ab_t, Ab_{t-1}, t)}{(1+\rho)^{(t-1990)}} \tag{6.1}$$

$$w.r.t. \quad (\theta_t - \theta_{1990}) \leq \Delta\theta_{\max} \tag{6.2}$$

with:

$f(\cdot)$: the abatement cost function,
Ab_t : the abatement rate at time t (% of baseline emissions),
ρ : the discount rate (5% / year),
θ_t : the global mean temperature rise at time t,
$\Delta\theta_{\max}$: the constraint on global mean temperature rise relative to 1990.

The carbon cycle (10.3) is taken from Nordhaus (12): it is a linear three-reservoir model which describes carbon accumulation and transportation between the atmosphere, the biosphere (oceanic and continental) and deep ocean. The model accounts for some inertia in natural processes. Related parameters (transfer coefficients and initial conditions) are given in the appendix.

$$M_{t+1} = C_{trans} M_t + \delta(1 - Ab_t) em_t \, u \tag{6.3}$$

with:

M_t : the carbon contents of each reservoir at time t, a column vector,
C_{trans} : the net transfer coefficients matrix, a 3x3 matrix,
δ : the time step of the model (10 years),
em_t : baseline carbon dioxide emissions at time t, exogenous (A1B scenario),
u : a column vector (1,0,0).

Lastly, variations in global mean temperature derive from a two-box climate model (11.4) describing the modification of the thermal equilibrium between atmosphere and ocean in response to enhanced greenhouse effect (carbon dioxide only) based on specifications close to Nordhaus's one (12). To improve the quality of the insight on the timing of abatement over the short run, we calibrated this model in such a way that it gives a better description of warming over forthcoming decades: we prioritize the description of the interaction between the atmosphere and the superficial ocean neglecting interactions with the deep ocean. A thorough description is provided in the appendix.

$$\theta_{t+1} = L(\theta_t, M_t). \tag{6.4}$$

The model defined by equations (10.1), (10.2), (10.3) and (11.4) can be run on a perfect information mode (*RESPONSE_T/c*) and on an uncertainty mode (*RESPONSE_T/s*). In the second option, uncertainty on climate sensitivity is discrete: we consider three possible states of the world (s) in which climate sensitivity may be $\{2.5\,^\circ\text{C};\ 3.5\,^\circ\text{C};\ 4.5\,^\circ\text{C}\}$ with the corresponding *ex ante* subjective probabilities (p_s) $\{1/6;\ 2/3;\ 1/6\}$. Information arrives at a fixed time in the future (t_{info}). The program solves a set of three parallel problems – three equations (10.3) and (11.4) representing three alternative states of the world; climate dynamics (11.4) is notably dependent on the value of climate sensitivity. The objective function (10.5) is re-specified as the minimization of expected costs of abatement paths:

$$\underset{Ab_t^s}{Min} \sum_s p_s \sum_{t=1990}^{2300} \frac{f(Ab_t^s Ab_{t-1}^s, t)}{(1+\rho)^{(t-1990)}}. \tag{6.5}$$

Environmental constraint (10.6) is rewritten to consider each state of the world:

$$w.r.t. \quad (\theta_t^s - \theta_{1990}) \leq \theta_{MAX}. \tag{6.6}$$

Additional constraints (10.7) are added to impose that, before the disclosure of information, decision variables be the same across all states of the world:

$$\forall t \leq t_{\text{info}},\ \forall\left(s,s'\right) \in S,\ Ab_t^s = Ab_t^{s'}. \tag{6.7}$$

Model solutions correspond technically to perfect information when $t_{\text{info}} = 1990$, imperfect information with learning when $1990 < t_{\text{info}} < 2300$, absolute uncertainty when $t_{\text{info}} = 2300$.

Let us start from a +2 °C target[12] with respect to 1990. It corresponds to an expected value of 500 ppm for GHGs concentration, actually shifting from a very

[12] This figure is circulated in many studies such as the Global Fast Track Assessment (13) where the additional number of people at risk of water shortage increases sharply once global mean temperature rise gets close to +2 °C. (14) also suggests that a +2 °C temperature increase dramatically reduces suitable areas for Robusta coffee in Uganda. Note that this target is less binding than EU long-term climate goal [15], amounting to a maximum +2 °C global mean temperature rise *wrt preindustrial level.*

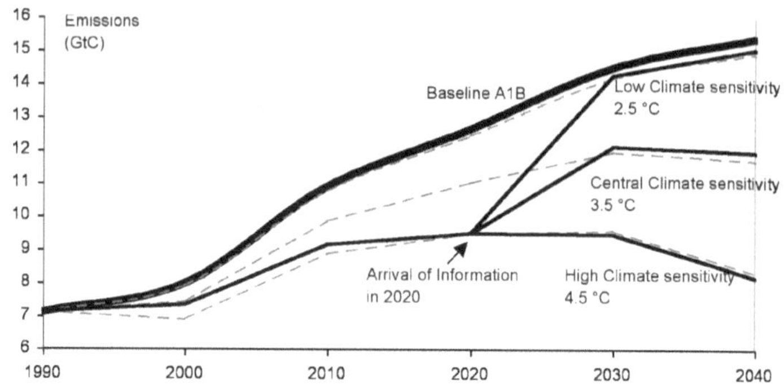

Fig. 6.1 Hedging strategies for a given +2 ° C temperature ceiling: with perfect information (grey dashed line) and with uncertainties (black continue line).

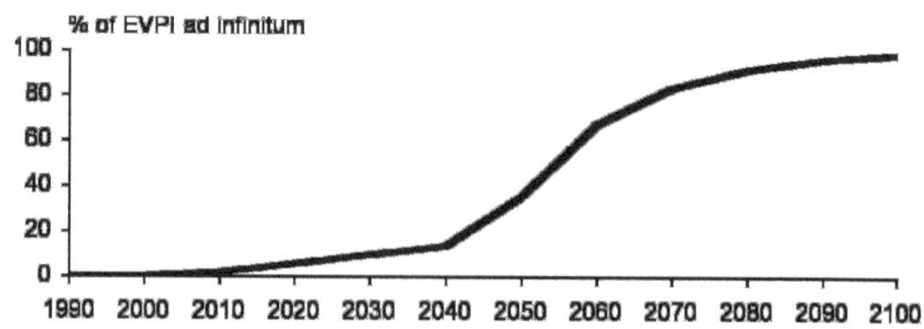

Fig. 6.2 Expected Value of Perfect Information (EVPI) with respect to the date of resolution of uncertainty on climate sensitivity. Information value raises brutally after 2040, that means there is a significant interest in revealing this value before this date (17).).

stringent 440 ppm when climate sensitivity is set to its upper value to a very lax 590 ppm when climate sensitivity is at its lower value. A +1 ° C and +3 ° C target would respectively lead to a 379– 448 ppm range (expected value: 408 ppm) and to a 515–780 ppm range (expected value: 617 ppm) for concentration ceiling.

It appears that for a +2 ° C target, and assuming that information on the value of climate sensitivity arrives in 2020, the earlier periods optimal emissions path is very close to the one consistent with the most pessimistic hypothesis about this value (figure 6.1). When compared with results obtained with GHGs concentration ceilings, the dominance of the worst case hypothesis is reinforced: pessimistic assumptions regarding climate sensitivity lead to a tighter constraint (440 ppm) and, more importantly, imply that the +2 ° C temperature ceiling is reached as early as 2050 in the baseline case. Consequently, the model accounts for the fact that any delay in climate policy will result in a costly acceleration of GHG abatement.

A good indicator of the environmental irreversibility is captured by the value of information on climate sensitivity. The Expected Value of Perfect Information

(EVPI) is classically the difference between the expected value of the objective function in the "Learn then Act" (climate sensitivity known from the outset and policy adopted consequently) and in the "Act then Learn" (a policy must be adopted before the value of this parameter is revealed) (16). Logically, the later the date of resolution of uncertainty the higher the EVPI. Before 2040 (figure 6.2), it increases linearly up to 13% of its final value and then sharply between 2040 and 2070 to reach 83% of this value. To give a comparative benchmark, the expected value of discounted abatement costs over the three states of the world in the Learn then Act hypothesis would amount almost to 52 percentage points in the same metrics. Such a high opportunity cost of knowing climate sensitivity before 2040 highlights the risk of postponing too much a serious hedging strategy in case of pessimistic prospects about the progress in scientific knowledge and public awareness of climate risks.

The main criticism addressed to a stochastic cost-effectiveness analysis is that it gives too high a weight to the tightest constraint. The set of probabilities can be indeed interpreted either in terms of subjective probabilities or in terms of shares in a population of subgroups advocating for a given constraint. In the latter case the program comes to find a compromise between competing views of the world. But a minority, say a fringe of 10% of the population, arguing for a 390 ppm target, would automatically exert a disproportionate influence on decision because costs of postponing action for this target tend towards infinity.

In practice though, faced with such a situation, societies would overshoot the ceiling at the risk of some damages admitting that a window of opportunity has been missed (18), rather than bear the social costs of an exaggerated deceleration of emissions. The necessity of examining such trade-off is the main argument for shifting from cost-effectiveness analysis to some form of cost-benefit approach.

6.5 The pure preference for the current climate regime: *RESPONSE_P*

As explained in section 7.2, the first form of cost-benefit analysis, consistent with an attitude of distrust regarding any numerical assessment of damages, considers a willingness to pay for mitigating climate change and a *pure preference for current climate regime* (PPCR).

Let $U(.)$ denote the utility function. C_t denotes current consumption level; climate change is expressed by global mean temperature rise, θ_t. We specify $U(\cdot)$ such as $U(C_t, \theta_t) = \ln(C_t)\left(\overline{\theta} - \theta_t\right)^{\beta}$ with *0< (1<)Ct* and $\theta_t < \overline{\theta}$, $0 < \beta < 1$. $\overline{\theta}$ denotes an absolute threshold beyond which climate change impacts would be overbearingly disruptive; we arbitrarily set this parameter to +4 °C (keeping in mind that such a warming on a global scale would imply, in some regions, a warming greater than +6 °C, that is to say tremendous local climatic shocks). With this specification, willingness to pay increases with the expected level of climate change. Moreover, the preservation of the current climate regime is treated as a superior good. This point can be quickly verified. Let $WTP(\theta)$ be the maximum amount of current income we are willing to pay to prevent a climate change of magnitude θ:

$$\ln(C)\left(\overline{\theta}-\theta\right)^{\beta}=\ln\left(C-WTP(\theta)\right)\left(\overline{\theta}\right)^{\beta}$$

leading to

$$WTP(\theta)=C-C^{\left(\frac{\overline{\theta}-\theta}{\overline{\theta}}\right)^{\beta}}.$$

Hence, marginal willingness to pay is:

$$\frac{\partial WTP(\theta)}{\partial\theta}=\frac{\beta\ln C}{\overline{\theta}}\left(\frac{\overline{\theta}-\theta}{\overline{\theta}}\right)^{\beta-1}C^{\left(\frac{\overline{\theta}-\theta}{\overline{\theta}}\right)^{\beta}}>0.$$

Therefore, *WTP(θ)* is a growing function of temperature change θ.

Let $\pi(\theta)$ denote the ratio between *WTP(θ)* and income. We have:

$$\pi(\theta)=\frac{WTP(\theta)}{C}=1-C^{\left(\frac{\overline{\theta}-\theta}{\overline{\theta}}\right)^{\beta}-1}$$

leading to

$$\frac{\partial\pi(\theta)}{\partial C}=-\left(\left(\frac{\overline{\theta}-\theta}{\overline{\theta}}\right)^{\beta}-1\right)C^{\left(\frac{\overline{\theta}-\theta}{\overline{\theta}}\right)^{\beta}-2}>0.$$

Thus, for the same climate change magnitude θ, $\pi(\theta)$ is an increasing function of income: climate protection is a superior good.

So far no opinion polls exist on the willingness to pay for climate stability; would they though, their results would be very sensitive to the political and/or media life cycles that determine the way information is conveyed to public opinion (19). A more secure approach is to reveal the implicit utility function behind figures circulating about the *reasonable* maximum value for temperature change (for example +2 °C in the Energy Modeling Forum ongoing round or for some NGOs). To do so, for each value of pure time preference (PTP), we can determine the elasticity of utility w.r.t. climate regime (β) that exactly balances the marginal welfare impacts of consumption and climate amenity value along the optimal abatement trajectories obtained for this target in the certainty case: practically, for this value of β, the marginal welfare impact of the consumption loss resulting from a tightening of the environmental objective from +2.05 °C to +1.95 °C is exactly compensated by the marginal welfare improvement due to lower temperatures. This procedure ensures consistency between claims for a given target and expectations on baseline emissions, abatement costs and climate sensitivity. For example, for a given abatement cost curve, a +2 °C ceiling implies higher mitigation costs under higher climate sensitivity. Sticking to this objective thus implies a higher WTP for climate protection than if one expects low climate sensitivity (see table 6.1).

An important feature of the new program which maximizes $U(\cdot,\cdot)$ without absolute constraint on the quality of the environment, is that an overshoot is now allowed in case of delayed action: this occurs if the cost of maintaining the temperature below the desired target is greater than the marginal WTP to avoid extra warming.

Table 6.1 Parameter β values in function of climate sensitivity and pure time preference.

Climate sensitivity	2.5 °C	3.5 °C	4.5 °C
PTP = 1%.year^{-1}	1.4 10^{-4}	7.7 10^{-4}	13.8 10^{-4}
PTP = 3%.year^{-1}	2.7 10^{-4}	16.4 10^{-4}	34.4 10^{-4}

In *RESPONSE_P/c* all equations and model specifications remain identical to *RESPONSE_T/c* except the objective function which is re-specified as the maximization of an intertemporal utility function (10.8):

$$\underset{Ab_t}{Max} \sum_{t=1990}^{2300} N_t \ln\left(c\frac{(Y_t - f(Ab_t, Ab_{t-1}, t))}{N_t}\right)\left(\overline{\theta} - \theta_t\right)^{\beta} e^{-\eta(t-1990)} \tag{6.8}$$

with:

N_t : the population level (source A1B),
Y_t : the gross world product (source A1B),
c : the propensity to consume (0.8),
β : the elasticity of utility w.r.t. climate regime (see table 1),
η : the pure time preference (1 or 3%/year, corresponding to 3 and 5%/year discount rate in *RESPONSE_T/c*).

In a deterministic mode, it is first remarkable that there is no overshoot beyond a +2 °C target for a 1% pure time preference, even in the most pessimistic value for climate sensitivity (results not shown here). A moderate overshoot (up to 0.15 °C) during 50 years[13] is found with a pure time preference as high as 3% (figure 6.3). Second, the model does not advocate lower abatement in the first periods: up to 2020 mitigation costs are twice as high as in a cost-efficiency framework. This paradox, noted by Hammitt [8], can be easily explained: in a cost-efficiency framework, agents give a high value to climate (the costate variable at a given point in time) only when the target is approached whereas in our *PPCR* approach, climate change is given a significant value by current generations. As time passes, future (and richer) generations give a higher value to it since it is a superior good.

Let us now turn to a situation where, given the mandate of staying below a +2 °C target for an expected +3.5 °C value of climate sensitivity, the central planner calibrates the β coefficient accordingly ($\beta = \beta_C = 16.4 \times 10^{-4}$) and considers the resulting utility function as expressing the real preferences of the population. But

[13] That figure should be considered cautiously because of the short-term calibration of the temperature model.

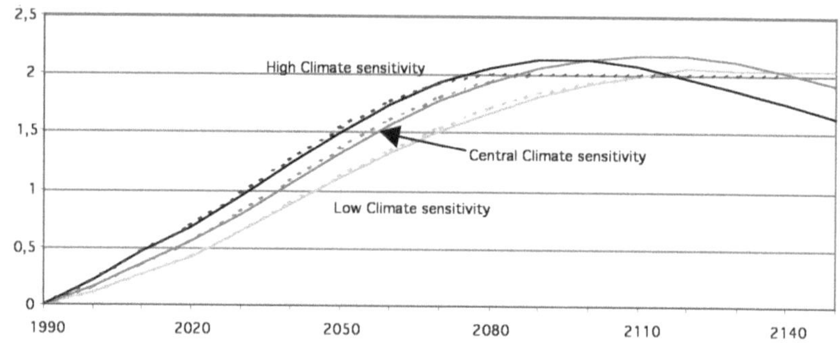

Fig. 6.3 Global mean temperature increase for 3 climate sensitivities: in a cost-benefit analysis with Pure Preference for Climate Current Regime based on a desired 2 ° C target (full curve) and in a cost-effectiveness analysis with a 2 ° C target (dotted curve).

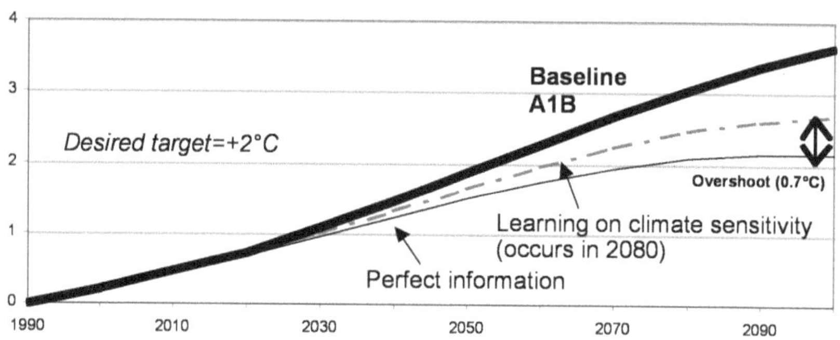

Fig. 6.4 Comparing the cost of misestimating belief on climate change damages and climate dynamics. In all cases, climate sensitivity has a high value. Global mean temperature increase for baseline case (bold black curve), for optimal strategy with perfect information (thin black curve) and for optimal strategy with learning on climate sensitivity (grey curve).

+3.5 ° C is only the mean of three possible values and as information arrives, climate sensitivity is set to its true value whereas the value of β is not revised. This is captured by *RESPONSE_P/s* in which uncertainty on climate sensitivity specifications is similar to *RESPONSE_T/s*.

RESPONSE_P/s parallels RESPONSE_P/c except its objective function (10.10) which we specify such as the maximization of expected utility across the probability distribution of the three possible states of the world:

$$\underset{Ab_t^s}{Max} \sum_{s}^{p_s} \sum_{t=1990}^{2300} N_t \ln\left(c\frac{(Y_t - f(Ab_t^s, Ab_{t-1}^s, t))}{N_t} \right) \left(\overline{\theta} - \theta_t^s\right)^{\beta^C} e^{-\eta(t-1990)}. \qquad (6.9)$$

Additional constraints (10.11) are added to impose that, before the disclosure of information, decision variables be the same across all states of the world:

$$\forall t \leq t_{\text{info}},\ \forall \left(s,s'\right) \in S,\ Ab_t^s = Ab_t^{s'}. \tag{6.10}$$

For a resolution of uncertainty as late as 2080, the optimal response leads to a +0.7 °C overshoot if climate sensitivity is finally +4.5 °C (dotted grey curve, figure 6.4). This has to be compared to the modest overshoot in the certainty case (+0.1 °C) (black thin curve, figure 6.4). However, this overshoot does not mean an absence of action: the simulation shows a very significant deviation from the global mean temperature increase in the baseline scenario (bold black curve, figure 6.4).

Consistently, under this pessimistic assumption regarding climate sensitivity, mitigation costs in the Kyoto commitment period are significantly lower in the learning case ($\beta = 16.4 \times 10^{-4}$) than in the perfect information case ($\beta = 34.4 \times 10^{-4}$) : 0.02% of GWP to be compared with 0.08% of GWP. Sensitivity tests about the date of arrival of information on climate sensitivity show that mitigation costs around 2010 are remarkably constant and that the learning date has no dramatic influence on the magnitude of the overshoot, which varies from +0.6 °C (early learning) to +0.7 °C (late learning). These results suggest that the difference between β values dominates uncertainty regarding climate sensitivity.

The key question remains the real magnitude of WTP and its variations amongst regions depending on preferences and income level. At a regional level, such a *PPCR* analysis would enable to scrutinize compensation schemes between countries necessary to reach a consensus on a global temperature target. Some regions might indeed be willing to adopt a very low temperature ceiling corresponding to a global constraint too tight to be agreed upon at an international level. Would this global constraint be slackened, these regions would understandably demand for compensations.

6.6 Key issues with the strong form of cost-benefit analysis

Some authors are reluctant to resort to a cost-benefit analysis because, they claim, discounting cannot but underestimate the value of environmental damages and thus jeopardize the welfare of future generations. This seems the case in the few existing cost-benefit analyses (see the EMF review (20)), which univocally advocate a slow departure from current emission trends unless 'bad news' regarding climate damages appears.

Contrary to a *PPCR* approach, where environmental variations affect welfare in the first periods, climate impacts occur only several decades after mitigation efforts are undertaken and, once discounted, marginal benefits of those actions are easily outweighed by their costs. This is the reason why a zero coefficient for pure time preference (PTP) has been argued (21). But this option faces serious problems. First, as shown by Koopmans (22), time consistent decision-making over infinite consumption plans requires a strictly positive *PTP*. In addition, introducing a zero or very low *PTP* in a growth model entails high savings and low consumption for the current (and poorest) generation. This is arguably not consistent with intergenerational equity.

This paper does not address the alternative proposals suggested in the literature to avoid the sacrifice of both current and future generations (e.g., (23)). Despite their interest, such proposals either raise serious dynamic consistency problems or do not change the response much for the early periods (24). We rather concentrate on the interplay between conventional discounting and expectations regarding the shape of the damage function, future economic growth and future emissions. To this aim, we introduce a zero *PPCR* ($\beta = 0$). Prior to discussing our results, which are mainly derived from numerical experiments, we present an analytical model to better understand the interplay between discounting and the shape of the damage function, and demonstrate the importance on short term response of three parameters other than the shape of damage curve and discount rate: growth assumptions and emissions scenario, short term climate response and abatement cost functions.

6.6.1 Interplay between the discount rate and the shape of the damage function

Let us start from a simple two-period decision model. At date t_1, a first decision is made to spend c_1 in abatement expenditures; in the same way, we spend c_2 in abatement expenditures at date t_2. Resulting damages, $D(c_1,c_2)$, befall at a posterior point in time. We define φ (the discount factor) as $\varphi = 1/(1+\rho)$, ρ being the social discount rate.

The planner's optimal abatement schedule is solution of the following cost minimization problem, with $n$the number of years between t_1 and t_2 and m the distance between t_1 and the time at which damages occur ($m > n$):

$$\underset{c_1,c_2}{Min}\ c_1 + \varphi^n c_2 + \varphi^m D(c_1,c_2).$$

The relationship between the slope of the damage function and the impact of the discount rate can be illustrated analytically by decomposing the damage function in two terms: an indicator of impacts $\theta(c_1,c_2)$ and a damage function per se $\Psi(\theta)$: $D(c_1,c_2) = \Psi[\theta(c_1,c_2)]$.

For illustrative purposes, let us assume that $\theta(\cdot,\cdot)$ has the following form, where α is strictly lower than unity to capture decreasing environmental return of abatement: $\theta(c_1,c_2) = ac_1^\alpha + bc_2^\alpha$.

The optimal abatement policy, (c_1^*,c_2^*) satisfies the following conditions:

$$\begin{cases} \alpha a (c_1^*)^{\alpha-1}\Psi_\theta(\theta^*) = -\varphi^{-m} \\ \alpha b (c_2^*)^{\alpha-1}\Psi_\theta(\theta^*) = -\varphi^{n-m} \end{cases}$$

where

$$\begin{cases} \Psi_\theta(\cdot) \equiv \frac{\partial \Psi}{\partial \theta} \\ \theta^* \quad = \theta\left(c_1^*, c_2^*\right) \end{cases}.$$

Under the above assumptions, a rapid calculus shows that the variation of optimal first-period abatement when the discount rate varies from ρ to ρ' is as follows (where "'" denotes the values of the variables for the alternative optimum):

$$\frac{c_1^{*'}}{c_1^*} = \left(\frac{1+\rho}{1+\rho'}\right)^{\frac{m}{1-\alpha}} \left(\frac{\psi_\theta\left(\theta^{*'}\right)}{\psi_\theta\left(\theta^*\right)}\right)^{\frac{1}{1-\alpha}}.$$

If damages are linear in the environmental indicator (i.e., if $\psi_\theta(\theta)$ is constant) then the variation of first-period abatement becomes:

$$\frac{c_1^{*'}}{c_1^*} = \left(\frac{1+\rho}{1+\rho'}\right)^{\frac{m}{1-\alpha}}.$$

If $\alpha = 1/3$ (which corresponds to quadratic marginal abatement costs) and $m =$ 100 years, a 1% increase of the discount rate implies a 76% decrease in first period marginal abatement costs.

But if marginal damages vary with the environmental indicator ($\Psi(\theta) = \theta^{-k}$), the variation of first period abatement becomes

$$\frac{c_1^{*'}}{c_1^*} = \left(\frac{1+\rho}{1+\rho'}\right)^{\frac{m}{1-\alpha}} \left(\frac{1+\xi\theta'^{\frac{-n\alpha}{1-\alpha}}}{1+\xi\theta^{\frac{-n\alpha}{1-\alpha}}}\right)^{\frac{k+1}{\alpha k+1}}$$

with $\xi = \left(\frac{b}{a}\right)^{\frac{1}{1-\alpha}}$. If k= 5 and b/a= 2 (technical change makes abatement twice less costly in the second period), the optimal first period abatement diminishes only by 17% when the discount rate rises by 1%. Even if k= 1 (quadratic damage function), first period abatement diminishes only by 45%. The impact of the discount rate on early decades abatement is thus strongly dependent on the interplay between the indicator of climate change and the damage function.

6.6.2 Importance of parameters other than the shape of damage curve

This critical role comes back again when comparing results from Dixit and Pyndick (25) on the one hand and Narain and Fisher (26) or (27) on the other hand. Comparing the environmental irreversibility effect and the investment irreversibility effect, the former, using a real option model, conclude to the dominance of the investment irreversibility effect in the case of a linear damage function, whereas the latter, including an avoidable climatic catastrophe in the analysis, find an opposite result. However, given the likely controversies about the shape of the damage function it would be misleading to focus on this sole parameter despite its critical significance. Three other key determinants of the timing of abatements are indeed of significance: (a) the underlying growth scenario which dictates the level of the discount rate and the emissions baseline; (b) the short term response of climate system to a given inflow of carbon; and (c) the abatement costs.

To demonstrate the importance of these determinants, we will introduce the following modifications in the DICE model (12), (28) while keeping its quadratic damages function of temperature rise[14]:

- We use the A1B SRES scenario as the baseline emissions (10.88 GtC and 12.64 GtC emissions in 2010 and 2020) instead of the DICE baseline which is very close to the B2 SRES scenario (8.78 GtC and 9.05 GtC emissions in 2010 and 2020, respectively).
- We modify the short term climate response ($\theta(c_1, c_2)$). DICE two-box climate model provides a fair description of long-term climate change but underestimates short term atmospheric temperature rise because of the specification of upper and lower compartments (atmosphere and superficial ocean, deep ocean). This is not the case with the climate model presented in appendix: though similar to the one in DICE, it has been calibrated so as to describe more precisely short-term climate change.
- We retain a marginal abatement cost curve as exposed in appendix. The specification is quadratic and accounts for socio-economic inertia. It leads to an equivalent burden for 2010 (0.35% of GWP as compared to 0.36% of GWP following DICE specifications) but with a moderately lower price of carbon: 60 \$/tC instead of 75 \$/tC.

Figure 6.5 demonstrates that changing the specification $\theta(c_1, c_2)$ or choosing an alternative emissions baseline raises abatement rates in 2015 from 5.6 to 7.2% (resp. 5.6 to 8.6%). When both effects are combined, the abatement rate variation is increased by 50% (from 5.6 to 8.6%). It is more than doubled (from 5.6 to 12.5%) if abatement costs are 20% lower.

These results do not pretend to be conclusive about the validity of the Kyoto Protocol. They simply underline that, even without singularity in damage functions, the optimal level of departure from current trends is sensitive to the description of short term climate response and emissions trends, in addition to the value of the discount rate.

---

[14] We used DICE99 version available at `http://www.econ.yale.edu/{\%}7Enordhaus/homepage/dice{\_}section{\_}IV.html`

Damage function is polynomial of degree 2. Both coefficients are positive so they do not allow for global benefits of climate change for low temperature change. Benchmark corresponds to a 1.5% GWP loss for a 2.5 °C global mean temperature rise. Furthermore, the argument of DICE-99 damages function is global mean temperature rise since 1900. To keep results comparable, we reformulated DICE including the following modification: the argument of the damages function becomes global mean temperature rise relative to the first period of the model (1995).

6.6.3 Interplay between the shape of damage curve and climate sensitivity

Let us now turn to the numerical analysis of the linkages between the timing of GHGs abatement, the value of the discount rate and that of the time derivative of damages which are critical for early decades actions (see section 6.1). This raises the question of singularities triggered by the interplay between climate change $\theta(.)$ and the responses of environmental and socio-economic systems $\psi(.)$.

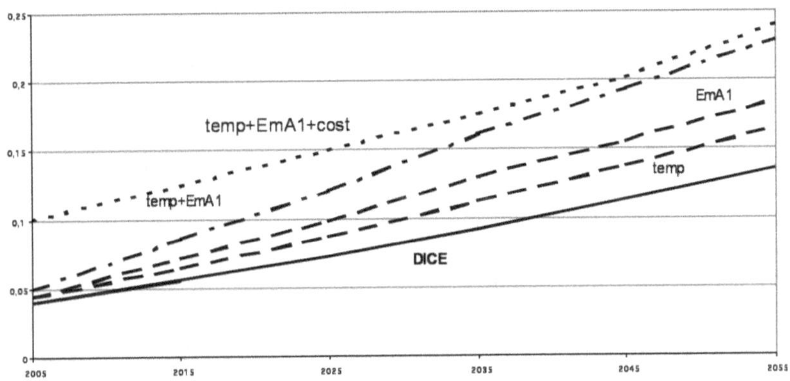

Fig. 6.5 Abatement rate for DICE and for DICE including the modified temperature model (temp), a different baseline (EmA1), both former modifications (temp+EmA1) and finally same than before with the new cost curve (temp+EmA1+cost).

Recently, concerns about such singularities have been evoked beyond environmentalist quarters[15]: "[My] biggest fear is that international policy is being made based on smooth climate change" (G. Yohe). It is hardly disputable that potential sources of abrupt impacts exist along the chain from global warming to changes in local ecosystems. Large scale catastrophic events are the most obvious examples: slow-down of the thermohaline circulation in the North Atlantic, transformation of monsoons patterns or of El Niño cycles. Local climate surprises may also be triggered by smooth evolutions as soon as a threshold is exceeded: for example coral reefs could experience severe bleaching episodes due to a warming of sea surface temperature.

But one major layer of uncertainty lies in the very translation from impacts to losses in social welfare. On the one hand, archaeologists [30] have established coincidences between sudden climate shifts and deep societal mutations; on the other hand, it can be argued that technologically advanced societies are far more resilient.

[15] The Boston Globe (12 December 2002) on the occasion of the publication announcement of the National Academies report "Abrupt climate change: inevitable surprises" (29).

But this response in turn shows that damages depend strongly on the mobilization of adaptation capacities, among which compensations between 'winners' and 'losers'. For example, variation of crops productivity, triggered by changes in temperature, CO_2 concentration, rainfall regime or soil degradation, will also depend upon the capacity to invest in water management systems in affected regions and/or to cover the basic needs of populations through an accessible world market. In the same way, higher frequency of extreme events may aggravate the vulnerability of countries with fragile socioeconomic systems; for example, the political disorganization in Guatemala cannot be fully isolated from the catastrophes that have been affecting this country in the past several years.

6.6.3.1 Smooth vs. threshold function: levels and rates of climate damages

A review of the main shortcomings of widely-used impact functions can be found in (2) (chap. XIX). In the perspective of analyzing short-term response to climate change, the main drawbacks of these functions can be exemplified with the DICE-94 damage function (28). This function is close to $a_1\theta^{a_2}$, where θ stands for global mean temperature rise since 1900 (benchmark estimate is a 1.33% GWP loss for a +3 °C global mean temperature increase; for $a_2 = 2$, we obtain $a_1 = 1.33/9$). Base value of $a_2 = 2$ has greatly influenced previous studies. But such a function has three intrinsic drawbacks:

- Climate surprises leading to high GWP losses can only be represented by adopting unrealistically high global mean temperature rise values. As an example, referring to DICE-94 damage function, the global mean temperature rise corresponding to a 10% GWP loss (which is higher than the economic shock of WWI) amounts to more than +8 °C.
- If a higher exponent is selected so as to lower the global mean temperature rise corresponding to this 10% GWP loss (for $a_2 = 4$, this rise is +5 °C), this leads to the paradoxical consequence that the larger the long-term damages, the smaller the short term ones (because of an increased convexity).
- Lastly, multiplying the scale parameter of the damage function (a_1) to get more realistic damages on the short term (without altering the convexity of the function) also quickly leads to unrealistic high damages on the longer term.

One technical option allowing to represent the episodes of very significant damages without assuming unrealistic temperature increases is the use of sigmoid-like functional forms (31). To carry out simulations comparable with our previous cost-effectiveness analysis, we set the middle of the threshold to +2 °C (with a transition range from 1.7 °C to 2.3 °C). The maximum damage plateaus at a 4% GWP loss.

To clarify the interplay of assumptions regarding the shape of damage functions with climate sensitivity, we compare in this section how, given the uncertainty on climate sensitivity, the timing of mitigation policies is affected by the specification of damage functions (threshold vs. quadratic). In the next section, we perform a set of complementary simulations where climate sensitivity is known (set to its central value) whereas damages are subject to beliefs on the occurrence of singular events.

6.6.3.2 Threshold vs. quadratic function under climate dynamics uncertainty: RESPONSE_D

Let us first note that, under assumption of singularities in the damage curve, climate sensitivity determines the period at which the time derivative of damages becomes higher than the discount rate. This is demonstrated in figures 6.6 and 6.7: the 4% GWP loss is reached in 2050 or 2100, depending upon assumptions on climate sensitivity.

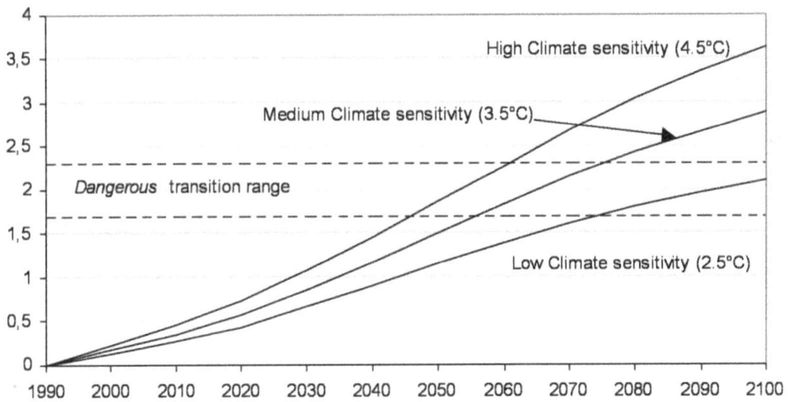

Fig. 6.6 Global mean temperature rise with respect to 1990, with A1B emission baseline and three values for climate sensitivity.

Let us now consider three possible states of the world (s) in which climate sensitivity may be $\{2.5\,^\circ\mathrm{C};\, 3.5\,^\circ\mathrm{C};\, 4.5\,^\circ\mathrm{C}\}$ with the corresponding ex ante subjective probabilities $(p_s) = \{1/6; 2/3; 1/6\}$. The resolution of the uncertainty may occur at different points in time during the 21st century (t_{info}). Damage functional forms are assumed to be known in each simulation and are either quadratic or sigmoid. They have been calibrated so that their total expected damages follow comparable trajectories in the reference case up to 2100. However, beyond 2100, quadratic damages are far higher than threshold ones; this has significant consequences on abatement pathways.

To analyze the impacts of these mechanisms on the optimal pathway, we performed numerical experiments based on the *RESPONSE_D/s* variant of our generic model, in which the objective function (10.12) is the maximization of the expected intertemporal (logarithmic) utility of income, Y, minus abatement costs, $f(.)$ and

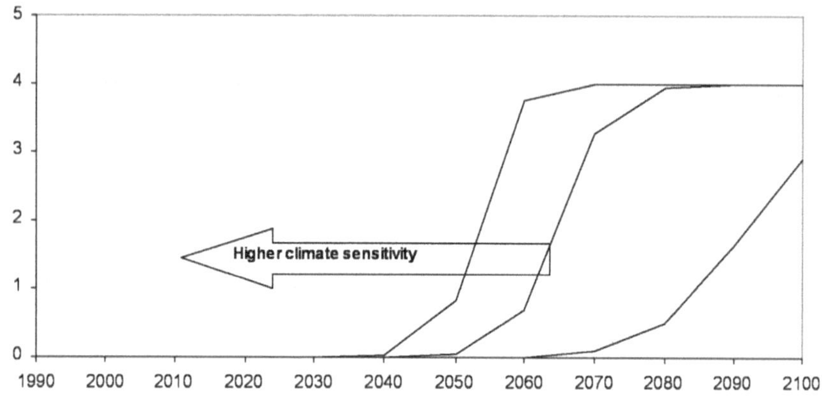

Fig. 6.7 Depending on the value of climate sensitivity, in the case of singularities around a +2 °C warming threshold, abrupt shifts (in the baseline case) occur sooner as climate sensitivity is higher.

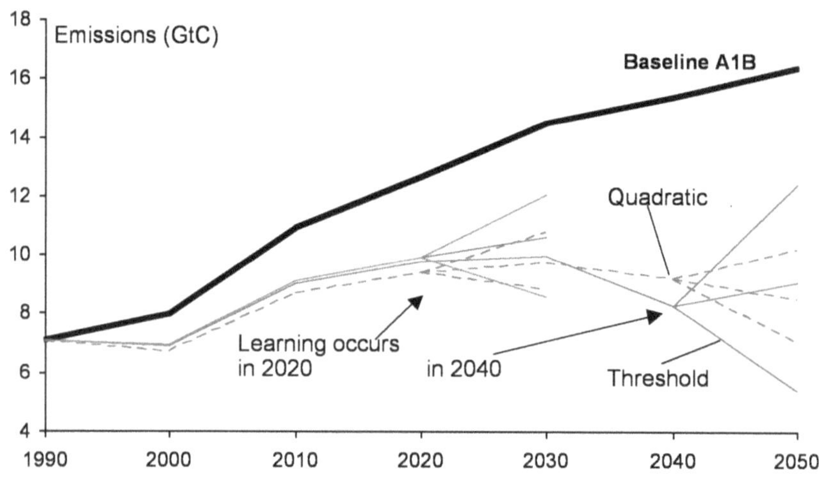

Fig. 6.8 Emissions path for quadratic (dashed grey curves) and threshold (full grey curves) damage functions and learning on climate sensitivity in 2020 or 2040.

resulting damages, $\Psi(\cdot)$:

$$\underset{Ab_t^s}{Max} \sum_s^{p_s} \sum_{t=1990}^{2300} N_t \ln\left(\frac{(Y_t - f(Ab_t^s, Ab_{t-1}^s, t) - \Psi(\theta_t^s, t))}{N_t}\right) e^{-\eta(t-1990)} \quad (6.11)$$

Damages functions are the following: quadratic damages function

$$\Psi(\theta_t^s, t) = a(\theta_t^s)^2 Y_t$$

threshold damages function

$$\Psi(\theta_t^s, t) = \left(\frac{d}{1 + \left(\frac{2-e}{e}\right)^{\left(\frac{K+Z-2\theta_t^s}{K-Z}\right)}} + b\theta_t^s\right) Y_t$$

with a =0.6%GWP, d=4% of GWP, e=0.01, Z=1.7 °C, K=2.3 °C, b=0.5 %GWP.

As in earlier versions, learning process is represented by the following constraint (10.13):

$$\forall t \leq t_{\text{info}}, \ \forall (s, s') \in S, \ Ab_t^s = Ab_t^{s'}. \quad (6.12)$$

Figure 6.8 shows the optimal emission paths for quadratic (dashed grey curves) and threshold damage functions (grey curves) when learning on climate sensitivity occurs in 2020 or 2040. No direct policy conclusion can be derived from the comparison between both emissions paths[16], since the ultimate damages levels for each shape are not equal. This explains why the abatement pathways are comparable in the early periods. The main information from this experiment is that, if information on climate sensitivity arrives later than 2030, threshold functions lead to higher abatement rates and that the value of information on climate sensitivity (figure 6.9) increases very significantly. This is due to the fact that one gets close to the threshold, a mechanism which does not appear with quadratic functions. This confirms Peck and Teisberg findings (32) that the value of information gets higher the more non-linear damages are. In policy terms, this confirms the existence of a window of opportunity, already found in the cost-efficiency analysis with temperature ceiling and in the PPCR approach.

6.6.3.3 Uncertainty regarding damage function when climate dynamics is known

To analyze the specific importance of uncertainty on the shape of damage function, we calibrated both specifications on the same arbitrary benchmark value: 1% GWP

[16] In particular the fact that optimal emissions paths are similar until 2030 should be considered as a calibration artefact.

loss for a 2 °C temperature increase. Climate sensitivity is assumed to be known and set to its central value (3.5 °C). Moreover, expected damages exhibit similar temporal trends at least during the first half of the current century.

This comes to respecifying *RESPONSE_D/s* objective function in the following manner. Ex ante subjective probabilities (p_s) are assigned to two states of the world (s): either damages functions are quadratic (Q) or they exhibit threshold (T). To reflect the diversity of beliefs, we have tested four sets of values for p_s: $\{p_Q = 1, p_T = 0\}$; $\{p_Q = 0.95, p_T = 0.05\}$; $\{p_Q = 0.5, p_T = 0.5\}$; $\{p_Q = 1, p_T = 0\}$. Objective function (10.14) is the maximization, for each set of subjective probabilities, of expected intertemporal (logarithmic) utility of income, Y, minus abatement costs, $f(.)$ and resulting damages, $\Psi(\cdot)$:

$$\underset{Ab_t^s}{Max}\sum_s^{p_s}\sum_{t=1990}^{2300} N_t \ln\left(\frac{(Y_t - f(Ab_t^s, Ab_{t-1}^s, t) - \Psi_s(\theta_t^s, t))}{N_t}\right) e^{-\eta(t-1990)} \tag{6.13}$$

with

$$\Psi_s(\theta, t) = \begin{cases} a\,(\theta_t^s)^2\, Y & s = Quadratic \\ \left(\dfrac{d}{1+\left(\frac{2-e}{e}\right)^{\left(\frac{K+Z-2\theta}{K-Z}\right)}}\right) Y_t & s = Threshold \end{cases}$$

with a = 0.25% of GWP, d = 3% of GWP, e = 0.01, Z = 1.7 °C, K = 2.3 °C, b = 0.5 %GWP.

As earlier, learning process is represented by the following constraint (6.14):

$$\forall t \le t_{\text{info}},\ \forall\left(s, s'\right) \in S,\ Ab_t^s = Ab_t^{s'}. \tag{6.14}$$

Because quadratic functions refer to ultimate damages far higher than threshold functions, abatement rates are similar in the early decades. This is why results in figure 6.10 show the same limitations as in figure 6.8. However despite this artifact, abatement pathways diverge significantly if information is disclosed after 2030. After this date, the optimal pathways are critically dependent on the subjective probability sets: it is remarkable however that a 5% subjective probability only for the threshold function (upper dotted line) leads to a significant departure from the quadratic case while a 50/50 distribution of probabilities leads to emissions pathway very close to the optimal pathway in case of early certainty about the existence of the threshold. Here again, the worst case hypothesis dominates the result.

6.7 Conclusions

In this paper, we have compared optimal climate policy in the short run under three different decision-making frameworks: cost-effectiveness with temperature ceiling objectives, cost-benefit analysis with pure preference for current climate regime and

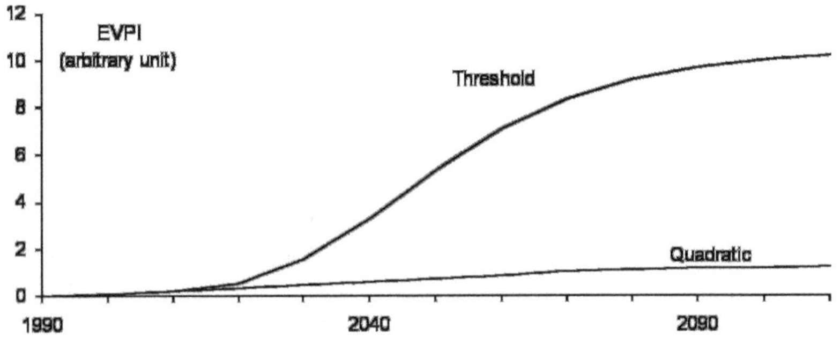

Fig. 6.9 Expected Value of Perfect Information on climate sensitivity for quadratic and threshold damage functions.

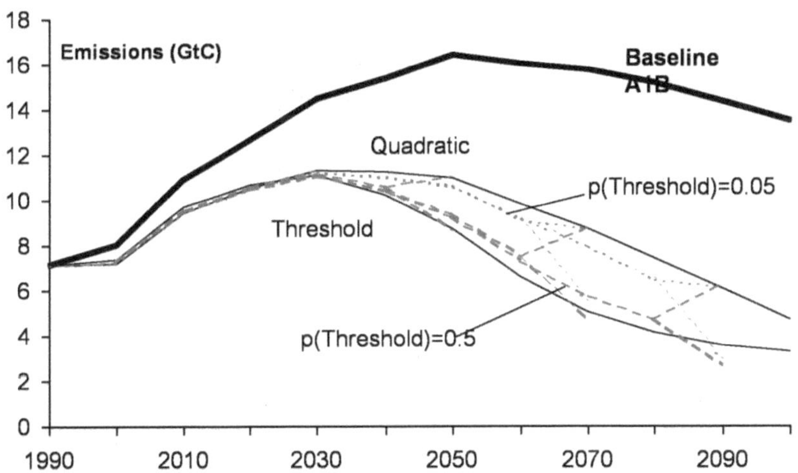

Fig. 6.10 Emissions paths for different subjective probabilities of the threshold function with learning. Threshold function probability: 0% for the upper thin black line; 5% for the dotted grey line, 50% for the dashed grey line; 100% for the lower thin black line. Learning occurs in 2040, 2060 and 2080.

full cost-benefit approach with monetized evaluations of impacts. Five key lessons on short term decision emerge from this analysis.

1. Given the cascade of uncertainty from emissions to damages, the difference between various decision-making frameworks appears to matter less than the difference between stochastic and non stochastic approach.
2. In a stochastic approach, it is not necessary to assume ultimate catastrophic impacts to conclude to the optimality of early GHGs abatements. Singularities in the damage curves are sufficient to draw such a conclusion mainly because they increase the role of the uncertainty on climate sensitivity. In a stochastic framework, with uncertainty about the shape of the damage curve, the choice of the optimal strategy is dominated by the likelihood of occurrence of function with singularities.
3. In addition to the shape of damage curves, the optimal timing of emissions abatement is very sensitive to the way the carbon cycle, the climate sensitivity and baseline emissions over the first decades are calibrated, and to the extent to which their intrinsic uncertainty is considered.
4. A window of opportunity exists in all decision-making frameworks, cost-benefit analysis with smooth damage curves excepted. The value of information is low in the first periods but increases drastically after 2020 to 2040. This time-horizon has to be compared with the fifty years necessary to change energy systems, and to the fact that, according to the climate models, clear signals may not emerge from the noise of climate variability before 2050.
5. The introduction of a pure preference for current climate allows for an overshoot of desired temperature (or concentration) targets without lowering the first period effort because it counterbalances the influence of discounting, all the more so as the environment is treated as a superior good.

The core difficulties remain: (a) the revelation of the pure preference for current climate regime (including its volatility due to the media life cycles), (b) the evaluation of the interplay between the various influences of climate change on the economy. Among these interplays we will insist, as an invitation to further thoughts, on the role of the inertia and of the geographical distribution of damages. One major source of singularity in damage curves comes indeed from the joint effect of uncertainty and the inertia of human systems: a two percent of GDP loss may either represent a benign shock when spread over a century or havoc when concentrated on five years (this is the cost of WW1 for France). Another related source of singularity is the propagation effect (climate refugees, for example), in case of uncompensated shocks at a local level.

Coping with these difficulties will confront the methodological difficulties of incorporating intrinsically controversial information at various spatial scales, including information derived from 'grass-root' case studies, into an integrated modeling framework. The increase of the size of the models to be mobilized will make all the more necessary the development of compact models of the sort used in this paper. Both mathematically controllable and flexible enough, they are an appropriate communication tool between scientific disciplines and between science and stakeholders in a process of public decision-making under scientific controversy.

Acknowledgements The authors wish to acknowledge financial support from the French Ministry of Ecology and Sustainable Development and French Interministerial Mission on Climate Change through the GICC Research Program (No.10/99) and from the European Union Fifth Framework Programme for Energy, Environment and Sustainable Development through the contract EVK2-2001-00156 (PRUDENCE). We would also like to thank Vincent Gitz, Frédéric Ghersi and an anonymous reviewer for helpful remarks and comments. The remaining errors are entirely the authors'.

Disclaimer.

The findings, interpretations and conclusions expressed in this paper are entirely those of the authors. They do not necessarily represent the view of the World Bank, its Executive Directors, or the countries they represent.

Appendix

A.1. Baseline growth scenario and exogenous related data (income and population)

All experiments are based on the SRES A1B scenario which has been computed by NIES (National Institute for Environmental Studies, Japan) with the AIM model (Asian Pacific Integrated Model) (33). We choose the A1B scenario because it corresponds to rather optimistic beliefs about the future. A1B is indeed the picture of a prosperous and generous world where economic growth is high with a considerable catch-up of developing countries while continuous structural change and rapid diffusion of more efficient technologies yield to decreasing GHGs emissions as soon as 2050. A1B is thus consistent with beliefs such as "*it is better to invest in R&D in the energy sector and/or research in climate change-related fields than to deep-cut fossil fuel emissions at once while alternative technologies are expensive and climate change consequences might prove ultimately benign*" or "*abatement opportunity cost is lower than that of fostering development in potential vulnerable regions*". It is therefore relevant to examine how statements like "*one should delay GHGs emissions reduction efforts*" are to be revised when using a proper precautionary approach.

A.2. Specification of abatement cost function

We use the following abatement cost function:

$$f(Ab_t, Ab_{t-1}, t) = \frac{1}{3} BK\, PT_t\, \gamma(Ab_t, Ab_{t-1})\; em_t\; (Ab_t)^3$$

with:

$f(Ab_t, Ab_{t-1}, t)$: total cost of mitigation measures at time t, (trillion US$),
BK : initial marginal cost of backstop technology (thousand US$/tC),
PT_t : technical change factor,
$\gamma(Ab_t, Ab_{t-1})$: socio-economic inertia factor,
em_t : baseline CO_2 emissions at time t(GtC),
Ab_t : abatement rate at time t(% of baseline emissions).

Under these specifications, marginal costs of abatement are convex (quadratic). This is consistent with assumptions by experts and the results of technico-economic models. Note that $f(\cdot)$does not allow for so-called no-regret potential.

BK stands for the initial marginal cost of backstop technology, i.e., the carbon free-technology which would enable to completely reduce GHGs emissions were it to be substituted to current existing energy systems. Its value depends on a set of assumptions regarding its nature (windpower, nuclear, ...), its development date, its penetration rate and technical change. Given our own assumptions on technical change, we retain an initial 1,100 US$/tC cost.

PTt captures the influence of autonomous technical change on abatement costs. It translates the decrease of the costs of carbon-free technology over time, but the improvement of energy intensity which is already taken into account in the baseline. We assume that the costs of abatement technologies decrease at a constant 1% per year rate but we assume costs cannot decrease beyond 25% of their initial values. PT_t thus take the form below (which leads to an ultimate cost of 275 US$.tC^{-1})

$$PT_t = 0.25 + 0.75e^{-0.01\delta t}$$

Where δ is the time step of the model (10 years).

$\gamma(Ab_t, Ab_{t-1})$captures the influence of socio-economic inertia as a cost-multiplier (transition costs between a more and a less carbon-intensive economic structure). $\gamma(\cdot)$ is a multiplicative index. It is equal to 1 (no additional costs) if abatement increases at a rate lower than a given threshold τ between two consecutive periods. But it increases linearly with the speed of variation of abatement rate when this rate is higher than τ, i.e., the annual turnover of productive capital below which mitigation policies do not lead to premature retirement of productive units. Here τ is set to 5% per year (average capital stocks turnover of 20 years).

$$\gamma(Ab_t, Ab_{t-1}) = \begin{cases} 1 & if\ \frac{Ab_t - Ab_{t-1}}{\delta\tau} \leq 1 \\ \frac{Ab_t - Ab_{t-1}}{\delta\tau} & otherwise \end{cases} .$$

A.3. Three-reservoir linear carbon-cycle model

We use the C-Cycle of Nordhaus (12), a linear three-reservoir model (atmosphere, biosphere+ surface ocean and deep ocean). Each reservoir is assumed to be homogenous (well-mixed in the short run) and is characterized by a residence time inside the box and corresponding mixing rates with the two other reservoirs (longer timescales). Carbon flows between reservoirs depend on constant transfer coefficients. GHGs emissions (CO_2 solely) accumulate in the atmosphere and they are slowly removed by biospheric and oceanic sinks.

The dynamics of carbon flows is given by is given by:

$$\begin{pmatrix} A_{t+1} \\ B_{t+1} \\ O_{t+1} \end{pmatrix} = C_{trans} \cdot \begin{pmatrix} A_t \\ B_t \\ O_t \end{pmatrix} + \delta(1 - Ab_t)em_t.u$$

with:

A_t: carbon contents of atmosphere at time t(GtC),
B_t: carbon contents of upper ocean and biosphere at time t(GtC),
O_t : carbon contents of deep ocean at time t(GtC),
C_{trans} : net transfert coefficients matrix,
u : column vector (1,0,0).

As such, the model has a built-in ten-year lag between CO_2 emissions and CO_2 accumulation in the atmosphere, which reflects the inertia in C-cycle dynamics. Nordhaus calibration on existing carbon-cycle models gives the following results (for a decadal time step):

$$C_{trans} = \begin{pmatrix} 0.66616 & 0.27607 & 0 \\ 0.33384 & 0.60897 & 0.00422 \\ 0 & 0.11496 & 0.99578 \end{pmatrix}$$

initial conditions (GtC):

$$C_{1990} = \begin{pmatrix} 758 \\ 793 \\ 19230 \end{pmatrix}.$$

The main criticism which may be addressed to this C-cycle model is that the transfer coefficients are constant. In particular, they do not depend on the carbon content of the reservoir (e.g., deforestation hindering biospheric sinks) nor are they influenced by ongoing climatic change (e.g., positive feedbacks between climate change and carbon cycle).

A.4. The reduced-form climate model

This model[17] is very close to Schneider and Thompson's two-box model (34). A set of two equations is used to describe global mean temperature variation (equation (A.2)) since pre-industrial times in response to additional human-induced forcing (equation (A.1)). More precisely, the model describes the modification of the thermal equilibrium between atmosphere and surface ocean in response to anthropogenic greenhouse effect. Calibration was carried out with H. Le Treut (IPSL) from data kindly provided by P. Friedlingstein (IPSL). All specifications correspond to decadal values, which is the time step of the model.

Radiative forcing equation:

$$F(t) = F_{2X} \frac{\log\left(\frac{M_t}{M_{PI}}\right)}{\log 2} \quad \text{(A.1)}$$

with

M_t : atmospheric CO_2 concentration at time t (ppm),
$F(t)$: radiative forcing at time t (W/m^2),
M_{PI} : atmospheric CO_2 concentration at pre-industrial times, set at 280 ppm,
F_{2x} : instantaneous radiative forcing for $2 \times M_{PI}$, set at $3.71 W/m^2$).

Temperature increase equation:

$$\left\{ \begin{bmatrix} \theta_{At}(t+1) \\ \theta_{Oc}(t+1) \end{bmatrix} = \begin{bmatrix} 1-\sigma_1(\lambda+\sigma_2) & \sigma_1\sigma_2 \\ \sigma_3 & 1-\sigma_3 \end{bmatrix} \begin{bmatrix} \theta_{At}(t+1) \\ \theta_{Oc}(t+1) \end{bmatrix} + \sigma_1 \begin{bmatrix} F(t) \\ 0 \end{bmatrix} \right. \quad \text{(A2)}$$

with

$\theta_{At}(t)$: global mean atmospheric temperature rise wrt preindustrial times ($^\circ$C),
$\theta_{Oc}(t)$: global mean oceanic temperature rise wrt preindustrial times ($^\circ$C),
λ : climate response parameter ($\mathrm{C^{-1}\,W\,m^{-2}}$),
σ_1 : transfert coefficient (set at $0.479\,\mathrm{C\,W^{-1}\,m^2}$),
σ_2 : transfert coefficient (set at $0.109\,\mathrm{C^{-1}\,W\,m^{-2}}$),
σ_3 : transfert coefficient (set at 0.131).

Climate sensitivity (T_{2x}) is given by $T_{2x} = F_{2x}/\lambda$. We assume that uncertainty is mainly due to uncertainty on (atmospheric) climate feedbacks process (represented by λ) rather than uncertainty on F_{2x}. A higher climate response parameter will lead to a low climate sensitivity. We explore three values for climate sensitivity and λ is set accordingly to F_{2x}/T_{2x} (see table 6.2).

---

[17] A more detailed description of the model and calibration process may be found in (17).

Table 6.2 Three values for climate sensitivity and λ.

State of the World	Low	Central	High
Climate sensitivity (T_{2x})	2.5 °C	3.5 °C	4.5 °C
Ex ante subjective probability (p_s)	1/6	2/3	1/6
λ	1.48	1.06	0.82

A.5. Numerical resolution

To avoid boundary effects, we did not specify terminal conditions in 2100 but set the time horizon of the model at 2300. All the models have been run under the GAMSMINOS non-linear solver. The model codes are available from the authors on request[18].

References

[1] United Nations Framework Convention on Climate Change (1992).

[2] J.J.McCarthy et al., eds., *Climate Change 2001: Impacts, Adaptation and Vulnerability. Contribution of Working Group II to the Third Assessment Report of the Intergovernmental Panel on Climate Change* (Cambridge University Press, Cambridge, UK & US, 2001).

[3] J.P. Bruce, H. Lee and E.F. Haites, eds., *Climate Change 1995: Economic and Social Dimensions of Climate Change. Contribution of Working Group III to the Second Assessment Report of the Intergovernmental Panel on Climate Change* (Cambridge University Press, Cambridge, UK & US, 1996).

[4] B. Metz et al., eds., *Climate Change 2001: Mitigation. Contribution of Working Group III to the Third Assessment Report of the Intergovernmental Panel on Climate Change* (Cambridge University Press, Cambridge, UK & US, 2001).

[5] L.B. Lave, Formulating greenhouse policies in a sea of uncertainties, *Energy Journal* 12(10.1) (1992) 9–21.

[6] R. Costanza et al., The value of the world's ecosystem services and natural capital, *Nature* 387 (1997) 253–260.

[7] M.A. Toman,Why not to calculate the value of the world's ecosystem services and natural capital, *Ecological Economics* 25(10.1) (1998) 57–60.

[8] J.K. Hammitt, Evaluation endpoints and climate policy: atmospheric stabilisation, benefit-cost analysis and near-term green-house gas emissions, *Climatic Change* 41 (1999) 447–468.

[18] Also available on the url `http://www.centre-cired.fr/IMG/pdf/These_PhilippeAmbrosi.pdf`

[9] M. Ha Duong, M. Grubb and J.C. Hourcade, Influence of socioeconomic inertia and uncertainty on optimal CO2-emission abatement, *Nature* 390 (1997) 270–274.
[10] T.M.L. Wigley, R. Richels and J.A. Edmonds, Economic and environmental choices in the stabilization of atmospheric CO2 concentrations, *Nature* 379 (1996) 240–243.
[11] J.T. Houghton et al., eds., *Climate Change 2001: the Scientific Basis. Contribution of Working Group I to the Third Assessment Report of the Intergovernmental Panel on Climate Change* (Cambridge University Press, Cambridge, UK & US, 2001).
[12] W. Nordhaus and R. Boyer, *Warming the World: Economics Models of Climate Change* (MIT Press, Cambridge, MA, USA, 1999).
[13] M. Parry et al., Millions at risk: defining critical climate change threats and targets, *Global Environmental Change* 11 (2001) 181–183.
[14] O. Simonett, Potential impacts of global warming: cases studies on climatic change, GRID-Geneva, Geneva, Switzerland (1989).
[15] EU, Communication on Community strategy on climate change: Council conclusions, Brussels (1996).
[16] A.S. Manne and R. Richels, *Buying Greenhouse Insurance: The Economic Cost of CO2 Emissions Limits* (MIT Press, Cambridge, MA, USA, 1992).
[17] Ph. Ambrosi, J.C. Hourcade and H. Le Treut, Incertitudes sur la sensibilité du climat: valeur de l'information et politique climatique optimale, in: *Modélisation intégrée, évaluation des risques climatiques et des politiques de précaution*, eds. J.C. Hourcade and H. Le Treut, Progamme GICC (Mate/MIES/Ademe), Rapport final (2002).
[18] J.C. Hourcade and T. Chapuis, No-regret potentials and technical innovation, *Energy Policy* 23(4/5) (1995) 433–445.
[19] G. Scimeni, Environmental strategies and anticipatory strategies, in: *Preventive Environmental Policy*, ed. C.E. Simonis (Campus, NY, USA, 1987).
[20] A.S. Manne, Hedging strategies for global carbon dioxide abatement: a summary of poll results, EMF-14 Sub group – Analysis for decision under uncertainty, Stanford University, Stanford, CA (1995).
[21] W.R. Cline, La lutte contre l'effet de serre, *Finance et développement* 30(10.1) (1996) 4–6.
[22] T.C. Koopmans, Stationary ordinal utility and impatience, *Econometrica* 28(10.2) (1960) 287–309.
[23] G. Chichilnisky, An axiomatic approach to sustainable development, *Social Choice and Welfare* 13(10.2) (1996) 231–257.
[24] F. Lecocq and J.C. Hourcade, Incertitude, irréversibilités et actualization dans les calculs économiques sur l'effet de serre, in: *Les enjeux économiques de l'effet de serre*, ed. R. Guesnerie (La Documentation Française, Paris, 2003) pp. 177–199.
[25] A.K. Dixit and R.S. Pyndick, *The Timing of Environmental Policy* (Princeton University Press, Princeton, NJ, USA, 1994).

[26] U. Narain and A.C. Fisher, *Global Warming, Endogenous Risk and Irreversibility* (Department of Agricultural and Resource Economics-UC Berkeley, CA, USA, 2002).
[27] Gjerde, J., S. Grepperud and S. Kverndokk, Optimal climate policy under the possibility of a catastrophe, *Resources and Energy Economics* 21 (1999) 289–317.
[28] W. Nordhaus, *Managing the Global Commons: the Economics of Climate Change* (MIT Press, Cambridge, MA, USA, 1994).
[29] R. Alley, ed., *Abrupt Climate Change: Inevitable Surprises* (National Academy Press, Washington, DC, 2002).
[30] H. Weiss and R.S. Bradley, What drives societal collapse?, *Science* (2001) 609–610.
[31] P. Dumas and M. Ha Duong, Non-linearities and uncertain climate impacts, in: *Energy Modeling Forum (EMF)*, Laxenburg, Austria (2001).
[32] S. Peck and T. Teisberg, The importance of nonlinearities in global warming damage costs, in: *Assessing Surprises and Nonlinearities in Greenhouse Warming* (RFF, Washington, DC, 1993).
[33] N. Nakicenovic, ed., *Special Report on Emissions Scenarios: a Special Report of Working Group III of the Intergovernmental Panel on Climate Change* (Cambridge University Press, UK & US, 2000).
[34] S.H. Schneider and S.L. Thompson, Atmospheric CO2 and climate: importance of the transient response, *Journal of Geophysical Research* 86 (1981) 3135–3147.

Chapter 7
A Stochastic Control/Game Approach to the Optimal Timing of Climate Policies

O. Bahn, A. Haurie and R. Malhamé

Abstract This chapter deals with an application of stochastic control or stochastic game methods to the design of optimal timing of climate policies. In the first part, we propose a stochastic control approach for a *cost-benefit* model that takes into account the uncertainty on the access to a backstop (clean) technology. In a second part, we show how this model can be extended to a game theoretic framework, assuming non-cooperative behavior of two groups of countries that are affected by climate change related damages induced by their joint greenhouse gas emissions. Finally we discuss the possibility of implementing successive control synthesis cycles preceded by learning cycles concerning climate sensitivity statistics.

7.1 Introduction

Our aim, in this chapter is to show how the methods and techniques of stochastic control could be used to address some fundamental issues in the design of efficient climate policies. Climate policies should aim, according to the United Nations Framework Convention on Climate Change (24), at stabilizing atmospheric greenhouse gas (GHG) concentrations "at a level that would prevent dangerous anthropogenic interference with the climate system". In an attempt to define what such a "dangerous" level could be, the European Council[1] has proposed to limit temperature increase (relative to pre-industrial levels) at 2°C and rate of temperature

Olivier Bahn
GERAD and MQG, HEC Montréal, Montréal, Canada; e-mail: olivier.bahn@hec.ca

Alain Haurie
GERAD-HEC, Montréal and ORDECSYS, Geneva, Switzerland; e-mail: alain.haurie@ordecsys.com

Roland Malhamé
GERAD and Ecole Polytechnique, Montréal, Canada; e-mail: roland.malhame@gerad.ca

[1] EU Council, 7631/04.

J.A. Filar and A. Haurie (eds.), *Uncertainty and Environmental Decision Making*, International Series in Operations Research & Management Science 138,
DOI 10.1007/978-1-4419-1129-2_7,

increase at 0.2°C per decade. In order to respect these temperature limits, optimal timing of GHG emissions should basically consider the so-called climate sensitivity, defined by the Intergovernmental Panel on Climate Change (14) as the equilibrium change in average surface warming resulting from a doubling of atmospheric (equivalent) carbon dioxide (CO_2) concentration. Irrespective of how critical climate sensitivity is to adequate climate policy making, its value remains quite uncertain. Another important source of uncertainty concerns the access to a "clean" technology. Indeed, the possibility to harness an energy form without GHG emissions would considerably help in tackling the climate change problem. However, in order to obtain this access to a "clean" production economy, an R&D activity must take place, and the time at which this technological breakthrough will intervene is also highly uncertain.

To address the climate sensitivity uncertainty, Ambrosi et al. (1) propose a stochastic framework[2]. In their model, the policy variables have no influence on the stochastic process which represents the knowledge of the true climate sensitivity. Therefore their approach is more related to the stochastic programming approach[3] proposed by Manne & Richels (17) or more recently by Bosetti et al. (5) than to the dynamic programming approach which is characteristic of the general stochastic control pradigm. The latter have also proposed a stochastic game formalism, with uncertainty not influenced by the decision variables of the players, to study the research and development (R&D) policy required for an economically viable climate change mitigation.

In this chapter we focus on the second source of uncertainty (access to a clean technology) and propose a stochastic control and stochastic game formalism adapted to the study of optimal timing of climate policies. Unlike the preceding stochastic programming based approaches, a true stochastic control framework allows for models where control decisions can affect the nature of uncertainty. As in our model proposed in (3), we consider a situation where policy variables (players decisions) have an influence on the stochastic process representing timing of access to a backstop (clean) technology that will sensibly reduce GHG emissions from economic production activities. More precisely, this stochastic process is represented as a controlled Markov jump process. The backstop technology could be for example *carbon capture and sequestration* (CCS) or even *controlled hydrogen fusion*: These technologies have attracted significant investments over time but they are not as yet commercially available. A transition to such a stage could usher the way into a cleaner economy whereby carbon-free electricity could partly be used to produce hydrogen in turn to be used as the main fuel throughout the economy.

Our model takes its inspiration from the seminal work of (19), (20) and (21), but introduces two important new features: (i) a description of a dual production economy called "dirty" and "clean", respectively emitting a high amount of carbon and using much less fossil fuels and (ii) a description of uncertainty related to access to the clean production technology under a form of controlled jump process. The

[2] This article is reproduced as one chapter of this book.

[3] Note that Chapter 2 by Labriet et al. is an application of stochastic programming to a large scale energy/environment model of the world.

stochastic control problem is formulated as a piecewise deterministic control system and the information structure of the game is the one introduced under the name of "piecewise open-loop" in (11) and exploited in (12) in a stochastic oligopoly model. In this chapter, uncertainty on climate sensitivity is modeled via an a-priori probability distribution and its average effect within a cost-benefit approach. Furthermore, an extension of the modeling framework to the case of two non-cooperative players is considered.

The chapter is organized as follows. In Section 7.2, we revisit the model proposed in (3) and we show how a cost-benefit analysis can be formulated, using a stochastic control paradigm, and what its implications are in terms of timing of the climate policy. Up to this point, a single decision maker would have been considered. In Section 7.3, we extend the previous model to a non-cooperative game setting. As an illustration, we consider a situation where the world is divided into two coalitions, say developing and developed countries, and propose a game model where the two players define non-cooperatively their climate policies all the while being affected by climate change related damages induced by their joint GHG emissions. In Section 7.4, we speculate on the possibility of refining climate sensitivity statistics based on observations of actual GHG emissions and average earth surface temperature rises, under the proposed control policy, with periodic updating of that control policy based on the new information as sufficient data is accumulated. Finally Section 7.5 summarizes our conclusion.

7.2 A stochastic control model for the optimal timing of climate policies

This section revisits the stochastic control model that we already proposed in (3) which will be run now with new parameters and assumptions concerning the economy. The economic model corresponds to an economic growth à la Ramsey (23) with two types of technological organization, a "carbon economy" where a high level of carbon emissions is necessary to obtain output and a more expensive "clean economy" where a much lower level of emissions is necessary to produce the economic good. More precisely, production occurs in these economies using three factors, labor, physical capital and fossil energy which generates carbon emissions. The resulting homogenous good can either be consumed or invested. The clean economy cannot initially contribute to production. For this to happen, a technological "breakthrough" in carbon-free technology is a precondition. However, investment into this technology will increase the probability that such a breakthrough takes place. Until then, the carbon-free related physical capital is treated as cumulative research and development (R&D) investment.

7.2.1 Variables

The following variables enter in the description of the economic model:

$C(t) \geq 0$: total consumption at time t, in trillions (10^{12}) of dollars;
$c(t)$: per capita consumption at time t, $c(t) = \frac{C(t)}{L(t)}$;
$E(t) \geq 0$: global yearly emissions of GHG at time t, in Gt (10^9 tons) carbon equivalent;
$E_1(t) \geq 0$: part of the emissions attributed to the carbon economy;
$E_2(t) \geq 0$: part of the emissions attributed to the carbon-free economy;
$\text{ELF}(t) \geq 0$: economic loss factor at time t due to climate changes, in %;
$I_i(t) \geq 0$: investment at time t in capital $i = 1,2$, in trillions of dollars;
$K_1(t) \geq 0$: physical stock of productive capital in the carbon economy at time t, in trillions of dollars;
$K_2(t) \geq 0$: physical stock of (R&D or productive) capital in the carbon-free economy at time t, in trillions of dollars;
$L(t) \geq 0$: labor (world population), in millions (10^6) persons;
$L_1(t) \geq 0$: part of the labor force allocated to the carbon economy;
$L_2(t) \geq 0$: part of the labor force allocated to the carbon-free economy;
t : running time, (unit = 10 years, in our discrete time formulation);
$\xi(t) \in \{0,1\}$: indicator variable for the carbon-free economy, where $\xi(t) = 0$ indicates that $K_2(t)$ is an R&D capital;
W : total discounted welfare;
$Y(t)$: economic output at time t, in trillions of dollars.

7.2.2 Economic dynamics

The implicit objective of the economy is to maximize a welfare criterion given by the integral over an infinite time horizon of a discounted utility from per capita consumption. The pure time preference discount rate ρ (set typically to 3% per year or 30% per decade) and the welfare criterion is given by:

$$W = \int_0^\infty e^{-\rho t} L(t) \log[c(t)]\, dt. \tag{7.1}$$

Total labor (L) is divided between labor allocated to the carbon economy (L_1) and labor allocated to the carbon-free economy (L_2):

$$L(t) = L_1(t) + L_2(t). \tag{7.2}$$

Total labor evolves according to the following law of motion:

$$\dot{L}(t) = g_L(t) L(t), \tag{7.3}$$

where the growth rate g_L decays exponentially with coefficient δ_{g_L} (set to 0.3):

$$\dot{g}_L(t) = -\delta_{g_L} g_L(t). \tag{7.4}$$

Yearly GHG emissions come from the carbon economy (emission level E_1) and the carbon-free economy (emission level E_2):

$$E(t) = E_1(t) + E_2(t). \tag{7.5}$$

GHG emissions accumulate in the atmosphere, which yields climate changes that trigger damages. Hence economic losses occur, measured through the economic loss factor ELF(t) applied on economic output (Y). The economic output is defined in the two economies by an extended Cobb-Douglas production function in three inputs, capital (K), labor (L) and energy (which use is measured through emission level E):

$$\begin{aligned} Y(t) &= A_1(t)K_1(t)^{\alpha_1}(\phi_1(t)E_1(t))^{\theta_1(t)}L_1(t)^{1-\alpha_1-\theta_1(t)} \\ &\quad + \xi(t)A_2(t)K_2(t)^{\alpha_2}(\phi_2(t)E_2(t))^{\theta_2(t)}L_2(t)^{1-\alpha_2-\theta_2(t)}. \end{aligned} \tag{7.6}$$

Notice that in Eq. (7.6), production from the carbon-free economy occurs only when $\xi(t) = 1$, namely when the technological breakthrough has occurred. The different terms in Eq. (7.6) are detailed below. Total factor productivity (A_i) evolves according to the following law of motion:

$$\dot{A}_i(t) = g_{A_i}(t)A_i(t) \quad i = 1,2, \tag{7.7}$$

where the growth rate g_{A_i} decays exponentially with coefficient $\delta_{g_{A_i}}$ (set to 0.2):

$$\dot{g}_{A_i}(t) = -\delta_{g_{A_i}} g_{A_i}(t) \quad i = 1,2. \tag{7.8}$$

Capital stock (K_i) evolves according to the choice of investment (I_i) and depreciation rate δ_{K_i} (set to 0.1) through the standard relationship given by:

$$\dot{K}_i(t) = I_i(t) - \delta_{K_i} K_i(t) \quad i = 1,2. \tag{7.9}$$

The elasticity of output α_i with respect to capital K_i is set to 0.3. The evolution of energy efficiency (ϕ_i) is given by:

$$\dot{\phi}_i(t) = g_{\phi_i}(t)\phi_i(t) \quad i = 1,2, \tag{7.10}$$

where the growth rate g_{ϕ_i} decays exponentially with coefficient $\delta_{g_{\phi_i}}$ (set to 0.3):

$$\dot{g}_{\phi_i}(t) = -\delta_{g_{\phi_i}} g_{\phi_i}(t) \quad i = 1,2. \tag{7.11}$$

Elasticity of output with respect to emissions (θ_i) evolves according to the following law of motion:

$$\dot{\theta}_i(t) = g_{\theta_i}(t)\theta_i(t) \quad i = 1,2, \tag{7.12}$$

where the growth rate g_{θ_i} decays exponentially with coefficient $\delta_{g_{\theta_i}}$ (set to 0.008):

$$\dot{g}_{\theta_i}(t) = -\delta_{g_{\theta_i}} g_{\theta_i}(t) \quad i = 1,2. \tag{7.13}$$

Remaining economic output (given climate change damages) is used for consumption (C), investment (I) and the payment of energy costs:

$$\mathrm{ELF}(t)Y(t) = C(t) + I_1(t) + I_2(t) + p_{E_1}(t)\phi_1(t)E_1(t) + p_{E_2}(t)\phi_2(t)E_2(t). \tag{7.14}$$

Energy prices (p_{E_i}) are inflated over time using the discount rate (ρ) as follows:

$$\dot{p}_{E_i}(t) = e^{\rho t} p_{E_i}(t) \quad i = 1,2. \tag{7.15}$$

Initial values

We assume that parameters in the above equations are initialized as follows[4]:

$L(0)$: initial value for population level; $L(0) = 6409$;
$g_L(0)$: initial value for growth rate of population; $g_L(0) = 0.08$;
$A_i(0)$: initial value for total factor productivity; $A_1(0) = A_2(0) = 0.0302$;
$g_{A_i}(0)$: initial value for growth rate of total factor productivity; $g_{A_1}(0) = g_{A_2}(0) = 0.08$;
$\phi_i(0)$: initial value for energy efficiency; $\phi_1(0)$=1.0; $\phi_2(0)$=5.0;
$g_{\phi_i}(0)$: initial value for growth rate of energy efficiency; $g_{\phi_1}(0) = g_{\phi_2}(0) = 0.15$ per decade;
$\theta_i(0)$: initial value for elasticity of output with respect to emissions; $\theta_1(0) = \theta_2(0) = 0.05$;
$g_{\theta_i}(0)$: initial value for growth rate of elasticity of output with respect to emissions; $g_{\theta_1}(0) = g_{\theta_2}(0) = 0.0$;
$p_{E_i}(0)$: initial price of energy input; $p_{E_1}(0) = 0.35$; $p_{E_2}(0) = 0.6$;

Asymptotic values

Asymptotic values of the parameters are given below:

$L(\infty)$: asymptotic value for population level; $L(\infty) = 8368$;
$A_i(\infty)$: asymptotic value for total factor productivity; $A_1(\infty) = A_2(\infty) = 0.045$;
$\phi_i(\infty)$: asymptotic value for energy efficiency; $\phi_1(\infty) = 1.65$; $\phi_2(\infty) = 8.24$;

[4] Mostly following the DICE model (20) and more precisely the DICE-2007 version; see: `http://nordhaus.econ.yale.edu/`.

7.2.3 Technological breakthrough dynamics

The possibility of access to a cleaner economy at time t is represented by a binary variable $\xi(t) \in \{0,1\}$ whereby the clean technology starts being available only when ξ switches from an initial value of 0 to 1. The initial value $\xi(0) = 0$ indicates that there is no access to the clean technology at the initial time. The switch to the value 1 occurs at a random time which is controlled through the global accumulation of R&D capital K_2. More precisely one introduces a jump rate function[5] $q_b(t, K_2(t))$ which is instrumental in determining the elementary probability of a switch around time t, given that no switch has occurred up to that time:

$$\mathrm{P}[\xi(t+dt) = 1 | \xi(t) = 0, K_2(t)] = q_b(t, K_2(t))\,dt + o(dt), \tag{7.16}$$

where

$$\lim_{dt \to 0} \frac{o(dt)}{dt} = 0.$$

We choose the following form for $q_b(t, K_2(t))$ which thanks to a square root function displays diminishing marginal returns of the R&D activity[6]:

$$q_b(t, K_2(t)) = \omega_b + \upsilon_b \sqrt{K_2(t)}, \tag{7.17}$$

where the parameters (ω_b, υ_b) are defined as follows:

ω_b : initial probability rate of discovery; $\omega_b = 0.05$;
υ_b : slope w.r.t $\sqrt{K_2(t)}$ of the probability rate of discovery; $\upsilon_b = 0.0019$.

Note that although the technological breakthrough is an abrupt occurrence, the diffusion of the clean technology into the economy would be gradual due to the dynamics of capital accumulations (see Eq. (7.9)).

7.2.4 GHG concentration dynamics

Let $M(t)$ denote the atmospheric concentration of GHG at time t, in GtC equivalent. The accumulation of GHG in the atmosphere is described by the following equation:

$$\dot{M}(t) = \beta E(t) - \delta_M (M(t) - M_p), \tag{7.18}$$

where the initial value of atmospheric concentration is $M(0) = 808.8$ GtC and where the different parameters are defined as follows:

β : marginal atmospheric retention rate; $\beta = 0.64$;

---

[5] We always assume the required regularity, for instance, by assuming continuous differentiability with respect to time and state.

[6] Notice that a simpler affine dependence relation was used in (3) which had some drawbacks since it allowed for very high investment in K_2 in order to force the access to the clean technology.

δ_M : natural atmospheric elimination rate; $\delta_M = 0.036$;
M_p : preindustrial level of atmospheric concentration; $M_p = 588.2$ GtC.

Equation (7.18) is a very simple representation of the GHG concentration dynamics yet consistent with the archetypal DICE model of Nordhaus (19). A computationally costly alternative would be to incorporate a more detailed representation of GHG dynamics, as proposed for instance by Manne *et al.* in their classical MERGE model (16), (18).

7.2.5 Climate change damages

In a cost-benefit approach, one must quantify somehow the damages related to climate change. To model these damages and their economic impacts, we follow the MERGE model, (18). In particular, we compute an economic loss factor due to climate changes at time t, appropriately modified to account for the a-priori probability density function f_0 of the climate sensitivity ω:

$$\mathrm{ELF}(t) = 1 - \left(\frac{M(t) - M_d}{\overline{cat}_{M_0} - M_d} \right)^2, \tag{7.19}$$

where M_d is the concentration level at which damages start to occur and where:

$$\overline{cat}_{M_0} = \mathrm{E}_{f_0} \left[cat_M(\omega) \right]. \tag{7.20}$$

cat_M is the sensitivity dependent "catastrophic" concentration level at which the entire production would be wiped out. For the illustrative purposes of this paper, $\overline{cat}_{M_0}$ is set roughly at 3.5 times the pre-industrial concentration level and M_d at 735.3 GtC (350 ppm). As seen in Eq. (7.14), economic loss is applied directly on economic output. Notice that within such framework, one aims at maximizing social welfare by curbing GHG emissions to a level where marginal abatement costs equal marginal benefits gained from avoiding climate change related damages.

7.2.6 Stochastic control model

Define the state variable $s = (\mathbf{K}, M, \xi)$ where $\mathbf{K} \in \mathbf{R}^{+2}$ represents the capital stocks, $M \in \mathbf{R}^+$ is the atmospheric GHG concentration and $\xi \in \{0, 1\}$ indicates the eventual availability of the advanced (clean) technology. At initial time $t^0 = 0$, the state $s^0 = (\mathbf{K}^0, M^0, \xi^0)$ is such that $K_2^0 \simeq 0$ and $\xi^0 = 0$ since the advanced technology is not yet available.

It will be convenient to introduce a special notation for the continuous state variable $x = (\mathbf{K}, M)$. Control variables are the emission rates $E_1(t), E_2(t)$, the investment rates $I_1(t), I_2(t)$ in the different types of capital and the labor allocation $L_1(t), L_2(t) = L(t) - L_1(t)$. Control variables are denoted $u = (\mathbf{E}, \mathbf{I}, \mathbf{L})$ where $\mathbf{E}(t) = (E_i(t))_{i=1,2}$, $\mathbf{I}(t) = (I_i(t))_{i=1,2}$, $\mathbf{L}(t) = (L_i(t))_{i=1,2}$. The dynamics of the state variable can be described by the following piecewise deterministic Markov process (PDMP) (8):

$$\dot{x}(t) = f^{\xi(t)}(t, x(t), u(t)). \tag{7.21}$$

One can then introduce the following state and control dependent instantaneous reward function[7]:

$$\mathscr{L}^{\xi}(t,x,u) = L \log\left(\frac{1}{L}\left[F^{\xi}(t,\mathbf{E},\mathbf{K},\mathbf{L}) - \sum_{i=1}^{2}(I_i + p_{E_i}\phi_i E_i)\right]\right), \tag{7.22}$$

where $F^{\xi}(t,\mathbf{E},\mathbf{K},\mathbf{L}) = \mathrm{ELF}(t)\, Y(t)$. The controls are subject to the constraints: $E_i(t) \geq 0$, $I_i(t) \geq 0$, $L_i(t) \geq 0$, $i = 1,2$ and $L_1(t) + L_2(t) = L(t)$. This is summarized in general notations by: $u(t) \in U(t)$, where $U(\cdot)$ is the admissible state dependent control set.

7.2.7 Single-jump piecewise deterministic control model

Let $\tau^0 = 0$ be the initial time and τ^1 be the jump time of the $\xi(\cdot)$ process indicating the time at which a technology breakthrough occurs. Denote s^0, s^1 the state observed at times τ^0, τ^1. A piecewise deterministic policy γ is a mapping that associates with τ^0 and s^0 a control $u_{\tau^0,s^0}(\cdot)$: $(\tau^0, \infty[\to U(\cdot)$ valid until occurrence of the system jump at some (random) time τ^1. At that point, the policy must be updated to $u_{\tau^1,s^1}(\cdot)$[8] according to the time τ^1 and the state s^1 reached at that time. This corresponds to the concept of piecewise deterministic control. Associated with a policy γ and an initial state s^0 one obtains an expected reward (welfare gain) defined by:

$$J(\gamma, s^0) = \mathrm{E}_{\gamma}\left[\int_0^{\infty} e^{-\rho t} \mathscr{L}^{\xi(t)}(t, x(t), u(t))\, dt\right], \tag{7.23}$$

where the expectation E_{γ} is taken with respect to the probability measure induced by the policy γ.

[7] To simplify the notations, $\mathscr{L}^{\xi(t)}(t,x(t),u(t))$ is denoted here $\mathscr{L}^{\xi}(t,x,u)$.

[8] For simplicity, in the remainder of this chapter, we shall drop the indices τ and s and write $u(\cdot)$ instead of $u_{\tau,s}(\cdot)$.

7.2.8 Dynamic programming

7.2.8.1 After the jump

From jump time τ^1 onwards, the carbon-free technology is available and hence the clean economy is productive. At time τ^1, for a given state $s^1 = (x^1, \xi^1) = (\mathbf{K}^2, M^2, 1)$, let us define the value function $V^1(\tau^1, x^1)$ as the solution to the optimization problem:

$$V^1(\tau^1,x^1) = \max_{u(\cdot)} e^{\rho\tau^1} \int_{\tau^1}^{\infty} e^{-\rho t} \mathscr{L}^1(t,x(t),u(t))\,dt \tag{7.24}$$

subject to the state equations:

$$\dot{x}(t) = f^1(t,x(t),u(t)); \quad u(t) \in U(t); \quad t \geq \tau^1; \quad x(\tau^1) = x^1. \tag{7.25}$$

7.2.8.2 At the initial time

At the initial time, the clean technology is not available and thus $\xi = 0$. Based on the dynamic principle, the value function starting at initial time τ^0(=0) and state $s^0 = (x^0, 0)$ must satisfy the following equation:

$$\begin{aligned} V^0(x^0) &= \max_{u(\cdot)} \mathrm{E}_{K_2(\cdot)} \Bigg[\int_0^{\tau^1} e^{-\rho t} \mathscr{L}^0(t,x(t),u(t))\,dt \\ &\quad + e^{-\rho\tau^1} V^1(\tau^1, x^1(\tau^1)) \Bigg] \end{aligned} \tag{7.26}$$

$$\begin{aligned} &= \max_{u(\cdot)} \Bigg[\int_0^{\infty} \Bigg[e^{-\int_0^s q_b(\alpha,K_2(\alpha))\,d\alpha} \\ &\quad q_b(s,x(s)) \int_0^s e^{-\rho t} \mathscr{L}^1(t,x(t),u(t))\,dt \Bigg] ds \\ &\quad + \int_0^{\infty} \left[e^{-\int_0^s q_b(\alpha,K_2(\alpha))\,d\alpha} q_b(s,K_2(s)) e^{-\rho s} \right] V^1(s,x(s))\,ds \Bigg] \end{aligned} \tag{7.27}$$

$$\begin{aligned} &s.t. \\ \dot{x}(t) &= f^0(t,x(t),u(t)); \quad u(t) \in U(t); \quad t \geq 0; \quad x(0) = x^0. \end{aligned}$$

In Eq. (7.26), the jump time τ^1 is stochastic. The associated jump rate is $q_b(t, K_2(t))$. Thus the expectation E_γ is replaced here by E_{K_2} to stress the fact that the effect of γ on τ^1 is mediated by K_2. One has denoted $s^1(\tau^1)$ the random state reached after the jump time. After integration by parts of the first term in Eq. (7.27), and if one assumes that the following constraint has to be satisfied by all admissible controls:

$$\lim_{s\to\infty} \left[e^{-\int_0^s q_b(\alpha,K_2(\alpha))\,d\alpha} \int_0^s e^{-\rho t} \mathscr{L}^1(t,x(t),u(t))\,dt \right] = 0,$$

one obtains the following equivalent infinite horizon deterministic control problem:

$$
\begin{aligned}
V^0(x^0) &= \max_{u(\cdot)} \int_0^\infty e^{-(\rho t+\int_0^t (q_b(s,K_2(s)))\,ds)} \left[\mathscr{L}^0(t,x(t),u(t))\right. \\
&+ \left. q_b(t,K_2(t))V^1(t,x(t))\right] dt \qquad (7.28)\\
&s.t. \\
\dot{x}(t) &= f^0(t,x(t),u(t)); \quad u(t)\in U(t); \quad t\geq 0; \quad x(0)=x^0 .
\end{aligned}
$$

7.2.8.3 Economic interpretation

Equation (7.28) corresponds to the formulation of the stochastic control problem (7.23) as an equivalent open-loop, infinite horizon, deterministic optimal control problem. The economic interpretation of the various terms in Eq. (7.28) follows. There is first an endogenous, state dependent (and thus indirectly control dependent), discount factor:

$$e^{-(\rho t+\int_0^t (q_b(s,K_2(s)))\,ds)} . \qquad (7.29)$$

The extended deterministic problem reward function is:

$$\mathscr{L}^0(t,x(t),u(t))+q_b(t,K_2(t))V^1(t,x(t)) . \qquad (7.30)$$

It takes into account the possible future access to an improved technology. Indeed, the dependence of jump rate q_b in Eq. (7.30) (as well as in Eq. (7.29)) on the control sub-state $K_2(t)$ captures the essence of the interaction of controls with uncertainty as represented by the access time to a clean technology. Finally, the decisions to abate (choice of $E_1(t),E_2(t)$) or to invest in dirty and clean technologies (choice of $I_1(t),I_2(t)$) and labor allocation $L_1(t)$ are then the result of a tradeoff between these different contributions to the reward.

7.2.9 Numerical method

We solve numerically the model described above to obtain some qualitative information on the solution of the optimal timing problem. The numerical method adopted consists in replacing the differential equations by difference equations with a 10-year time step, solving the equivalent deterministic control problems using mathematical programming techniques and approximating the value functions in a class of log-linear functions that are akin to the Cobb-Douglas function frequently used in economics.

7.2.9.1 Discrete time formulation

To solve numerically this control problem we use a discrete time/finite difference reformulation, with time steps of 10 years, of the differential equations entering the model. Our model (7.1) – (7.20) is discretized accordingly. As an illustration, we provide the discrete time reformulation of selected equations. Social welfare (W) is computed as the sum over T 10-year periods of the discounted utility (from per capita consumption) with discount rate $dr(t)$:

$$W = \sum_{t=0}^{T-1} 10dr(t)L(t)\log[c(t)]. \tag{7.31}$$

Eq. (7.9) describing the evolution of capital stock (K_i) is rewritten as follows:

$$K_i(t+1) = 10I_i(t) + (1-\delta_K)^{10}K_i(t) \quad i = 1,2. \tag{7.32}$$

Finally Eq. (7.18) characterising the accumulation (M) of GHG in the atmosphere becomes:

$$M(t+1) = 10\beta\,(E_1(t)+E_2(t)) + (1-\delta_M)M(t) + \delta_M M_p. \tag{7.33}$$

7.2.9.2 Approximation of the value functions

The numerical solution which is based on dynamic programming involves the following steps:

Step 1; expressing V^1: For a given grid of initial values (state space and jump time), solve the deterministic control problem which corresponds to Eqs. (7.24)-(7.25), using a discrete time approximation, with time steps of 10 years. The problem is formulated using the GAMS modeling language (6) and the nonlinear optimization problem is solved using (10). Then, based on a sufficient set of grid point evaluations, use log-linear least squares to fit a product form ($aK_1{}^bK_2{}^cM^d$) to the functions $V^1(\tau^1,\cdot)$.

Step 2; expressing V^0: For the given initial state, solve the deterministic optimal control problem which corresponds to Eq. (7.28), where the approximation of $V^1(t,x(t))$ obtained at Step-1 is used in the formulation of the reward function.

7.2.9.3 Class of approximating functions

The value function $V^1(t,x(t))$ is approximated via a class of functions of the form:

$$a(\tau^1)K_1(t)^{b(\tau^1)}K_2(t)^{c(\tau^1)}M(t)^{d(\tau^1)}$$

which closely resemble extended Cobb-Douglas functions, standard in economics. They should be suitable for describing the interplay of economic aggregates at work here. Note that coefficients a, b, c and d are functions of the jump time τ^1 at which the value function must be evaluated. This is consistent with the non-stationary characteristics of process dynamics (Eqs. (7.1) – (7.20)).

7.2.9.4 Grid specification

The goal is to produce an analytical expression for an approximation of the value function V^1. This analytical expression will enter into the definition of the reward function in the deterministic equivalent infinite horizon optimal control computation in Eq. (7.28). Therefore, it is important that the coefficients in this approximate function be computed on the basis of grid point evaluations which are likely to lie in the neighborhood of the optimal control trajectory. The grid is iteratively made more relevant by producing Monte Carlo simulations of optimally controlled trajectories computed on the basis of initial analytical value function expressions, which are then be updated on the basis of the new states recorded at jump times.

7.2.10 Numerical results

In this section, we compare the results of the optimal control policy before the jump occurs (scenario labeled "stoch. control") to the optimal controls that would result under different deterministic scenarios whereby technology is either available at the outset (scenario labeled "K2 now") or will never become available on the control horizon (scenario labeled "only K1"). In addition, we report on a deterministic counterfactual baseline (denoted "BaU"), where one does not suffer from climate change related damages[9], and thus where one does not impose any GHG emission control policy.

7.2.10.1 GHG emission and concentration paths

Figure 7.1 reports on GHG emission trajectories, that adjust to limit climate change related damages, and on the resulting concentration paths. In both scenarios "*only* K1" and "K2 *now*", emissions drop by around 18% the first 10 years. This corresponds to the "sudden discovery" that climate damages have been occurring between M_d (350 ppm) and $M(0)$ (385 ppm); see Eq. (7.19). Then, in the "*only* K1" scenario, emissions are progressively reduced, but concentration keeps slowly increasing to reach a level of around 1160 GtC (around 550 ppm) in the 23nd century. Whereas in the "K2 *now*" scenario, emissions are initially slightly higher as

[9] Namely, $ELF(t)$ is set to 1 in Eq. (7.19).

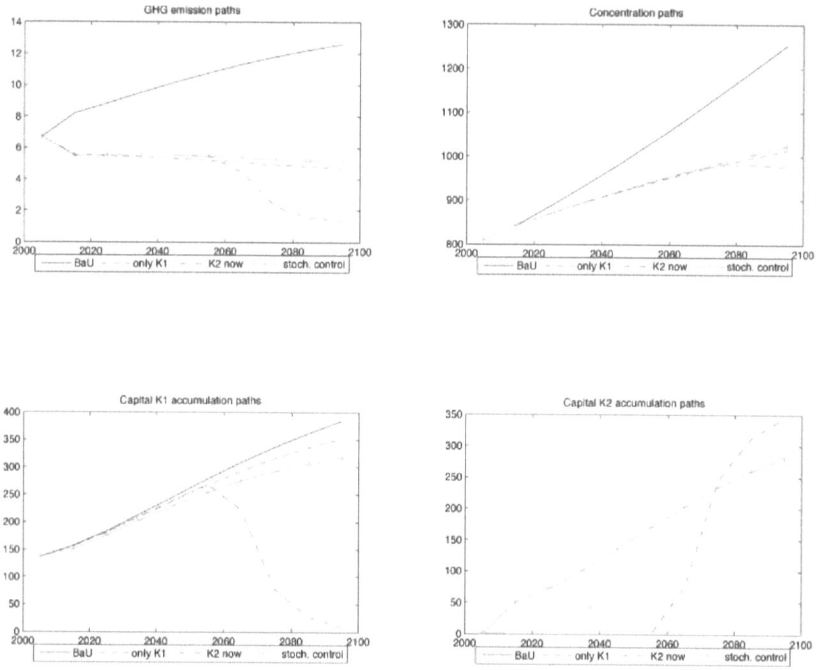

Fig. 7.1 Results for the cost-benefit analysis

the availability of the clean technology enables one to reach later a lower emission level. Consequently, concentration decreases from 2085 on to reach a lower level of around 910 GtC (430 ppm) in the 23nd century. In other words, less abatement possibilities leads one to target a higher concentration level in scenario "*only* K1", and thus to bear a higher damage cost.

In the stochastic control scenario, emissions follow roughly the "*only* K1" path, although slightly higher. Here one aims at maintaining production levels as close as possible to the "*only* K1" path, and thus one resorts to using more energy as a production factor to compensate for the relative decrease of dirty capital K_1; see Section 7.2.10.2 below. Consequently, concentration reaches slightly higher levels than in this latter scenario.

7.2.10.2 Capital K_1 accumulation paths

As reported in the previous section, taking into account climate change damages leads one to aim for a long-run concentration level of around 550 ppm in the "*only* K1" scenario, and around 430 ppm in the "K2 *now*" scenario. To achieve these levels, GHG emissions must be curbed from the baseline path, and since the (carbon) economy has to use less (fossil) energy, economic output is reduced; namely GDP losses occur. In the "*only* K1" scenario, where the clean technology is not available, accumulation of capital K_1 is reduced (by about 14%, compared to the BaU case) to adjust to the lower production levels. When the clean technology is available ("K2 *now*" case), the previously mentioned GDP losses are mitigated through the introduction of the clean technology (see Section 7.2.10.3 below) and a phasing out of capital K_1 occurs, by 2065.

Recall that, following Eq. (7.14), available production is used for consumption, investment and the payment of energy costs. In the stochastic control scenario, contrary to the "*only* K1" scenario, investment is also channeled into the R&D capital K_2; see again Section 7.2.10.3 below. Consequently, fewer investment possibilities remain for productive capital K_1 whose accumulation is less than in the "*only* K1" scenario.

7.2.10.3 Capital K_2 accumulation paths

Figure 7.1 reports finally on the accumulation of capital K_2.

Neither in the baseline, where it is not needed, nor in the "*only* K1" scenario, where it is not available, does capital K_2 accumulate. In the "K2 *now*" scenario, the accumulation of clean capital simply (inverse) mirrors the phasing out of K_1.

In the stochastic control scenario, a steady accumulation of R&D capital K_2 takes place to hasten the expected technological breakthrough.

7.3 A stochastic game model

In this section, we extend our previous models to a stochastic game framework whereby 2 groups of countries interact non-cooperatively in the definition of a global climate policy.

7.3.1 International dimension of climate policies

It is well recognized that, in order to be effective, climate policies must have an international dimension. Indeed, greenhouse gas (GHG) emissions from emerging economies have caught up the levels emitted by industrialized economies; see for instance (13). Consequently, a climate policy aiming at stabilizing GHG concentration at a "sustainable" level must involve all major emitters, and that includes an equal number of developing and developed countries, (4). However in each country environmental protection may have a different importance in the overall political agenda, as distinct countries perceive differently the level of gravity associated with environmental degradation. Furthermore, each country may be faced with different developmental issues and may be equipped with different technological means. In particular, the ability in each country to limit over time GHG emissions using efficient carbon-free technologies remains uncertain as it depends on the country's current level of technological development and of national R&D policies yet to be defined.

This situation (of shared international responsibilities but different possibilities and willingness to act) calls for a non-cooperative game paradigm to study how countries interact to define a global climate policy. In an effort to account for the inherent uncertainty in so far as the possibility of a technological breakthrough in clean technologies is concerned, we propose a stochastic game model where two groups of countries, corresponding to the major GHG emitters, interact non-cooperatively in the definition of a global climate policy but share the consequences of global environmental damage.

7.3.2 Variables

We select the following variables, where $j = 1,2$ is the index of each of the 2 regions and t the model running time:

$C(j,t)$: total consumption in region j at time t, in trillions (10^{12}) of dollars;
ELF(t): economic loss factor due to climate changes at time t, in %;
$E_1(j,t)$: yearly emissions of GHG (in Gt–10^9 tons–carbon equivalent) in the carbon economy of region j at time t;
$E_2(j,t)$: yearly emissions of GHG in the clean economy of region j at time t, in GtC;
$I_i(j,t)$: investment in capital $i = 1,2$ in region j at time t, in trillions of dollars;
$K_1(j,t)$: physical stock of productive capital in the carbon economy of region j at time t, in trillions of dollars;
$K_2(j,t)$: physical stock of productive capital in the clean economy of region j at time t, in trillions of dollars;
$L_1(j,t)$: part of the (exogenously defined) labor force $L(j,t)$ of region j allocated at time t to the carbon economy, in millions (10^6) of persons;
$L_2(j,t)$: part of the labor force of region j allocated at time t to the clean economy, in millions of persons;
$M(t)$: atmospheric concentration of GHG at time t, in GtC equivalent;
$WRG(j)$: discounted welfare of region j;
W: total discounted welfare;
$Y(j,t)$: economic output of region j at time t, in trillions of dollars.

7.3.3 Equations

In each region j, social welfare (WRG) is given by the accumulation of a discounted utility obtained from available per capita consumption with discount rate $\rho(j)$:

$$WRG(j) = \int_0^\infty e^{-\rho(j)t} L(j,t) \log[C(j,t)/L(j,t)]\, dt. \tag{7.34}$$

Total labor (L) is divided between labor allocated to the carbon economy (L_1) and labor allocated to the clean economy (L_2):

$$L(j,t) = L_1(j,t) + L_2(j,t). \tag{7.35}$$

Capital stock (K_i) evolves according to the choice of investment (I_i) and depreciation rate δ_K as follows:

$$\dot{K}_i(j,t) = I_i(j,t) - \delta_K K_i(j,t) \quad i = 1,2. \tag{7.36}$$

Economic output (Y) occurs in the two economies according to an extended Cobb-Douglas production function in three inputs, capital (K), labor (L) and energy (which use is measured through emission level E):

$$\begin{aligned} Y(j,t) = {} & A_1(j,t) K_1(j,t)^{\alpha_1(j)} (\phi_1(j,t) E_1(j,t))^{\theta_1(j,t)} L_1(j,t)^{1-\alpha_1(j)-\theta_1(j,t)} \\ & + \xi(t) A_2(j,t) K_2(j,t)^{\alpha_2(j)} (\phi_2(j,t) E_2(j,t))^{\theta_2(j,t)} L_2(j,t)^{1-\alpha_2(j)-\theta_2(j,t)}, \end{aligned} \tag{7.37}$$

where A_i is the total factor productivity in the carbon (resp. clean) economy (when $i = 1$, resp. $i = 2$), α_i is the elasticity of output with respect to capital K_i, ϕ_i is the energy conversion factor for emissions E_i and θ_i is the elasticity of output with respect to emissions E_i. Notice that in Eq. (7.37) production from the carbon-free economy occurs only when $\xi(t) = 1$, namely when a technological breakthrough has occurred. Economic output is used for consumption (C), investment (I) and the payment of energy costs:

$$\begin{aligned} \mathrm{ELF}(t) Y(j,t) = {} & C(j,t) + I_1(j,t) + I_2(j,t) + p_{E_1}(j,t) \phi_1(j,t) E_1(j,t) \\ & + p_{E_2}(j,t) \phi_2(j,t) E_2(j,t), \end{aligned} \tag{7.38}$$

where p_{E_i} is the energy price in the carbon (resp. clean) economy (when $i = 1$, resp. $i = 2$). Accumulation M of GHG in the atmosphere evolves according to:

$$\dot{M}(t) = \beta \sum_{j=1}^{2} (E_1(j,t) + E_2(j,t)) - \delta_M (M(t) - M_p), \tag{7.39}$$

where β is the marginal atmospheric retention rate, δ_M the natural atmospheric elimination rate and M_p the preindustrial level of atmospheric concentration. Finally, increasing atmospheric GHG concentrations yield economic losses (due to climate changes) that reduce the available consumption. As before, we assume that the economic loss factor ELF is defined by:

$$\mathrm{ELF}(t) = 1 - \left(\frac{M(t) - M_d}{\overline{cat}_{M_0} - M_d} \right)^2, \tag{7.40}$$

where M_d is the concentration level at which damages start to occur and $\overline{cat}_{M_0} = \mathrm{E}_{f_0}[cat_M(\omega)]$ with cat_M being a "catastrophic" concentration level depending on the climate sensitivity ω with an a-priori probability density function f_0. For simplicity, we assume that parameter M_d and the vision on the climate sensitivity uncertainty (and hence on parameter $\overline{cat}_{M_0}$) are respectively the same for each player.

7.3.4 Uncertain technological breakthrough

Let us recall that the access to a cleaner economy at time t is represented by a binary variable $\xi(t) \in \{0,1\}$ whereby the clean technology starts being available only when ξ switches from an initial value of 0 to 1. The initial value $\xi(0) = 0$ indicates that there is no access to the clean technology at the initial time. The switch to the value 1 occurs at a random time which is influenced by the global accumulation of R&D capital $\tilde{K}_2$ where:

$$\tilde{K}_2(t) = \sum_{j=1}^{2} K_2(j,t). \tag{7.41}$$

The jump rate function is defined as:

$$q_b(t, \tilde{K}_2(t)) = \omega_b + \upsilon_b \sqrt{\tilde{K}_2(t)}, \tag{7.42}$$

where ω_b is the initial probability rate of discovery and υ_b the slope w.r.t $\tilde{K}_2(t)$ of the probability rate of discovery.

7.3.5 Controlled stochastic game

Define the state variable process:

$$s(\cdot) = ((\mathbf{K}(j,\cdot) : j = 1,2), M(\cdot), \xi(\cdot)), \tag{7.43}$$

where $\mathbf{K}(j,\cdot) \in \mathbb{R}^{+2}$ represents the evolution of capital stocks in region j, $M(\cdot) \in \mathbb{R}^+$ is the evolution of atmospheric GHG concentration, and $\xi(\cdot)$ that takes its value in $\{0,1\}$ indicates the eventual availability of the advanced (clean) technology. At initial time $t^0 = 0$, the state $s(0) = s^0 = (\mathbf{K}^0, M^0, \xi^0)$ is such that $K_2^0 \simeq 0$ and $\xi^0 = 0$ since the advanced technology is not yet available. For ease of notations we have denoted $\mathbf{K}(\cdot) := ((\mathbf{K}(j,\cdot) : j = 1,2)$, the vector of capital stocks for all the regions.

It will be convenient to introduce a special notation for the continuous state variable $x = (\mathbf{K}, M)$. Control variables are the emission rates $E_1(j,t), E_2(j,t)$, the investment rates $I_1(j,t), I_2(j,t)$ in the different types of capital and the labor allocation $L_1(j,t), L_2(j,t) = L(j,t) - L_1(j,t)$. Control variables are denoted $u = (\mathbf{E}, \mathbf{I}, \mathbf{L})$ where $\mathbf{E}(t) = (E_i(j,t))_{i,j=1,2}$, $\mathbf{I}(t) = (I_i(j,t))_{i,j=1,2}$, $\mathbf{L}(j,t) = (L_i(j,t))_{i,j=1,2}$. The dynamics of the state variable can be described by the following PDMP:

$$\dot{x}(t) = f^{\xi(t)}(t, x(t), u(t)). \tag{7.44}$$

Given a state variable x and a control variable u, the instantaneous utility of consumption for each region is determined. Therefore one can introduce the reward function[10]:

$$\mathscr{L}^{\xi}(j,t,x,u) = L(j,t) \log\left(\frac{1}{L(j,t)}\left[\mathrm{ELF}(t) F^{\xi(t)}(j,t,\mathbf{E}(j,t),\mathbf{K}(j,t),\mathbf{L}(j,t)) - \sum_{i=1}^{2}(I_i(j,t) + p_{E_i}\phi_i E_i(j,t))\right]\right), \tag{7.45}$$

where $F^{\xi(t)}(j,t,\mathbf{E}(j,t),\mathbf{K}(j,t),\mathbf{L}(j,t)) = Y(j,t)$. The controls are subject to the constraints: $E_i(j,t) \geq 0$, $I_i(j,t) \geq 0$, $L_i(j,t) \geq 0$, $i = 1,2$ and $L_1(j,t) + L_2(j,t) \leq L(j,t)$, $j = 1,2$. This is summarized in general notations by: $u(t) \in U(t)$.

7.3.6 Game in normal form

We represent here a stochastic game, played according to a *piecewise open-loop information structure*. This class of strategies has been first introduced in (11) for stochastic games played over piecewise deterministic systems. It has been used in (12) to represent oligopoly competition.

7.3.6.1 Piecewise open-loop strategies

Let $\tau^0 = 0$ be the initial time and τ^1 be the jump time of the $\xi(\cdot)$ process indicating the time at which a technology breakthrough occurs. Denote s^0, s^1 the state observed at times τ^0, τ^1. A strategy γ_j for Player j is a mapping that associates with τ^0 and s^0 a control $u_{j,\tau^0,s^0}(\cdot)$: $(\tau^0, \infty[\to U(j,\cdot)$ that will be used by Player j until occurrence of the system jump at some (random) time τ^1. At that point, the policy must be updated to $u_{j,\tau^1,s^1}(\cdot)$[11] according to the time τ^1 and the state s^1 reached at that time. This corresponds to the concept of piecewise deterministic control. Associated with

[10] To simplify notations, $\mathscr{L}^{\xi(t)}(j,t,x(t),u(t))$ is denoted here $\mathscr{L}^{\xi}(j,t,x,u)$.

[11] For simplicity, in the remainder of the chapter, we shall drop the indices τ and s and write $u(j,\cdot)$ instead of $u_{j,\tau,s}(\cdot)$.

a policy vector $\gamma = \{\gamma_j : j = 1,2\}$ and an initial state s^0 one obtains an expected reward (welfare gain) for Player j defined by:

$$J(j,\gamma,s^0) = \mathrm{E}_\gamma \left[\int_0^\infty e^{-\rho(j)t} \mathscr{L}^{\xi(t)}(j,t,x(t),u(t))\,dt \right], \tag{7.46}$$

where the expectation E_γ is taken with respect to the probability measure induced by the policy γ. Such a strategy is admissible if it generates a control for each player that satisfies almost surely all the constraints.

7.3.6.2 Nash equilibrium

A strategy γ^* is a Nash equilibrium if the following inequality holds:

$$J(j,\gamma^*,s^0) \geq J(j,(\gamma^{*-j},\gamma_j),s^0) \tag{7.47}$$

for all initial state s^0 and Player j, where (γ^{*-j},γ_j) stands for the policy vector where Player $\iota \neq j$ uses strategy γ_ι^* whereas Player j uses any admissible strategy γ_j.

7.3.7 Dynamic programming

In a Piecewise open-loop information structures, Player controls are adapted at jump times, observing the system state, right after the jump. The Nash equilibrium solution can thus be characterized by using a dynamic programming argument for the discrete event system which is obtained by observing this control system and deciding on the subsequent control policy only at jump times. In our case there are only two jump times, the initial time and the time when the breakthrough occurs.

7.3.7.1 After the jump

Assume that the jump occurs at time τ^1. From time τ^1 onwards, the carbon-free technology is available. At time τ^1, for a given state $s^1 = (x^1,\xi^1) = (\mathbf{K}^1,M^1,1)$, the value function $V^1(j,\tau^1,x^1)$ is defined as the payoff to Player j in the equilibrium solution to an open-loop differential game. In this equilibrium solution each player solves the following optimal response problem:

$$\begin{aligned} V^1(j,\tau^1,x^1) = \max_{u(j,\cdot)} \Big[& e^{\rho(j)\tau^1} \int_{\tau^1}^\infty e^{-\rho(j)t} \mathscr{L}^1(j,t,x(t),[u(j,t),u^*(\iota,t)])\,dt : \\ & \text{given } u^*(\iota,\cdot);\iota \neq j \Big], \end{aligned} \tag{7.48}$$

subject to the state equations:

$$\dot{x}(t) = f^1(t,x(t),[u(j,t),u^*(\iota,t)]); \qquad (7.49)$$
$$[u(j,t),u^*(\iota,t)] \in U(t); \quad t \geq \tau^1; \quad x(\tau^1) = x^1,$$

where $u^*(\iota,\cdot)$ is the equilibrium control of the other player.

7.3.7.2 At the initial time

At initial time the discrete state is $\xi(0) = 0$, meaning that one does not have access to the clean technology. The stochastic control problem describing the optimal response of player j to the equilibrium control of the other player ι can be written as follows:

$$\begin{aligned} V^0(j,x^0) &= \max_{u(j,\cdot)} \mathrm{E}_{K_2(\cdot)} \left[\int_0^{\tau^1} e^{-\rho(j)t} \mathscr{L}^0(j,t,x(t),[u(j,t),u^*(\iota,t)])\,dt \right. \\ &\quad \left. + e^{-\rho(j)\tau^1} V^1(j,\tau^1,x^1(\tau^1)) : \text{ given } u^*(\iota,\cdot); \iota \neq j \right] \\ s.t.& \\ \dot{x}(t) &= f^0(t,x(t),[u(j,t),u^*(\iota,t)]); \\ &\quad [u(j,t),u^*(\iota,t)] \in U(t); \quad t \geq 0; \quad x(0) = x^0. \end{aligned}$$

The deterministic equivalent infinite horizon differential game problem is given by:

$$\begin{aligned} V^0(x^0) &= \max_{u(j,\cdot)} \int_0^{\infty} e^{-(\rho(j)t + \int_0^t q_b(s,\tilde{K}_2(s))\,ds)} \left[\mathscr{L}^0(j,t,x(t),[u(j,t),u^*(\iota,t)]) \right. \\ &+ \left. q_b(t,\tilde{K}_2(t)) V^1(j,t,x(t)) : \text{ given } u^*(\iota,\cdot); \iota \neq j \right] dt \qquad (7.50) \\ s.t.& \\ \dot{x}(t) &= f^0(t,x(t),[u(j,t),u^*(\iota,t)]); \quad u(t) \in U(t); \quad t \geq 0; \quad x(0) = x^0. \end{aligned}$$

7.3.8 Solving the deterministic equivalent dynamic games

We have seen how, in a piecewise open-loop information structure, the characterization of equilibrium solutions could be obtained through a sequence of solutions of deterministic differential games. Finding an equilibrium solution in a differential game is not an easy task. It can be handled through the implementation of nonlinear complementarity algorithm, like the one implemented in the PATH solver (9) or it can be obtained via the implementation of a heuristic cobweb search of a fixed point of the optimal response function of the two players. For simplicity of notations denote:

$$\psi(j,\tau^1,s^1,u(\cdot)) = e^{\rho(j)\tau^1} \int_{\tau^1}^{\infty} e^{-\rho(j)t} \mathscr{L}^1(j,t,x(t),u(j,t))\,dt$$

where $x(\cdot)$ is the state trajectory resulting from the choice of control

$$u(\cdot) = (u(1,\cdot), u(2,\cdot))$$

and initial state $s^1 = (x^1, 1)$. Let us take any weighting $r = (r_j > 0 : j = 1, 2)$ and define the following function:

$$\theta(\tau^1, s^1, u(\cdot), v(\cdot); r) = \sum_{j=1}^{2} r_j \psi(j, \tau^1, s^1, (u(\iota, \cdot), v(j, \cdot)), \tag{7.51}$$

where, as before, $\iota = 2$ if $j = 1$ and $\iota = 1$ if $j = 2$ and where $v(\cdot) \in U(\cdot)$. Now define the generalized reply mapping as the point to set map:

$$\Theta(u(\cdot)) = \left\{ \tilde{v}(\cdot) : \theta(\tau^1, s^1, u(\cdot), \tilde{v}(\cdot); r) = \max_{v(\cdot) \in U(\cdot)} \theta(\tau^1, s^1, u(\cdot), v(\cdot); r) \right\}. \tag{7.52}$$

A fixed point $u^*(\cdot) \in \Theta(u^*(\cdot))$ is a Nash equilibrium for this game. To compute such a fixed point we implement the following cobweb method:

Let $G \subset \mathbf{R}^5$ be a grid for initial value $s^1 = (K_1^1(j), K_2^1(j), M^1)$ and let $\mathscr{T}$ be a set of initial times.

Step-0 For each $\tau^1 \in \mathscr{T}$ and each $s^1 \in G$

Step-1 find $u^*(\cdot) = \text{argmax}_{u(\cdot)} \left[\sum_{j=1}^{2} r_j \psi(j, \tau^1, s^1, u(\cdot)) \right]$

Step-2 take $u(\cdot) = u^*(\cdot)$; find

$$v^*(\cdot) = \text{argmax}_{v(\cdot)} \left[\sum_{j=1}^{2} r_j \psi(j, \tau^1, s^1, (u(\iota, \cdot), v(j, \cdot)) \right]$$

Step-3 take $u(\cdot) = v^*(\cdot)$; find new

$$v^*(\cdot) = \text{argmax}_{v(\cdot)} \left[\sum_{j=1}^{2} r_j \psi(j, \tau^1, s^1, (u(\iota, \cdot), v(j, \cdot)) \right]$$

Step-4 repeat step-3 until $\max_{v(\cdot)} \left[\sum_{j=1}^{2} r_j \psi(j, \tau^1, s^1, (u(\iota, \cdot), v(j, \cdot)) \right]$ does not change significantly.

Step-5 record for each $j = 1, 2$ the value $V^1(j, \tau^1, x^1) = \psi(j, s, u^*(\cdot))$ where $s = (x^1, 1)$, $x^1 \in G$.

Step-6 Adjust by LSE an analytical form for the function $V^1(j, \tau^1, x^1)$, $x \in \mathbf{R}^5$.

A similar method is used to compute the Nash-equilibrium of the equivalent deterministic differential game at initial time.

7.3.9 Discrete time formulation

As in Section 7.2.9, we use a discrete time version of the model to compute numerically an approximate solution. In particular, social welfare ($WRG(j)$) of Player j is now given by the sum:

$$WRG(j) = \sum_{t=0}^{T-1} 10dr(j,t)L(j,t)\log[C(j,t)/L(j,t)]. \tag{7.53}$$

Capital stock (K_i) evolves through investment (I_i) and depreciation rate δ_K as follows:

$$K_i(j,t+1) = 10I_i(j,t) + (1-\delta_K(j))^{10}K_i(j,t) \quad i = 1,2. \tag{7.54}$$

Finally, accumulation M of GHG in the atmosphere evolves according to:

$$M(t+1) = 10\beta \sum_{j=1}^{2}(E_1(j,t)+E_2(j,t)) + (1-\delta_M)M(t) + \delta_M M_p. \tag{7.55}$$

7.3.10 Numerical results

The numerical solution of this stochastic game equilibrium problem involves "deep computing". Indeed, we must solve an equilibrium economic growth problem formulated as a deterministic differential game at a large number of initial states. If the grid contains 5 points for each of the 5 state variables, we obtain $5^5 = 3125$ initial state values. Since our method of computing an equilibrium is a cobweb / Gauss-Seidel one, it requires on average 10 optimizations to reach the equilibrium. So we have 30000 solutions of a problem which is a Ramsey-like optimal economic growth problem. This has to be repeated for a sequence of initial times.

Clearly, this procedure would benefit a lot from a parallel computing implementation. In Figure 7.2, we report on some preliminary computational results obtained on a single machine, for a model which has sightly different parameters in the production function[12].

Figure 7.2 shows an interesting behavior at the equilibrium. As usual, the Nash equilibrium solution is quite detrimental to the environment: GHG emission path reaches high values and concentration increases rapidly. Player B, representing the emerging economy, invests heavily in the dirty capital and let Player A, representing the more advanced economy, do the R&D investment until 2040. From this date on, Player B joins the race for discovery of the clean technology substitute, and eventually overtakes Player A in R&D investments.

[12] Basically, these numerical results were obtained under an simplifying assumption of a stationary value function V^1.

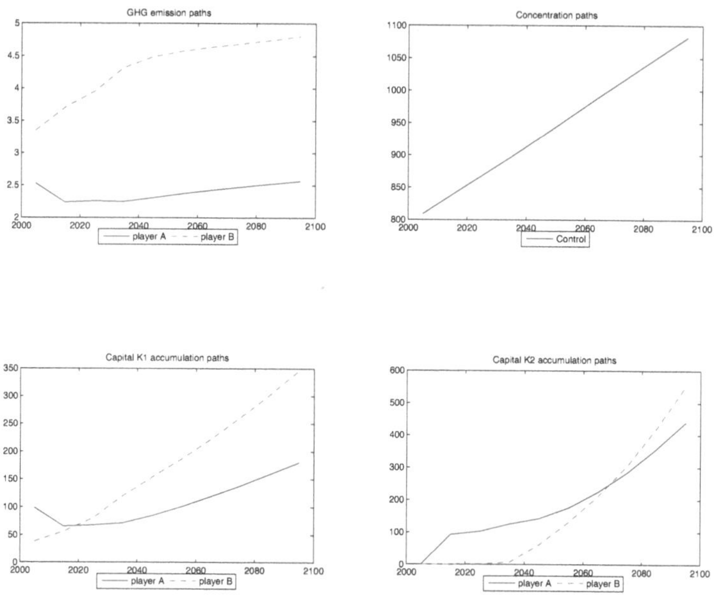

Fig. 7.2 Results for the dynamic game framework

7.4 Toward a model of learning about climate sensitivity

An issue of importance is the recognition that learning about climate sensitivity occurs all the while during which control is exerted in the climate system. Furthermore, ability to learn about climate sensitivity may increase with damages done to the environment under the form of increased GHG concentrations. This means that the control problem has an inherent dual nature to it, i.e., no separation exists between ability to learn and actual controlled state, contrary to the case of linear stochastic systems; see (7). Such dual control problems (where learning can be better achieved at the cost of some temporary degradation of control quality) are notoriously difficult to solve optimally. Instead, we propose to circumvent this difficulty by recognizing that learning about climate sensitivity is a slow process; thus, if one waits for successive periods of say 50 years before acknowledging the new data acquired over past trajectories and integrating it within a Bayesian framework to improve the climate sensitivity statistical distribution (see (2) for example), this should not constitute too much of an under utilization of information. As a result, we anticipate a scheme whereby control policies are synthesized over an infinite horizon, based on a priori sensitivity information, but are actually applied only for the next fifty years, past which a reassessment of climate sensitivity statistics is carried out. In doing so, one could rely on a random diffusion model relating earth surface mean temperature to GHG concentrations and climate sensitivity (based on (22) for example) which would be driven by the GHG resulting from the climate policies actually applied for the past 50 years, and the output of which would be made to match the actual mean earth surface observed temperature fluctuations during that period. At the end of this learning phase, one would produce an improved statistical distribution for climate sensitivity and use this latest information to launch another control synthesis cycle over infinite horizon cycle. Thus in summary, we propose control rule updates on fixed, say fifty year (or less) periods, at the end of each a sensitivity statistical distribution update is carried out.

7.5 Conclusion

In this research, dealing with the optimal timing of climate policies under uncertainty, we have used three different types of stochastic control or stochastic game paradigms. In the first model, reported in (3), we considered both the uncertainty on climate sensitivity and on the time of access to a carbon-free technology, in a cost-effectiveness framework, using a piecewise deterministic control approach. Using a dynamic programming method applied at jump times of the uncertainty process, we obtained some qualitative insight on a climate policy which corresponds largely to the implementation of a precautionary principle. The second type of model, which has been presented in this chapter, is a cost-benefit model, also using the piecewise deterministic control framework, but with only one type of uncertainty, namely the time to access the carbon-free technology. The third type of model is a stochastic game, based on the cost-benefit model, but with two groups of countries in competition to share the burden of climate change. We characterized a Nash-equilibrium solution for this game and compared it with a globally optimal solution. This chapter has shown how uncertainty could be introduced in economic growth models with climate change effects. It has also shown the insights that can be gained from this consideration of uncertain events, in particular in the domain of R&D and technology development. The solution of equilibrium conditions in the game formulation poses a formidable numerical challenge, even for this relatively small and compact model.

Acknowledgements The authors thank Laurent Drouet and Julien Thénié, from ORDECSYS, for their help in the implementation of the numerical method to solve the game model.

References

[1] P. Ambrosi, J.-C. Hourcade, S. Hallegatte, F. Lecocq, P. Dumas and M. Ha Duong, Optimal control models and elicitation of attitudes towards climate damages, *Environmental Modeling and Assessment*, Vol. 8, pp. 133–148, 2003.

[2] J.D. Annan and J.C. Hargreaves, Using multiple observationally-based constraints to estimate climate sensitivity, *Geophysical Research Letters*, Vol. 33, L06704, 2006.

[3] O. Bahn , A. Haurie and R. Malhamé, A stochastic control model for optimal timing of climate policies, *Automatica*, Vol. 44, pp. 1545–1558, 2008.

[4] K.A. Baumert, T. Herzog and J. Pershing., *Navigating the Numbers: Greenhouse Gas Data and International Climate Policy*, World Resources Institute, 2005.

[5] V. Bosetti, C. Carraro, M. Galeotti, E. Massetti and M. Tavoni, WITCH: A world induced technical change hybrid model, *The Energy Journal*, Special Issue on Hybrid Modelling of Energy Environment Policies: Reconciling Bottom-up and Top-down, pp. 13–38, 2006.

[6] A. Brooke, D. Kendrick, A. Meeraus and R. Raman, *GAMS, A Users Guide*, GAMS Development Corporation, 2005.

[7] P.E. Caines, *Linear Stochastic Systems*, John Wiley & Sons, Inc., New York, NY, 1987.

[8] M. Davis, *Markov Models and Optimization*, Chapman and Hall, London, 1993.

[9] S.P. Dirkse and M.C. Ferris, The PATH solver: A non-monotone stabilization scheme for mixed complementarity problems, *Optimization Methods and Software*, Vol. 5, pp. 123–156, 1995.

[10] General Algebraic Modeling System, *GAMS, The Solver Manuals*, GAMS Development Corporation, 2005.

[11] A. Haurie, Piecewise deterministic differential games, In: T. Başar and P. Bernhard, Editors, *Differential Games and Applications*, Lecture Notes in Control and Information Sciences, Springer-Verlag, 1989.

[12] A. Haurie and M. Roche, Turnpikes and computation of piecewise open-loop equilibria in stochastic differential games, *Journal of Economic Dynamics and Control*, Vol. 18, pp. 317–344, 1994.

[13] IEA, *Key World Energy Statistics*, International Energy Agency, 2008.

[14] IPCC, *Climate Change 2007: The Physical Science Basis*, Contribution of Working Group I to the Fourth Assessment Report of the Intergovernmental Panel on Climate Change, Cambridge University Press, Cambridge, U.K., 2007.

[15] N. Nakicenovic and R. Swart, *Special report on emissions scenarios*, Cambridge University Press, Cambridge, U.K., 2000.

[16] A.S. Manne, R. Mendelsohn and R. Richels, MERGE: A model for evaluating regional and global effects of GHG reduction policies, *Energy Policy*, Vol. 23, pp. 17–34, 1995.

[17] A.S. MANNE AND R.G. RICHELS, *Buying Greenhouse Insurance: The Economic Costs of CO_2 Emission Limits*, The MIT Press, Cambridge, 1992.

[18] A.S. MANNE AND R.G. RICHELS, MERGE: An integrated assessment model for global climate change, In: R. Loulou, J.-P. Waaub and G. Zaccour, Editors, *Energy and Environment*, GERAD 25th Anniversary Series, pp. 175–189, Springer, 2005.

[19] W.D. NORDHAUS, An optimal path for controlling greenhouses gases, *Science*, Vol. 258, pp. 1315–1319, 1992.

[20] W.D. NORDHAUS, *Managing the Global Commons: The Economics of Climate Change*, MIT Press, Cambridge, Mass., 1994.

[21] W.D. NORDHAUS AND Z. YANG, RICE: A regional dynamic general equilibrium model of alternative climate change strategies, *American Economic Review*, Vol. 86, pp. 741–765, 1996.

[22] J.D. PELLETIER, The power-spectral density of atmospheric temperature from time scales of 10-2 to 106 yr, *Earth and Planetary Science Letters*, Vol. 158, pp. 157–164, 1998.

[23] F. RAMSEY, A mathematic theory of saving, *Economic Journal*, Vol. 38, pp. 543–549, 1928.

[24] UNITED NATIONS, Framework Convention on Climate Change, *International Legal Materials*, Vol. 31, pp. 849–873, 1992.

Chapter 8
Precautionary Effect and Variations of the Value of Information

M. De Lara and L. Gilotte

Abstract Consider an agent taking two successive decisions under uncertainty. After his first decision, a signal is revealed providing information about the state of nature and the second decision is taken accordingly. Suppose that the agent is an expected utility maximizer. The *precautionary effect* holds when, in the prospect of future information, his optimal initial decision is less than without (no signal). Indeed, the first decision is usually a scalar representing consumption, so that precaution is identified with less consumption. We introduce the *second-period value of information* as a function of the initial decision and show that, when this function is decreasing, the precautionary effect holds true. More generally the condition enables the comparison of optimal decisions related to different signals, not necessarily comparable. It also ties together and clarifies many conditions for the precautionary effect that are scattered in the environmental economics literature. A practical illustration with Nordhaus' DICE model is provided.

Key words: Value of information, uncertainty, learning, precautionary effect.

JEL Classification: D83

8.1 Introduction

Consider an agent taking two successive decisions to maximize his expected utility under uncertainty. After his first decision, a signal is revealed that provides information about the state of nature. The observation of the signal allows the decision-

Michel De Lara
Université Paris–Est, CERMICS, 6–8 avenue Blaise Pascal, 77455 Marne la Vallée Cedex 2, France, e-mail: delara@cermics.enpc.fr

Laurent Gilotte
CIRED, 45 bis avenue de la Belle Gabrielle, Nogent-sur-Marne, France, e-mail: gilotte@centre-cired.fr

J.A. Filar and A. Haurie (eds.), *Uncertainty and Environmental Decision Making*,
International Series in Operations Research & Management Science 138,
DOI 10.1007/978-1-4419-1129-2_8,

maker to revise his prior and the second decision is taken accordingly. Assuming that the first decision is a scalar representing consumption, the *precautionary effect* holds when initial consumption is less in the prospect of future information than without (no signal).

The example above is a stereotype of sequential decisions problems with learning were focus is put on comparison of the optimal initial decisions with different information structures. For instance, should we aim at more reductions of current greenhouse gases emissions today whether or not we assume some future improvement of our knowledge about the climate? Economic analysis has identified effects that go in opposite directions and make the conclusion elusive. This article proposes a simple and general relation between the value of future information on risks and the initial degree of precaution taken in a sequential decisions setting.

Seminal literature in environmental economics ((2), (18; 19)) focused on the irreversible environmental consequences carried by the initial decision and showed that the possiblity of learning should lead to less irreversible current decisions ('irreversibility effect'). Henry; Arrow and Fisher consider additive separable preferences, and so do (14), (12), (17). Epstein in (10) studies a more general nonseparable expected utility model, and derives a condition that identifies the direction of the irreversibility effect. Yet, the conditions under which this result holds are difficult to connect to the primitive utility model. Further contributions have insisted on the existence of an opposite economic irreversibility since environmental precaution imply sunk costs that may constrain future consumption ((21), (25), (13)).

Risk neutral preferences are studied in (28) for a global warming model. Assuming time separability of preferences, (15) and (9) examine risk averse preferences. Gollier, Jullien, and Treich identified conditions on the second-period utility function for the possibility of learning to have a precautionary effect with and alternatively without the irreversibility constraint.

The driving idea for linking the effect of learning and the value of information is the observation that, once an initial decision is made, the value of information can be defined as a *function* of that decision.[1] Hints towards this can be found in (6), (17) or (16), but the functional dependance is either non explicit or limited to a binary decision set. More recently, (26) defined, for a specific model of climate-change, the value of information as a function of the greenhouse gases concentration. In a special case, he found that, when this value of information (after the initial decision) is a monotone function of the pollution stock, then the optimal emission levels with and without information can be ordered.

We show that this result is very general and ties together different pieces of the literature on uncertainty and irreversibility. We define the value of information measured after an initial commitment is made and this *second-period value of infor-*

[1] After the paper was accepted for publication, we discovered that Jones and Ostroy define the value of information in this way in their paper (20) where they formalize the notion of flexibility in a sequential decision context, and relate its value to the amount of information an agent expects to receive. However, we feel that the different approach, the extensions, the examples treated, the numerical application, and the validity of our results for non necessarily finite random variables may still be of interest.

mation is now a *function of the initial decision.* We shall prove that, whenever this function is monotone, we are able to identify the problem with the lowest optimal initial decision. Our result requires the existence of a maximum but does not rely on any concavity assumptions. It is generalized to the value of exchanging one information structure for an other.

Section 8.2 presents a general analytical framework of sequential decisions under uncertainty, and gives our main result. It connects the comparison of the optimal initial decisions (with learning and without learning) to the variations of the value of information, considered as a function of the initial decision. In section 8.3, the main result is shown to tie together several former results of the literature. Section 8.4 uses Nordhaus' DICE model to provide a practical illustration. Section 8.5 concludes.

8.2 Effect of learning and second-period value of information

In this section, we provide a sufficient condition for the presence of learning to affect in a predictable way the optimal initial decision.

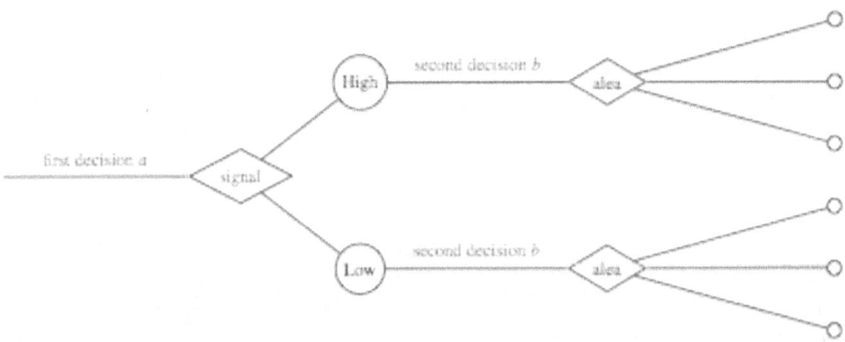

Fig. 8.1 Decision with learning; agent takes decision a; a signal is revealed; agent takes decision b accordingly.

Consider an utility maximiser who takes two successive decisions and solves[2]

$$\max_{a\in\mathbb{I}} \mathbb{E}\Big[\max_{b\in\mathbb{B}(a)} \mathbb{E}\big[U\big(a,b,X\big) \mid Y\big]\Big] . \tag{8.1}$$

The second decision b is taken knowing the value of a signal Y, as pictured in Figure 8.1. We assume that the initial decision a is a scalar belonging to an interval $\mathbb{I}$ of $\mathbb{R}$; decision b can be a vector and more generally belongs to a set $\mathbb{B}(a)$ which may

[2] We shall always assume that, for the problems we consider, the *sup* is attained and we shall use the notation *max*.

depend[3] on a ($\mathbb{B}(a)$ is a subset of a fixed larger set $\mathbb{B}$). Uncertainty is represented by states of nature ω belonging to a sample space Ω equipped with a prior $\mathbb{P}$, and by a random variable $X : \Omega \to \mathbb{X}$. Partial information on X is provided by means of a signal, that is, by a random variable $Y : \Omega \to \mathbb{Y}$. Appropriate measurability and integrability conditions are supposed to be satisfied[4].

Notice that we do not assume that the random variables X and Y are discrete.

8.2.1 Second-period value of the information

The evaluation of expected utility right after the first decision a has been taken is conditional on the signal Y and defined as follows:

$$\mathbb{V}^Y(a) := \mathbb{E}\big[\max_{b\in\mathbb{B}(a)} \mathbb{E}[U(a,b,X) \mid Y]\big] . \tag{8.2}$$

With this notation, the program of the informed agent is $\max_a \mathbb{V}^Y(a)$, while the non-informed agent solves $\max_a \mathbb{V}^0(a)$. The signal 0 returns zero for any state of nature, hence represents no information.

The program of the decision-maker who expects to receive information can now be rewritten as

$$\max_a \mathbb{V}^Y(a) = \max_a \{\mathbb{V}^0(a) + (\mathbb{V}^Y(a) - \mathbb{V}^0(a))\} . \tag{8.3}$$

stating that his objective is the same as the uninformed decision maker *plus* the expected value that is gained from receiving information after the initial decision a has been made. This value,

$$\Delta\mathbb{V}^Y(a) := \mathbb{V}^Y(a) - \mathbb{V}^0(a) , \tag{8.4}$$

is recognized therefore as the *second-period value of information* and depends on the initial decision. When future information is expected, the optimal initial decision must achieve a trade-off: it can be suboptimal from the point of view of the non-informed decision maker but compensates for this by an increase in the second-period value of information. This last notion allows to state our main result (see also (20)).

Proposition 1. *Assume that the programs* $\max_a \mathbb{V}^Y(a)$ *and* $\max_a \mathbb{V}^0(a)$ *have unique optimal solutions* $\overline{a}^Y$ *and* $\overline{a}^0$. *Whenever the second-period value of the information is a decreasing function of the initial decision, namely*

$$\Delta\mathbb{V}^Y : a \mapsto \mathbb{V}^Y(a) - \mathbb{V}^0(a) \text{ is decreasing,} \tag{8.5}$$

[3] This may materialize 'irreversibility' of the initial decision.

[4] The utility function U is defined on $\mathbb{I} \times \mathbb{B} \times \Omega$. We shall assume that $\mathbb{B}$ and Ω are measurable spaces, that U is measurable and bounded (weaker assumptions ensuring integrability are possible, too).

then, the effect of learning is precautionary *in the sense that the optimal initial decision is lower with information than without:*

$$\overline{a}^Y \leq \overline{a}^0 . \tag{8.6}$$

The proof is given in Appendix 8.6.1. This result has several extensions.

1. The above result, and all the extensions below, are not limited to the comparison of informed versus non-informed agents. If Y and Y' are two signals, such that $\Delta\mathbb{V}^{YY'}(a) := \mathbb{V}^Y(a) - \mathbb{V}^{Y'}(a)$ [5] is decreasing with a, then $\overline{a}^Y \leq \overline{a}^{Y'}$. The signal Y' need not be more or less informative than Y.
2. The signal need not be exogenous, but may be endogenous as follows. Active learning (or dependent learning) takes place when the initial decision can modify the signal before the second decision: $Y = Y(a)$. Our result remains valid with $\Delta\mathbb{V}^Y(a) := \mathbb{V}^{Y(a)}(a) - \mathbb{V}^0(a)$. Rouillon (26) studies a model of active learning in climate change economics and uses the variations of the value of information to conclude about the irreversibility effect.
3. Proposition 1 also holds under the weaker assumption that $\Delta\mathbb{V}^Y(a)$ is decreasing when $a < \overline{a}^Y$ (respectively increasing when $a > \overline{a}^Y$).
4. If the programs $\max_a \mathbb{V}^Y(a)$ and $\max_a \mathbb{V}^0(a)$ do not have unique optimal solutions, denote by $\arg\max_a \mathbb{V}^Y(a)$ and $\arg\max_a \mathbb{V}^0(a)$ the sets of maximizers. If $\Delta\mathbb{V}^Y(a)$ is decreasing, the upper bounds of these sets can be ranked: $\sup\arg\max_a \mathbb{V}^Y(a) \leq \sup\arg\max_a \mathbb{V}^0(a)$. If $\Delta\mathbb{V}^Y(a)$ is strictly decreasing, we obtain that $\sup\arg\max_a \mathbb{V}^Y(a) \leq \inf\arg\max_a \mathbb{V}^0(a)$ that is, $\overline{a}^Y \leq \overline{a}^0$ for any $\overline{a}^Y \in \arg\max_a \mathbb{V}^Y(a)$ and any $\overline{a}^0 \in \arg\max_a \mathbb{V}^0(a)$.

8.2.2 An example with quadratic costs

Consider the utility $U(a,b,X) = U_1(a) + U_2(b) - XD(a+b)$. In this model (see (7, p.229)), $a \in [0, a^\sharp]$ and $b \in [0, b^\sharp]$ are proxies for greenhouse gases emissions, U_1 and U_2 are utilities, $X \geq 0$ is an uncertain factor multiplying damages $D(a+b)$ where $a+b \geq 0$ is a proxy for final concentration due to accumulation of emissions. The damages D and the utility U_2 are supposed to be quadratic, while no assumption is made on U_1. This gives[6]

$$U(a,b,X) = U_1(a) + (b^\sharp)^2 - (b^\sharp - b)^2 - X(a+b)^2 , \tag{8.7}$$

where $0 \leq b \leq b^\sharp$. Under the assumption that the level of damages is moderate in the sense that $X < a^\sharp / b^\sharp$ where $0 \leq a \leq a^\sharp$, the following optimal solution for b is an interior solution and we have $\mathbb{V}^Y(a) = U_1(a) + (b^\sharp)^2 + \mathbb{E}[\max_{0 \leq b \leq b^\sharp} (-(b^\sharp - b)^2 -$

[5] The value of having information Y rather than Y', or the value of substituting Y for Y'.

[6] $U_2(b) = (b^\sharp)^2 - (b^\sharp - b)^2$ is increasing concave for $0 \leq b \leq b^\sharp$.

$\mathbb{E}[X \mid Y](a+b)^2)] = U_1(a) + (b^\sharp)^2 - (a+b^\sharp)^2\mathbb{E}[\frac{\mathbb{E}[X|Y]}{1+\mathbb{E}[X|Y]}]$. Thus, the second-period value of information is

$$\Delta\mathbb{V}^Y(a) = \mathbb{V}^Y(a) - \mathbb{V}^0(a) = (a+b^\sharp)^2\Big(\frac{\mathbb{E}[X]}{1+\mathbb{E}[X]} - \mathbb{E}[\frac{\mathbb{E}[X \mid Y]}{1+\mathbb{E}[X \mid Y]}]\Big) . \tag{8.8}$$

This is an increasing function in the domain $0 \le a \le a^\sharp$. Indeed, the coefficient of $(a+b^\sharp)^2$ is nonnegative by Jensen Theorem since $x \mapsto x/(1+x)$ is concave.[7] Thus, by Proposition 1, $\overline{a}^Y \ge \overline{a}^0$: one does more emissions with learning (in the expectation of information) than without learning, within this quadratic toy model where the magnitude of damages is assumed to remain moderate.

8.2.3 Comparison of initial and second-period values of information

Assume now that the decision-maker does not expect to receive future information. If offered to be able to observe signal Y in the future, how much welfare will he be ready to give up in exchange for this future signal? This amount will depend on whether the offer comes before or after he has commited to his initial decision. For an early offer, before the initial decision is taken, the early value of future information is

$$\overline{\Delta\mathbb{V}}^Y := \max_a \mathbb{V}^Y(a) - \max_a \mathbb{V}^0(a) . \tag{8.9}$$

For a late offer, after the initial decision has been committed, the decision-maker would be ready to give up the second-period value of information $\Delta\mathbb{V}^Y(a)$, which depends on what initial decision has been actually committed. Depending on the initial decision, second-period value of information can turn out to be greater or lower than the early value of information.

We have $\overline{\Delta\mathbb{V}}^Y = \mathbb{V}^Y(\overline{a}^Y) - \mathbb{V}^0(\overline{a}^0)$, by definition of the argmax $\overline{a}^Y$ and $\overline{a}^0$. Since $\mathbb{V}^Y(\overline{a}^0) \le \mathbb{V}^Y(\overline{a}^Y)$ and $\mathbb{V}^0(\overline{a}^0) \ge \mathbb{V}^0(\overline{a}^Y)$, we obtain the inequalities

$$\Delta\mathbb{V}^Y(\overline{a}^0) \le \overline{\Delta\mathbb{V}}^Y \le \Delta\mathbb{V}^Y(\overline{a}^Y) . \tag{8.10}$$

This comparison conveys the trade-off made by a decision-maker when he chooses his initial decision, depending on whether he expects or does not expect to observe a signal in the future. A decision maker expecting to observe a signal chooses an initial decision that increases the value of information above its early value. Conversely, a decision maker who does not expect to receive information selects a decision that brings the value of future information lower than what he would have been ready to pay early on, if offered to.

[7] One may also note that since in a game against nature the value of future information is nonnegative, $\Delta\mathbb{V}^Y(a) \ge 0$, then the coefficient of $(a+b^\sharp)^2$ is nonnegative.

Note also that the above comparison generalizes the relation between the initial value of information and the quasi-option value given by (17), who defines option value as $\mathbb{V}^Y(\overline{a}^Y) - \mathbb{V}^Y(\overline{a}^0)$ for a family of problems where $\mathbb{V}^Y(\overline{a}^0) = 0$.

8.3 Value of information as a key to the literature

A goal of the literature on irreversibility and uncertainty consists in identifying hypotheses or conditions under which it is possible to compare efficient decisions made with different information structures. Two kinds of conditions can be examined. A first thread follows (10) and concentrates on determining the direction of the effect of learning for all possible pairs of signals where one is more informative than the other. As noted in (28), this restricts the conclusion to limited classes of problems, for example those later identified by (15). An other thread looks for specific problems where some kind of comparison is possible though Epstein's conditions do not apply, as in (28).

In the above literature, the ranking of optimal initial decisions is obtained through conditions which are expressed by means of one of the two solutions. We show here how such conditions may be bypassed by making use of the monotonicity of the value of information.

8.3.1 Epstein (1980)

Epstein's Theorem in (10) is stated for discrete random variables X and Y. The random variable X is supposed to take its value in $\{x_1,\dots,x_m\}$. Any distribution ρ on $\{x_1,\dots,x_m\}$ is identified with an element of the simplex $\mathcal{S}^{m-1}$. Following Epstein, let us define for all $\rho \in \mathcal{S}^{m-1}$

$$J(a,\rho) := \sup_{b\in\mathbb{B}(a)} \mathbb{E}_\rho\left[U(a,b,x)\right] = \sup_{b\in\mathbb{B}(a)} \int_{\mathbb{X}} U(a,b,x)\rho(dx)\,. \tag{8.11}$$

Consider two signals Y and Y', taking values from the finite set $\mathbb{Y} = \{y_1,\dots,y_n\}$. Suppose that Y is more informative[8] than Y'. Suppose also that $J(a,\rho)$ is concave and differentiable in a, and that the programs $\max_a \mathbb{V}^Y(a)$ and $\max_a \mathbb{V}^{Y'}(a)$ have unique optimal solutions $\overline{a}^Y$ and $\overline{a}^{Y'}$.

Epstein's Theorem states that, if $\frac{\partial J}{\partial a}(\overline{a}^Y,\rho)$ is concave (resp. convex) in ρ, then $\overline{a}^Y \leq \overline{a}^{Y'}$ (resp. $\geq$). If $\frac{\partial J}{\partial a}(\overline{a}^Y,\rho)$ is neither concave nor convex, the ranking is ambiguous.

[8] A signal Y is said to be *more informative* than a signal Y' if the σ-field $\sigma(Y) := Y^{-1}(\mathcal{Y})$ contains $\sigma(Y')$. When Y' takes value in a separable complete metric space, a result due to Doob (8, Ch. 1, p. 18) states that it is equivalent to say that Y' is a measurable function of Y, namely $Y' = f(Y)$ where $f : (\mathbb{Y},\mathcal{Y}) \to (\mathbb{Y}',\mathcal{Y}')$.

Epstein's Theorem and our result are related by the following formulas. Denote by ν and ν' the unconditional distributions of the signals Y and Y' on $\mathbb{Y}$. For each $y \in \mathbb{Y}$, the conditional distribution of X knowing $Y = y$ belongs to the simplex $\mathscr{S}^{m-1}$ and we denote it by $\mathbb{P}_X^{Y=y}$. We have:

$$\begin{aligned}
\mathbb{V}^Y(a) &= \mathbb{E}_{\mathbb{P}}\big[\max_{b\in\mathbb{B}(a)} \mathbb{E}_{\mathbb{P}}[U\big(a,b,X\big) \mid Y]\big] \text{ by definition (8.2)} \\
&= \sum_{i=1}^{n} \nu\{y_i\} \max_{b\in\mathbb{B}(a)} \mathbb{E}_{\mathbb{P}_X^{Y=y_i}}[U\big(a,b,x\big)] \text{ by using the conditional distribution} \\
&= \mathbb{E}_\nu\big[\max_{b\in\mathbb{B}(a)} \mathbb{E}_{\mathbb{P}_X^Y}[U\big(a,b,x\big)]\big] \text{ where } \mathbb{E}_\nu[\phi(\mathbb{P}_X^Y)] = \sum_{i=1}^{n} \nu\{y_i\}\phi(\mathbb{P}_X^{Y=y_i}) \\
&= \mathbb{E}_\nu\big[J(a,\mathbb{P}_X^Y)\big] \text{ by (8.11).}
\end{aligned}$$

We deduce that[9]

$$\frac{d\Delta\mathbb{V}^{YY'}(a)}{da} = \mathbb{E}_\nu[\frac{\partial J}{\partial a}(a,\mathbb{P}_X^Y)] - \mathbb{E}_\nu[\frac{\partial J}{\partial a}(a,\mathbb{P}_X^{Y'})]\,. \tag{8.12}$$

Recall that if Y is more informative[10] than Y', then $\mathbb{E}_\nu[\phi(\mathbb{P}_X^Y)] \geq \mathbb{E}_{\nu'}[\phi(\mathbb{P}_X^{Y'})]$ for all convex function ϕ on the simplex (3). Thus, the Epstein's Theorem assumption that $\frac{\partial J}{\partial a}(\overline{a}^Y,\rho)$ is concave in ρ implies that $\frac{d\Delta\mathbb{V}^{YY'}}{da}(\overline{a}^Y) \leq 0$. In other words, the value of future information is locally decreasing (or stationary) at the initial optimal decision for the more informed agent.

Epstein's Theorem may be elaborated upon when restating that, under its assumptions, the value of future information is locally decreasing at $\overline{a}^{Y'}$, and the more informed agent takes a more precautionary decision $\overline{a}^Y \leq \overline{a}^{Y'}$.

Note that Epstein's proof assumes concavity of $J(a,\rho)$ in a, which allows to have a local assumption at $\overline{a}^Y$. In contrast, our main result makes a global assumption on the monotonicity of $a \mapsto \Delta\mathbb{V}^{YY'}(a)$. However, we need not have concavity, differentiability nor discrete random variables.[11]

8.3.2 Freixas and Laffont (1984)

Freixas and Laffont also study in (14) the monotonicity of $a \mapsto \Delta\mathbb{V}^{YY'}(a)$ for the additive separable case where $U(a,b,x) = U_1(a) + U_2(b,x)$. However, they do not provide the interpretation of $\Delta\mathbb{V}^{YY'}(a)$ in terms of value of substituting information structures. Their proofs depend on quasi-concavity assumptions.

---

[9] All appropriate regularity and integrability assumptions are supposed to hold true.

[10] See footnote 8

[11] The derivation of the formula $\mathbb{V}^Y(a) = \mathbb{E}_\nu\big[J(a,\mathbb{P}_X^Y)\big]$ remains valid when X and Y are random elements taking values from a Borel space (4).

Kolstad obtains necessary and sufficient conditions for a problem in (21) which is actually a sub-case of Freixas and Laffont though this does not appear at first glance from his notations but has to be derived from his hypotheses.

8.3.3 'All or nothing' decision set

The seminal literature, as well as more recent contributions, often considers linear dynamics and utilities — which imply all or nothing decisions — or hinges directly on a binary decision set (see for instance (2), (18), (16), (11) and (19, part 2)). With a binary decision set, the monotonicity of the value of information becomes trivial. Moreover, the direction of variation is easily determined under the hypothesis of total irreversibility, *i.e.* when one of the two possible initial decisions affects the second period utility so that it does not depend any longer on the second period decision. This is for example the case with the model of (2).

8.3.4 Ulph and Ulph (1997)

Ulph and Ulph developed a simple model of global warming where Epstein's conditions cannot apply. They proposed a specific condition that implies the irreversibility effect. We show that their assumptions imply the monotonicity of the second-period value of information and can be generalized to any information structure.

The model examined in (28) is $U(a,b,x) = U_1(a) + U_2(b-a) - xD(b)$, with an irreversibility constraint $\mathbb{B}(a) = [a, +\infty[$. In this model, b is a proxy for the atmospheric concentration of greenhouse gases and a and $b-a$ are proxies for greenhouse gases emissions. X is an uncertain nonnegative factor multiplying damages $D(b)$ resulting from the accumulation of greenhouse gases in the atmosphere. Functions U_1 and U_2 are supposed strictly concave increasing, and D strictly convex increasing. With our notations, their Theorem 3 states that if $(\overline{a}^0, \overline{b}^0)$ is such that the irreversibility constraint bites ($\overline{b}^0 = \overline{a}^0$), then $\overline{a}^Y \leq \overline{a}^0$.

Two features are essential to this result. First, the assumption that the optimal policy, $\overline{b}^0 = 0$, is a corner solution in the second period. Second, the shape of the payoff, which is linear in the random variable. In fact, in Ulph and Ulph's model, the condition $\overline{b}^0 = 0$ implies that the second-period value of any signal Y is a decreasing function for $a \geq \overline{a}^0$ (the proof[12] is given in Appendix 8.6.2). As a consequence, $\overline{a}^Y \leq \overline{a}^0$ for any signal Y.

[12] The intuition for monotonicity is as follows. The condition $\overline{b}^0 = \overline{a}^0$ implies that, when no information is available, it is optimal to cut emissions to zero in second period if the greenhouse gases concentration is above a certain threshold. Conversely, when information is obtained, it might open the opportunity to emit. The value of the information is then equal to the benefit of additional emissions in second period minus the expected additional damages. From the envelope theorem, these expected additional damages are strictly increasing at the margin for a small increase of concen-

8.4 Numerical application to a climate change model

We shall now provide a numerical illustration of the above approach with a stochastic version of the integrated-assessment model DICE 98 (23; 24). Uncertainty enters the model *via* the *climate sensitivity* which is now a random variable X with values 2.5 °C, 3.5 °C and 4.5 °C. Perfect information about the climate sensitivity is assumed to arrive in second period, which begins here in 2040. The model is a stochastic optimal-growth model of the world economy, where the discounted expected value of logarithmic utility from consumption is maximized. Decision b is a vector that defines the rates of investment and the rates of emissions allowance [13] from and after 2040. Decision a is the rate of emission allowances in 2030-2039 and is used to define the investment and emissions trajectory from now until 2039[14] .

The detailed climate-economy equations are slightly changed from the original version of DICE. Following (1), the temperature equation is an updated calibration and a threshold damage function replaces the original quadratic one. The full description for the original DICE model can be found in (23) or (24).

The Figure 8.2 plots the expected value of information as a function of the initial decision. Available initial decisions range from no effort until 2039 (100% emissions allowance) to targeting the maximum effort in 2039 (0% allowances). Three cases are presented corresponding to three different probability distributions for X: optimistic case (2.5 °C [2/3], 3.5 °C [1/3] and 4.5 °C [0]), centered case ([1/3,1/3,1/3]) and pessimistic case ([0,1/3,2/3]).

In all cases, the prospect of learning the true value of X in 2040 is an opportunity to allow initially more emissions than in the never-learn situation ($\overline{a}^Y > \overline{a}^0$). This is usually found in the empirical litterature (23), (28), (1). As far as we know, empirical models like this one remain out of bounds for the existing analytical literature about irreversibility, learning and climate change, which would offer no interpretation for such results. However, in the three cases of Figure 8.2, the direction of the learning effect can be related to the monotonicity of the expected value of information, which is strictly increasing with initial emissions allowances.

8.5 Conclusion

This article puts forward the relevance of the concept of value of information for the comparison of sequential decision problems. The difference between value of future information before and after an initial decision is taken was made explicit. The

---

tration, whereas benefits do not directly depend on concentration. As a consequence, the value of information decreases.

[13] The rate of allowance is one minus the rate of abatement of greenhouse gases emissions. This formulation is taken in order to keep the same notations as in Ulph and Ulph, where the control decision is the level of emissions rather than the level of abatement.

[14] The parameterization of the trajectories before 2039 is made in order to reproduce the optimal decisions in the DICE model under under different scenarios for the climate sensitivity.

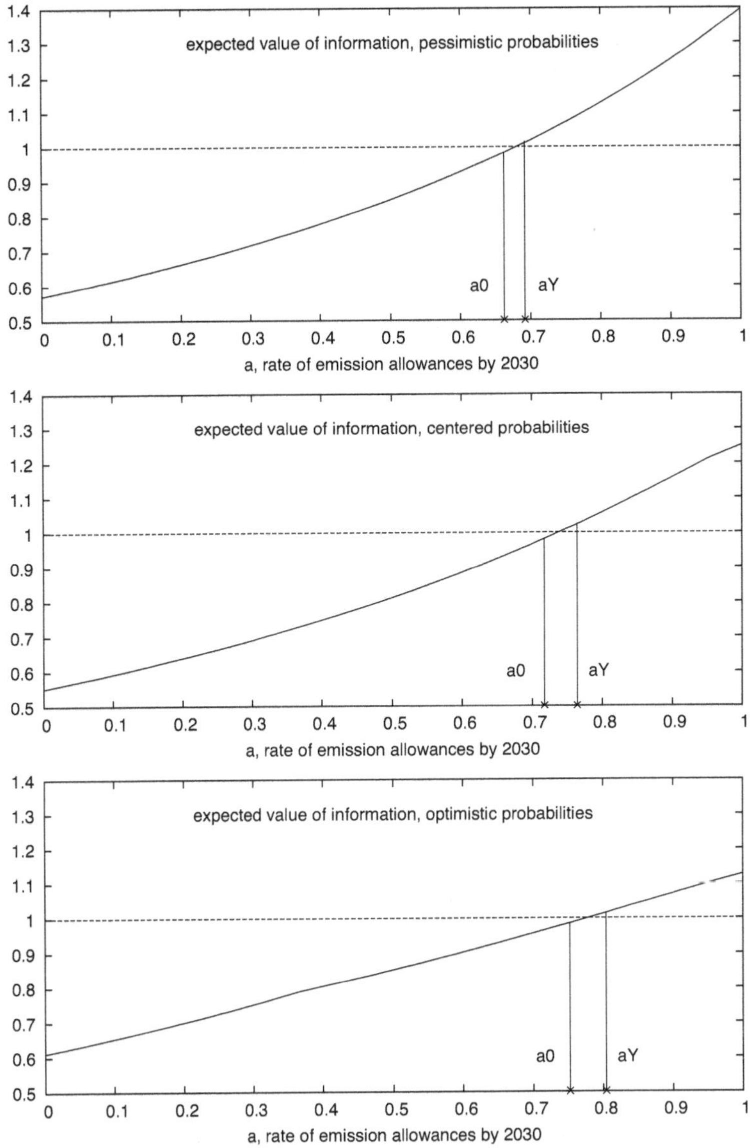

Fig. 8.2 Expected value of information in 2040 as a function of initial policy.
Initial policy is summarized by a, the rate of emissions allowance in 2030–2039. In each case, expected value of information increases with the emission allowances: the more we emit before 2040, the more we are willing to pay for information on the climate in 2040. As a consequence, the optimal initial policy with future learning, $\overline{a}^Y$, allows more emissions than the optimal initial policy without future learning, $\overline{a}^0$. The expected value of information has been normalized with $\overline{\Delta\mathbb{V}}^Y$, the early value of future of information (i.e. before any decision is made). Notice that this normalization is different in each case. It also illustrates that $\overline{\Delta\mathbb{V}}^Y$ is bounded above by the expected value of information after decision $\overline{a}^Y$ and bounded below by the expected value of information after decision $\overline{a}^0$.

second-period value of information should be viewed as a function of the initial decision and its monotonicity is sufficient for making a conclusion about the direction of the learning effect (precautionary or not). Interestingly, many of the conditions given in the literature about the irreversibility effect can be related to this monotonicity. On the one hand, our condition of monotonicity is a global one – not requiring to know one optimal solution in advance – while the above conditions are generally local – expressed at one of the optimal solutions. On the other hand, our framework requires less assumptions: no concavity or quasi-concavity assumptions, but only existence of a maximum; no optimal initial decision need to be known; the random variables and sample space are not necessarily discrete; the signals are not necessarily comparable. In particular, since it avoids standard, restrictive assumptions on the utility and on the environment (like linear evolution or scalar dimension), it is general enough to be applied to the study of numerical integrated assessment models, like DICE (24). Formally, it is not restricted to environmental problems.

Acknowledgements Research reported in this paper has been supported by a grant from the Institut Français de l'Energie (IFE). Special thanks to anonymous referees.

8.6 Appendix

8.6.1 General results on comparison of arg max

We recall here some results on comparisons between the arg max of two optimization problems. They may be seen as particular instances of results from a general theory with supermodular functions or functions with increasing differences, as developed in (27).

Proposition 2. *Let* $\mathbb{D} \subset \mathbb{R}$*, and let* $g : \mathbb{D} \to \mathbb{R}$ *and* $h : \mathbb{D} \to \mathbb{R}$ *be two functions having domain* $\mathbb{D}$*. We introduce the following set of maximizers*

$$\mathbb{D}_g := \arg\max_{u \in \mathbb{D}} g(u) \quad \textit{and} \quad \mathbb{D}_{g+h} := \arg\max_{u \in \mathbb{D}} (g+h)(u) \ ,$$

that we suppose non empty: $\mathbb{D}_g \neq \emptyset$ *and* $\mathbb{D}_{g+h} \neq \emptyset$*.*

1. *If h is strictly increasing on* $]-\infty, \sup \mathbb{D}_g] \cap \mathbb{D}$*, then*

$$\sup \mathbb{D}_g \leq \inf \mathbb{D}_{g+h} \, .$$

2. *If h is increasing on* $]-\infty, \sup \mathbb{D}_g] \cap \mathbb{D}$*, then*

$$\sup \mathbb{D}_g \leq \sup \mathbb{D}_{g+h} \, .$$

3. *If h is strictly decreasing on* $[\inf \mathbb{D}_g, +\infty[\cap \mathbb{D}$*, then*

$$\sup \mathbb{D}_{g+h} \leq \inf \mathbb{D}_g .$$

4. *If h is decreasing on $[\inf \mathbb{D}_g, +\infty[\cap \mathbb{D}$, then*

$$\inf \mathbb{D}_{g+h} \leq \inf \mathbb{D}_g .$$

Proof. We prove the first statement, the others being minor variations.

Let $u_g^\sharp \in \mathbb{D}_g$. For any $u \in \mathbb{D}$, we have $g(u) \leq g(u_g^\sharp)$. For any $u \in]-\infty, u_g^\sharp[\cap \mathbb{D}$, we have $h(u) < h(u_g^\sharp)$ if h is strictly increasing. Thus

$$u \in]-\infty, u_g^\sharp[\cap \mathbb{D} \Rightarrow g(u) + h(u) < g(u_g^\sharp) + h(u_g^\sharp) .$$

We conclude that $\mathbb{D}_{g+h} \subset [u_g^\sharp, +\infty[$, so that

$$\mathbb{D}_{g+h} \subset \bigcap_{u_g^\sharp \in \mathbb{D}_g} [u_g^\sharp, +\infty[= [\sup \mathbb{D}_g, +\infty[,$$

which proves that $\sup \mathbb{D}_g \leq \inf \mathbb{D}_{g+h}$.

8.6.2 Extension of Ulph and Ulph's result (1997)

Without learning, the second optimal decision $\overline{b}^0 = \overline{a}^0$ solves Euler's equation $U_2'(0) - \mathbb{E}[X]D'(\overline{a}^0) \leq 0$. Since D' is increasing, we deduce that $U_2'(0) - \mathbb{E}[X]D'(a) \leq 0$, for all $a \geq \overline{a}^0$. Hence, by Euler's condition, the irreversibility constraint bites, and we deduce that $\mathbb{V}^0(a) = \mathbb{E}[U_2(0) - XD(a)] = -\mathbb{E}[X]D(a)$ for all $a \geq \overline{a}^0$. Without learning, we obtain

$$\frac{d}{da} \max_{b \geq a} \mathbb{E}[U_2(b-a) - XD(b) \mid Y] = -\mathbb{E}[X \mid Y]D'(\overline{b}(a,Y))$$

by the Danskin theorem (see (5)), where $\overline{b}(a,Y) \geq a$ achieves the maximum. Since $X \geq 0$ and $-D'$ is decreasing, we obtain that $\frac{d}{da} \max_{b \geq a} \mathbb{E}[U_2(b-a) - XD(b) \mid Y] \leq -\mathbb{E}[X \mid Y]D'(a)$. By taking the expectation, this gives $\frac{d}{da}\mathbb{V}^Y(a) \leq -\mathbb{E}[X]D'(a)$. Since D is increasing, then $-D' \leq 0$, and we conclude that $\frac{d}{da}\Delta\mathbb{V}^Y(a) \leq 0$: the value of future information is decreasing for all a greater than $\overline{a}^0$.

References

[1] Philippe Ambrosi, Jean-Charles Hourcade, Stéphane Hallegatte, Franck Lecocq, Patrice Dumas, and Minh Ha-Duong. Optimal control models and elicitation of attitudes towards climate change. *Environmental Modeling and*

Assessment, 8(3):135–147, 2003. Special Issue on Modeling the economic response to global climate change.

[2] K. J. Arrow and A. C. Fisher. Environmental preservation, uncertainty, and irreversibity. *Quarterly Journal of Economics*, 88:312–319, 1974.

[3] Zvi Artstein. Gains and costs of information in stochastic programming. *Annals of Operations Research*, 85:128–152, 1999.

[4] D. P. Bertsekas and S. E. Shreve. *Stochastic Optimal Control: The Discrete-Time Case*. Athena Scientific, Belmont, Massachusets, 1996.

[5] Frank H. Clarke. *Optimization and nonsmooth analysis*, volume 5 of *Classics in Mathematics*. SIAM, 1990.

[6] Jon M. Conrad. Quasi-option value and the expected value of information. *Quarterly Journal of Economics*, 94:813–820, 1980.

[7] M. De Lara and L. Doyen. *Sustainable Management of Natural Resources. Mathematical Models and Methods*. Springer-Verlag, Berlin, 2008.

[8] C. Dellacherie and P.A. Meyer. *Probabilités et potentiel*. Hermann, Paris, 1975.

[9] Louis Eeckhoudt, Christian Gollier, and Nicolas Treich. Optimal consumption and the timing of the resolution of uncertainty. *European Economic Review*, 49(3):761–773, April 2005.

[10] L. G. Epstein. Decision making and temporal resolution of uncertainty. *International Economic Review*, 21:269–283, 1980.

[11] Anthony C. Fisher. Investment under uncertainty and option value in environmental economics. *Resource and Energy Economics*, 22:197–204, 2000.

[12] Anthony C. Fisher and W. M. Hanemann. Quasi option value: Some misconceptions dispelled. *Journal of Environmental Economics and Management*, 14:183–190, 1987.

[13] Anthony C. Fisher and Urvashi Narain. Global warming, endogenous risk, and irreversibility. *Environmental and Resource Economics*, 25:395–416, 2003.

[14] Xavier Freixas and Jean-Jacques Laffont. The irreversibility effect. In Marcel Boyer and R. Kihlström, editors, *Bayesian Models in Economic Theory*, chapter 7. North-Holland, Amsterdam, 1984.

[15] Christian Gollier, Bruno Jullien, and Nicolas Treich. Scientific progress and irreversibility: an economic interpretation of the "precautionary principle". *Journal of Public Economics*, 75:229–253, 2000.

[16] Minh Ha-Duong. Quasi-option value and climate policy choices. *Energy Economics*, 20:599–620, 1998.

[17] W. M. Hanemann. Information and the concept of option value. *Journal of Environmental Economics and Management*, 14:183–190, 1989.

[18] C. Henry. Investment decisions under uncertainty: The "irreversibility effect". *American Economic Review*, 64(6):1006–1012, 1974.

[19] C. Henry. Option values in the economics of irreplaceable assets. *Review of Economic Studies*, 41:89–104, 1974.

[20] Robert A Jones and Joseph M Ostroy. Flexibility and uncertainty. *Review of Economic Studies*, 51(1):13–32, January 1984.

[21] Charles D. Kolstad. Fundamental irreversibilities in stock externalities. *Journal of Public Economics*, 60:221–233, 1996.
[22] Alan S. Manne and Richard Richels. *Buying Greenhouse Insurance: The Economic Cost of CO_2 Emissions Limits*. MIT Press, 1992.
[23] W. D. Nordhaus. *Managing the Global Commons*. MIT Press, Cambridge, 1994.
[24] William D. Nordhaus and Joseph Boyer. *Warming the World: Economics Models of Global Warming*. MIT press, 2000. URL `http://www.econ.yale.edu/~nordhaus/homepage/dice\_section\_I.html`.
[25] Robert S. Pindyck. Irreversibilities and the timing of environmental policy. *Resource and Energy Economics*, 22:233–259, 2000.
[26] Sébastien Rouillon. Catastrophe climatique irréversible, incertitude et progrès de la connaissance. *Revue Économique*, 52(1):61–90, January 2001.
[27] D. M. Topkis. *Supermodularity and Complementarity*. Princeton University Press, Princeton, N.J., 1998.
[28] A. Ulph and D. Ulph. Global warming, irreversibility and learning. *The Economic Journal*, 107(442):636–650, 1997.

Chapter 9
Comparative Forecasting and a Test for Persistence in the El Niño Southern Oscillation

Belinda A. Chiera, Jerzy A. Filar, Daniel S. Zachary, Adrian H. Gordon

Abstract We present an analysis of two separate single-indicator forecasting methods for the El Nino Southern Oscillation based on oscillation persistence. We use the Southern Oscillation Index (SOI) to produce short term 5 month forecasts and a Bayesian approach to explore SOI persistence, with results compared to a benchmarking Taylor Series expansion. We find signal persistence is important when forecasting more than a few months and the models presented may provide a relatively simple approach to environmental risk forecasting in situations where the underlying phenomenon exhibits substantial persistence.

9.1 Introduction

The El Niño Southern Oscillation (ENSO) is a global phenomenon manifesting itself in fluctuations of the ocean and atmosphere, and is widely believed to be the main source of interannual climate variability (36). The quasi-periodic timing of ENSO events sees 12-to-18 month periods of either warm or complementary cold states of ENSO, which, when particularly strong, have the potential to cause catastrophic climatic conditions. The flow-on effects of these conditions are far reaching; weather sensitive industries including agriculture, commercial fishing, construction, and tourism are disrupted, making ENSO-related economic losses estimated in the

Belinda A. Chiera
University of South Australia, Australia, e-mail: belinda.chiera@unisa.edu.au

Jerzy A. Filar
University of South Australia, Australia, e-mail: jerzy.filar@unisa.edu.au

Daniel S. Zachary
Centre de Ressources des Technologies pour l'Environnement (CRTE), Luxembourg e-mail: dan.zachary@tudor.lu

Adrian H. Gordon (posthumous)

J.A. Filar and A. Haurie (eds.), *Uncertainty and Environmental Decision Making*, International Series in Operations Research & Management Science 138,
DOI 10.1007/978-1-4419-1129-2_9,

billions, with the potential for damage now in the trillions (4). On the flip side of the same coin, ENSO has also proven economically beneficial for the very same industries (4). Environmental concerns however are not as fortunate. In particular water conservation is of great importance to many countries facing severe drought conditions while others have to ward off the advent of flooding. Even higher are the unquantifiable losses incurred by ENSO: human losses as well as those suffered by communities without the benefit of precautionary notices are often the hardest hit. For these reasons, a mechanism to provide long lead times used to develop strategies for dealing with the anticipated impacts of ENSO are necessary to mitigate adverse impacts as well as to enhance the benefits where possible.

Providing ENSO forecasts has become a regular service of a number of leading global climate centres, with a number of these centres forming an important component of current day policy making (see for instance (20; 3)). The models used to generate the forecasts are quite complex, and include a number of interrelated ocean-atmosphere dynamics, with probability distributions often accompanying these forecasts (3). However as these dynamics are still poorly understood, it is still questionable whether the trade-off between model complexity and manageability translates into forecast quality.

Probabilistic forecasting has long been considered in the literature for a range of climatic conditions. If the dynamics of the system are not well understood, as is the case with ENSO, it is perhaps more reasonable to utilise a simplified probabilistic forecast. In particular, a forecast based on the properties of the ENSO signal alone may further aid in providing enhanced forecast quality in exchange for relatively little overhead. A probabilistic learning methodology, such as Bayesian-based modelling and forecasting, may prove useful in this instance.

In this paper we propose to model ENSO using a Bayesian forecast based on the Southern Oscillation Index (SOI), a signal which has long been used to characterise ENSO (34). Our technique, which we call the Bayesian Binary Tree (BBT) model, possesses a distinctive feature in that it is tailored to model ENSO when exhibiting two distinct phases, namely the El Niño and La Niña (respectively warm and cold) events. The BBT model is developed on the basis of an appropriate digitisation of the SOI into a binary signal, rather than the SOI itself, which has an aperiodic and oscillatory nature (16). A further advantage of digitising the SOI data is that the *persistence* of the ENSO, as captured by the SOI, is clearly visible in the binary signal and can be exploited to develop Bayesian forecasts that take into account the historical behaviour of the SOI.

We accompany the BBT model with an adjoining methodology used to forecast the original SOI data. Specifically, we develop a method for the probabilistic reconstruction of the forecast's binary signal into a forecast of the SOI index. We further supplement these SOI forecasts with a benchmarking procedure based on a Taylor series-type approximation of the original SOI signal.

To date, there are comparatively few applications of the Bayesian methodology to the El Niño Southern Oscillation (ENSO) phenomenon. Previously, there has been some research into identifying the dynamical regimes of ENSO using a stochastic Bayesian methodolgy (35), however, that approach uses statistical inverse modelling to reproduce key statistical characteristics of a coupled tropical Pacific model, which differs widely from the aims of this paper. In (18) a Bayesian approach for modelling the apparent teleconnection between ENSO and the Indian summer monsoon rainfall is presented. In this instance, a Bayesian dynamic linear model is developed with the aim of capturing the dynamic relationship between large-scale circulation indices and monthly variation of the Indian summer monsoon rainfall.

Other earlier works include that of (28) which analysed ENSO under an extended framework umbrella encompassing meteorology, plant science and economics, by utilising Bayesian decision analysis techniques. In (26) the authors describe the climatic system in the tropical Indian and Pacific basins using a statistical technique called Bayesian oscillation patterns, however, that work was focused on forecasting zonal wind and sea surface temperature in the Niño3 region rather than ENSO itself. Finally, the method given in (6) is the closest in spirit to the analysis we will consider here, as empirical forecasts were used in conjunction with Bayesian modelling techniques, however, in this instance the authors were concerned with developing a Bayesian methodology for combining empirical forecasts with raw, bias uncorrected coupled model ensemble forecasts, to produce skillful ENSO interval forecasts. These outcomes differ substantially from the aims of the work presented here in that firstly, in (6) SST Niño-3.4 index forecasts are used whereas we are interested in forecasting the oscillatory nature of the SOI, and secondly, there is no discussion of predicting the persistency of ENSO, rather the interest was solely focused on predicting the indices themselves.

This contribution is outlined as follows. In Section 9.2 we present a short analysis of the digitised SOI data, focusing on the pertinent characteristics of the resulting binary signal, namely persistence and the predictability barrier. In Section 9.3 we characterise the binary signal into *Up* and *Down* events and derive the BBT model, which then forms the basis for forecasting the original SOI data in Section 9.4. In Section 9.5 we present a benchmark against the forecasts made using the BBT model. We follow this with a comparison of the predictability power of the TSF and BBT indicators in Section 9.6 and discuss the results of the forecasting powers of these indicators in Section 9.7.

9.2 Analysis of the SOI data series

In what follows, we discuss the Southern Oscillation Index data series and the prevalence of persistent behaviour in this index.

9.2.1 The SOI data series

We obtained data for the Troup Southern Oscillation Index (Troup SOI) from the Australian Bureau of Meteorology (3). The Troup SOI is defined as the standardised anomaly of the Mean Sea Level Pressure (MSLP) difference measured at Tahiti and Darwin, calculated according to Troup (34; 3)

$$SOI = 10 \cdot \frac{P_d - \overline{P}_d}{\sigma_{P_d}} \tag{9.1}$$

where for any given month, P_d is the difference between the average Tahiti MSLP and Darwin MSLP, $\overline{P}_d$ is the long term average of P_d and σ_{P_d} is the long term standard deviation of P_d. The scaling factor of 10 is used merely as a convention. The index is interpreted such that a Troup SOI of -10, for example, means the SOI is 1 standard deviation on the negative side of the long-term mean, whereas a SOI of $+10$ means the SOI is 1 standard deviation on the positive side of the long term mean, and so forth. In terms of the physical interpretation of ENSO, a strongly negative SOI value is indicative of an El Niño event

whereas a strongly positive SOI value is representative of a La Niña event.

The monthly Troup SOI data ranges from 1876 – 2008. However, for the analysis presented here, we will use the monthly index values from 1900 – 2008 due to reliability issues. A plot of the monthly Troup SOI from January 1900 - December 2008 is shown in Figure 9.1. Seemingly, the SOI-based El Niño Southern Oscillation exhibits aperiodic and chaotic oscillatory dynamic behaviour, corroborating the results given in (16). An analogy given in (10) likened the SOI to a forced oscillation with damping in a noisy environment. At first sight, the SOI data in Figure

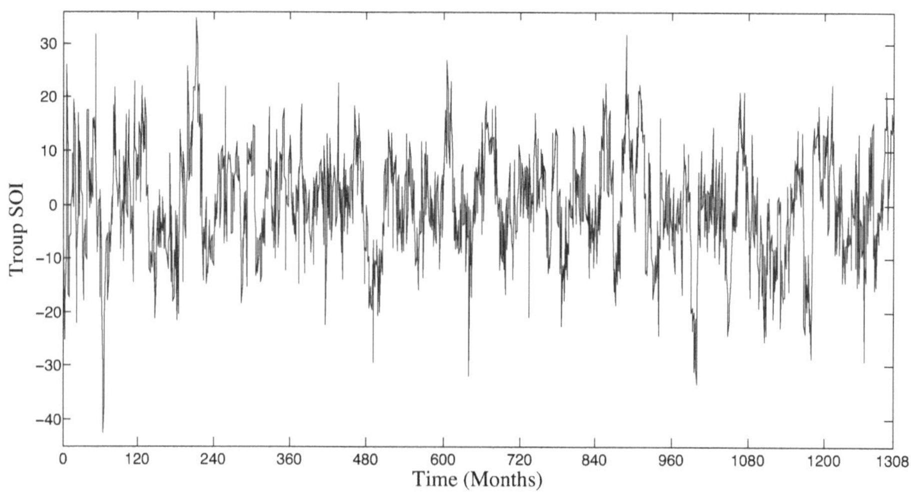

Fig. 9.1 The monthly Troup SOI data, corresponding to the time period January 1900 - December 2008.

9.1 suggest that devising a model that adequately simulates the raw signal is an extremely difficult undertaking. However, an exploratory analysis of the data reveals a number of interesting characteristics pertaining to some level of *persistence* in the index values, a characteristic which we propose to model first in place of the actual index values themselves. In particular, the SOI data show that once an oscillation cycle begins, that is when either the SOI is in a positive or negative state for a small number of consecutive months, there is a tendency to remain in that particular state for several months further.

Specifically, the underlying dynamical forces driving El Niño, although unknown in detail, display a probability in persistence, as shown in Figure 9.2, in which the probability distribution for persistent events (– •), derived from the SOI data, overlays that of the Binomial probability distribution (—). We have selected a Binomial distribution with probability $p = 0.5$ and the number of repetitions as depicted along the x-axis, where repetitions are being used to denote monthy persistence. Since $p = 0.5$ there is an equal probability of being in either a positive or negative state. We see from Figure 9.2 that the SOI-based probability distribution (–•) is

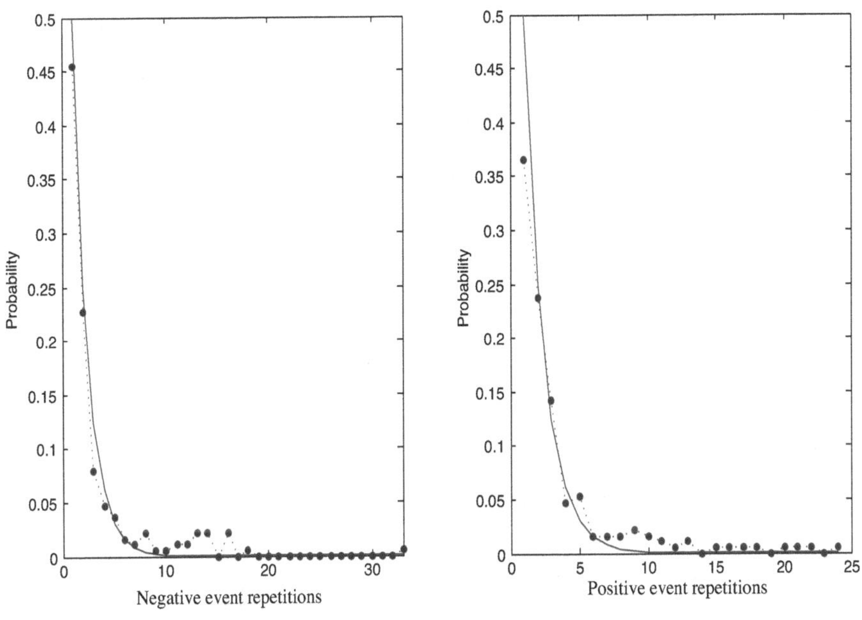

Fig. 9.2 SOI probability distributions (– •) compared to the Binomial distribution with probability $p = 0.5$ (—).

clearly augmented behaviour over the Binomial distribution (—), particularly when considering persistence of length between 10-20 months for negative values of the SOI and 8-20 months for positive values of the SOI. In both cases the probability of persistence decreases to zero by 20 months, suggesting an upper limit on realistic empirically-based predictions of ENSO. From Figure 9.2 we suggest a forecast horizon of length between $10-20$ months. Such an empirical finding is supported by (3; 15) and an ENSO forecasting predictability limit of 13 months has also previously been suggested (32).

In order to highlight the persistence properties of the SOI data, which are otherwise obfuscated when using the original signal, we instead digitise the data into a ± 1 signal. Here $+1$ is representative of a positive (La Niña) event and -1 a negative (El Niño) event. Figure 9.3 shows the digitised monthly data, from which we see it is easier to discern periods of persistence than when considering the original data in Figure 9.1. Note that in this case persistence is indicated by the lighter bands whereas periods of rapid oscillation between the two states are represented by the darker bands. We can readily see from the digitised data there are a number of persistence periods of varying lengths (measured in months) which we will use to form the basis of the Bayesian forecasting method that follows.

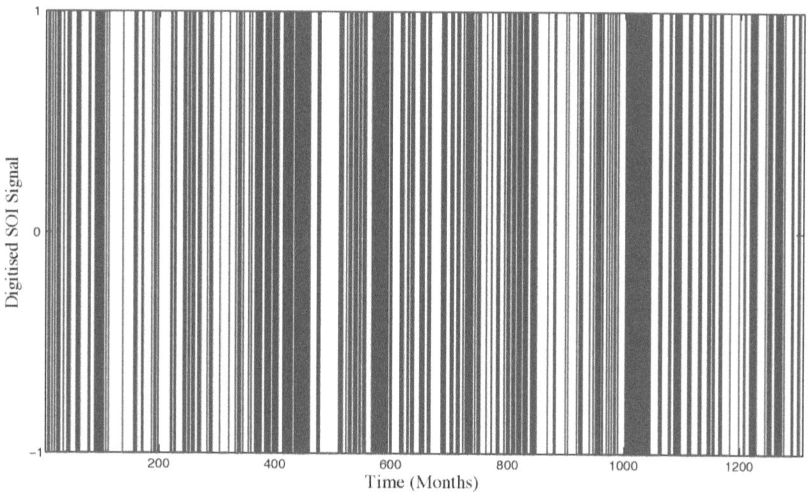

Fig. 9.3 The digitised monthly Troup SOI data, corresponding to the time period January 1900 - December 2008. Positive index values are assigned a value of +1 and negative index values are assigned a value of -1. The light bands highlight persistence and the darker bands highlight oscillation between the two states.

9.3 The Bayesian Binary Tree Model

In this section we develop the first of two forecasting models of the SOI data. Our first model is a probabilistic forecasting model based on a Bayesian analysis of the digitised SOI signal, a method originally explored in (5). Once the digitised signal has been forecast we will convert this forecast to values of the Troup index. An attractive advantage of a Bayesian model over other probability based models (for example Markov Chain models), is that a Bayesian model is a learning algorithm in that it is continually updated with new information about the data series in question. We can exploit this property by noting from the SOI probability distributions (Figure 9.2) that the probability of observing, for example, a fifth positive event, given four consecutive positive events have already occurred, will almost certainly be different to the probability of observing a sixth positive event directly following five consecutive positive events. It is precisely this type of probabilistic persistence which we will use to forecast ENSO episodes using the Bayesian methodology.

We begin by introducing the concept of *Up* and *Down* events in the ENSO oscillation cycle. If the digitised SOI data takes on the value $+1$, then we refer to the state of the oscillation as being *Up*, denoted by $\mathscr{U}$, and if the digitised value is -1 the oscillation is in a *Down* state, denoted by $\mathscr{D}$. If data are set aside from the original series to provide 'historical' information of the SOI data, we can use counts of *Up* and *Down* events to compute probabilistic persistence of the SOI in the following way.

We first define a counting scheme over the historical data where we count the number of *Up* episodes such that

$$
\begin{aligned}
\mathscr{U}_1 &= \text{Number of Up episodes of length 1} \\
\mathscr{U}_2 &= \text{Number of Up episodes of length 2} \\
\vdots \;\; & \;\; \vdots \\
\mathscr{U}_{\mathbf{u}} &= \text{Number of Up episodes of length } \mathbf{u}
\end{aligned}
$$

where $\mathbf{u}$ is the length of the longest recorded $\mathscr{U}$-episode. We then similarly define for *Down* episodes

$$
\begin{aligned}
\mathscr{D}_1 &= \text{Number of Down episodes of length 1} \\
\mathscr{D}_2 &= \text{Number of Down episodes of length 2} \\
\vdots \;\; & \;\; \vdots \\
\mathscr{D}_{\mathbf{d}} &= \text{Number of Down episodes of length } \mathbf{d}
\end{aligned}
$$

where $\mathbf{d}$ is the length of the longest recorded $\mathscr{D}$ episode. We can also define the total number of *Up* and *Down* episodes as

$$
\begin{aligned}
\mathscr{U}_T &= \sum_{i=1}^{\mathbf{u}} \mathscr{U}_i \\
\mathscr{D}_T &= \sum_{i=1}^{\mathbf{d}} \mathscr{D}_i
\end{aligned}
$$

respectively, which naturally partitions the historical portion of the data series into a total of $\mathscr{U}_T + \mathscr{D}_T$ episodes.

We now introduce conditional probabilities based on these counts. We can define

$$\alpha_j = P(\text{observing a } (j+1)^{th} \text{ Up event} \mid j \text{ consecutive Up events}).$$

Similarly

$$\beta_j = P(\text{observing a } (j+1)^{th} \text{ Down event} \mid j \text{ consecutive Down events}).$$

Both of these conditional probabilities have natural estimators, that is

$$\hat{\alpha}_j = \frac{\sum_{i=j+1}^{\mathbf{u}} \mathscr{U}_{\mathrm{i}}}{\sum_{i=j}^{\mathbf{u}} \mathscr{U}_{\mathrm{i}}}, \tag{9.2}$$

$$\hat{\beta}_j = \frac{\sum_{i=j+1}^{\mathbf{d}} \mathscr{D}_{\mathrm{i}}}{\sum_{i=j}^{\mathbf{d}} \mathscr{D}_{\mathrm{i}}}. \tag{9.3}$$

For example, if we have already observed three consecutive *Up* events, then using (9.2) we could compute

$$\hat{\alpha}_3 = \frac{\mathscr{U}_4 + \mathscr{U}_5 + \mathscr{U}_6 + \ldots \mathscr{U}_{\mathbf{u}}}{\mathscr{U}_3 + \mathscr{U}_4 + \mathscr{U}_5 + \ldots \mathscr{U}_{\mathbf{u}}}$$

and analogously, using (9.3) for three consecutive *Down* events we have

$$\hat{\beta}_3 = \frac{\mathscr{D}_4 + \mathscr{D}_5 + \mathscr{D}_6 + \ldots \mathscr{D}_{\mathbf{d}}}{\mathscr{D}_3 + \mathscr{D}_4 + \mathscr{D}_5 + \ldots \mathscr{D}_{\mathbf{d}}}.$$

We can also describe the future binary trajectories of the digitised SOI data graphically, based on these conditional probabilities, as shown in Figure 9.4. In this example a sequence of j consecutive 1s, denoted $\mathscr{U}_j$, has just been observed and the probabilistic trajectories of the data are shown in the tree, with the relevant probabilities on the edges of the tree resulting in the sequence shown at each node.

Finally, we can assess the validity of the proposed Bayesian model using the linear half-Brier Skill Score (2) which can be applied to forecasts of dichotomous events. The half-Brier skill score essentially determines whether the Bayesian model is more skilled than an uninformative forecast where all probabilities in question equal 0.5. The half-Brier skill score is defined as

$$\text{Skill} = 1 - \frac{1}{N}\sum_{i=1}^{N}(\delta_i - \hat{p}_i)^2$$

where $\delta_i = 1$ if the event in question occurred and 0 otherwise; $\hat{p}_i$ is the estimated probability of being in state i and N is the total number of observations. This score takes values in the range $[0,1]$ where 1 represents perfect forecasting. An uninformative forecast, with all $\hat{p}_i = 0.5$ will yield a skill score of 0.75.

The half-Brier skill score for the BBT when we are interested in forecasting an *Up* event is 0.9135 when using all but the last 6 months of the data, and is 0.9131 when we are interested in forecasting a *Down* event. In both cases the skill scores suggest the BBT is sufficiently more skilled than an uninformative forecast.

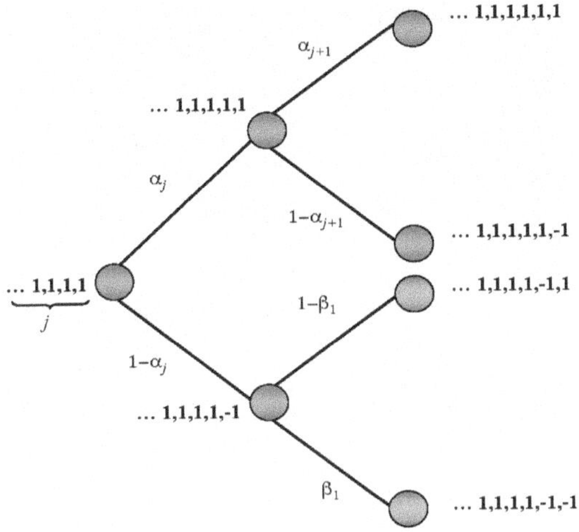

Fig. 9.4 The Bayesian Binary Tree for the digitised data for the case where a $\mathscr{U}_j$ sequence has been observed, with the tree marking out the probabilistic evolution of the sequence from the jth step onwards.

9.4 Forecasting SOI from the BBT Model

Although the probabilistic forecasting of the digitised signal appears more readily feasible than forecasting the Southern Oscillation index, ultimately it is precisely this index we are required to forecast. In this section we outline one possible method to reconstruct the SOI data series from the forecast digitised data.

The reconstruction method utilises the *Up* and *Down* episode histories from the untransformed SOI data. We first define the quantities $\bar{x}_{m,k}$ and $\bar{y}_{n,k}$ where

- $\bar{x}_{m,k}$ is the average SOI value for month k in an *Up* cluster of length m, where $k \in \{1,\ldots,m\}$, $m = 1,\ldots,\mathbf{u}$; and
- $\bar{y}_{n,k}$ is the average SOI value for month k in a *Down* cluster of length n, where $k \in \{1,\ldots,n\}$, $n = 1,\ldots,\mathbf{d}$

where a cluster is an uninterrupted series of *Up* or *Down* events.

We also need to take into account the possibility that our reconstruction may occur part-way through either an *Up* or *Down* episode, and we subsequently define the sub-totals $\mathscr{U}_{j\cdot}$ and $\mathscr{D}_{j\cdot}$ as

$$\mathscr{U}_{j\cdot} = \sum_{k>j} \mathscr{U}_k,$$
$$\mathscr{D}_{j\cdot} = \sum_{k>j} \mathscr{D}_k.$$

A natural estimate of the positively-valued $(j+1)^{st}$ SOI can then be given by

$$\begin{aligned}\hat{x}_{j+1} &= \frac{\mathscr{U}_{j+1}}{\mathscr{U}_{j\cdot}}\overline{x}_{j+1,j+1} + \frac{\mathscr{U}_{j+2}}{\mathscr{U}_{j\cdot}}\overline{x}_{j+2,j+1} + \cdots + \frac{\mathscr{U}_{\mathbf{u}}}{\mathscr{U}_{j\cdot}}\overline{x}_{\mathbf{u},j+1} \\ &= \sum_{k=1}^{\mathbf{u}-j} \frac{\mathscr{U}_{j+k}}{\mathscr{U}_{j\cdot}}\overline{x}_{j+k,j+1} \end{aligned} \tag{9.4}$$

for $j = 1,2,\ldots(\mathbf{u}-1)$. Analogously, the estimate of a negatively-valued $(j+1)^{th}$ SOI is given by

$$\hat{y}_{j+1} = \sum_{k=1}^{\mathbf{d}-j} \frac{\mathscr{D}_{j+k}}{\mathscr{D}_{j\cdot}}\overline{y}_{j+k,j+1}, \tag{9.5}$$

where $j = 1,2,\ldots(\mathbf{d}-1)$. We can associate these estimates with the nodes in our Bayesian Binary Tree in Figure 9.4 to produce the revised tree shown in Figure 9.5.

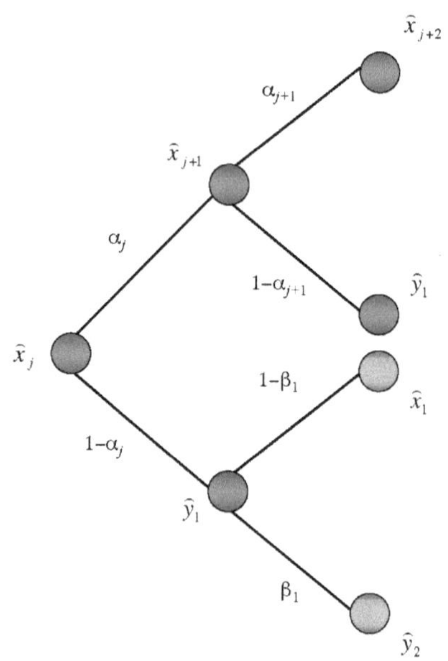

Fig. 9.5 The revised Bayesian Binary Tree containing the estimates $\hat{x},\hat{y}$ of the SOI data and their connection to the conditional probabilities α_j and β_j. The tree marks out the probabilistic evolution of the sequence from the jth step onwards.

We are now in a position to estimate the k-month forecast of the untransformed SOI data. We propose to do so with the help of three possible estimators:

$$\hat{z}(t+k) = E[SOI\ (t+k)] \tag{9.6}$$
$$\hat{z}_{\mathscr{U}}(t+k) = E[SOI\ (t+k)|\ \mathscr{U}] \tag{9.7}$$
$$\hat{z}_{\mathscr{D}}(t+k) = E[SOI\ (t+k)|\ \mathscr{D}] \tag{9.8}$$

where $\hat{z}(t+k)$ is the SOI forecast in k months' time. The first estimator (9.6) is self-explanatory, the second (9.7) refers to the expected value of SOI in month $(t+k)$ given we believe its value will be positive at that time, and the third estimator (9.8) refers to the expected value of SOI in month $(t+k)$ given we believe it will be negative. In this case, the Bayesian Binary Tree will have 2^k branches corresponding to sequences $\sigma = (\sigma_1, \dots \sigma_k)$, $\sigma_k \in \{-1, +1\}$ of all possible up/down signatures of the SOI signal at time instants $t_1, \dots t_k$.

Each of these branches has a probability $P(\sigma)$ where

$$P(\sigma) = \prod_{r=1}^{k} P(\sigma_r),$$

where each $P(\sigma_r)$ is a probability of type either α_j or β_j as in the BBT of Figures 9.4 and 9.5. Now let

$$\hat{z}_{t+i}(\sigma) = \begin{cases} \hat{x}_{t+i} & \text{if } \sigma_i = +1 \\ \hat{y}_{t+i}, & \text{if } \sigma_i = -1 \end{cases},$$

for $i = 1, 2, \dots k$, where $\hat{x}_{t+i}$ and $\hat{y}_{t+i}$ are calculated recursively using either (9.4) or (9.5). Hence a natural estimate $\hat{z}(t+k)$ is given by

$$\hat{z}(t+k) = \sum_{\sigma} P(\sigma) \hat{z}_{t+i}(\sigma).$$

Next, we partition the 2^k branches of the BBT into two sets:

$$A_k^+ = \{\sigma | \sigma_k = +1\}$$
$$A_k^- = \{\sigma | \sigma_k = -1\}$$

where we recall that $\sigma = (\sigma_1 \dots \sigma_k)$. Note that A_k^+ and A_k^- each contain 2^{k-1} branches and constitute the events that our forecast predicts a positive (respectively, negative) SOI value at time $(t+k)$. According to our BBT approach, we can now estimate the probabilities of A_k^+ and A_k^- by

$$P(A_k^+) = \sum_{\sigma \in A_k^+} P(\sigma)$$

and

$$P(A_k^-) = \sum_{\sigma \in A_k^-} P(\sigma).$$

Each $\sigma \in A_k^+$ the conditional probability of σ given A_k^+ can be given by

$$\gamma^+(\sigma) = \frac{P(\sigma)}{P(A_k^+)}$$

and analogously for each $\sigma \in A_k^-$

$$\gamma^-(\sigma) = \frac{P(\sigma)}{P(A_k^-)}.$$

Before we estimate $\hat{z}_{\mathscr{U}}(t+k)$ and $\hat{z}_{\mathscr{D}}(t+k)$ we need to revisit (9.4) and (9.5) to take into account the path defined by σ, as well as the history of the sequence preceding σ. This will determine whether the last month of a path $\sigma \in A_k^+$ is the $(j+1)^{st}$ month of an *Up* episode for $j \geq 1$, in which case (9.4) can be applied directly, or whether this is the 1st month of an *Up* episode ($j = 0$). We express this distinction as

$$\hat{x}_{t+k}(\sigma) = \begin{cases} \sum_{k=1}^{\mathbf{u}} \frac{\mathscr{U}_k}{\mathscr{U}_{1\cdot}} \overline{x}_{k,1} & \text{for the } 1^{st} \text{ month of an } Up \text{ episode,} \\ \sum_{k=1}^{\mathbf{u}-\mathrm{j}} \frac{\mathscr{U}_{j+k}}{\mathscr{U}_{j\cdot}} \overline{x}_{j+k,j+1} & \text{for the } j+1^{st} \text{ month of an } Up \text{ episode.} \end{cases} \tag{9.9}$$

and analogously for $\sigma \in A_k^-$

$$\hat{y}_{t+k}(\sigma) = \begin{cases} \sum_{k=1}^{\mathbf{u}} \frac{\mathscr{D}_k}{\mathscr{D}_{1\cdot}} \overline{y}_{k,1} & \text{for the } 1^{st} \text{ month of a } Down \text{ episode,} \\ \sum_{k=1}^{\mathbf{u}-\mathrm{j}} \frac{\mathscr{D}_{j+k}}{\mathscr{D}_{j\cdot}} \overline{y}_{j+k,j+1} & \text{for the } j+1^{st} \text{ month of a } Down \text{ episode.} \end{cases} \tag{9.10}$$

Using (9.9) and (9.10), natural estimates of $z_{\mathscr{U}}(t+k)$ and $z_{\mathscr{D}}(t+k)$, as defined by (9.7) and (9.8), are

$$\hat{z}_{\mathscr{U}}(t+k) = \sum_{\sigma \in A_k^+} \gamma^+(\sigma)\hat{x}_{t+k}(\sigma)$$

$$\hat{z}_{\mathscr{D}}(t+k) = \sum_{\sigma \in A_k^-} \gamma^-(\sigma)\hat{y}_{t+k}(\sigma).$$

Note that

$$\hat{z}(t+k) = P(A_k^+)\hat{z}_{\mathscr{U}}(t+k) + P(A_k^-)\hat{z}_{\mathscr{D}}(t+k). \tag{9.11}$$

9.5 A benchmark forecast using a Taylor Series

We now develop the second SOI forecasting method with the purpose of providing a benchmark against forecasts made using the BBT model. There are two important characteristics that must be realised in a benchmark forecast algorithm, namely: (1) simplicity; and (2) absence of any Bayesian weighting. An Mth-order Taylor series algorithm substitutes for both of these characteristics. Taylor series methods have been used in signal processing (12) and a variation of signal processing, in the form of time series analysis, has also been used to predict ENSO (13).

We construct an Mth-order Taylor Series Forecast (TSF)

$$S^T(\mathrm{t}+\mathrm{k}) = \sum_{\mathrm{m}=0}^{\mathrm{M}} \frac{\mathrm{k}^{\mathrm{m}} \mathrm{SOI}(\mathrm{t})^{(\mathrm{m})}}{\mathrm{m}!}, \tag{9.12}$$

using the standard expression where $SOI(t)$ is the measured value for the chosen month $t \in \{1, \ldots, 1308\}$ and $S^T(\mathrm{t}+\mathrm{k})$ is the $t+k^{th}$-month TSF-based SOI forecast.

A numerical approximation to the mth derivative $SOI(t)^{(m)}$ can be readily obtained using a difference approximation in steps of size 1 month:

$$\widehat{SOI}(t)^{(m)} = \sum_{p=0}^{m} (-1)^p \binom{m}{p} SOI(t-p), \tag{9.13}$$

where $\binom{m}{p}$ is the binomial coefficient.

We can quantify the proficiency of a TSF using a difference indicator over the full range of monthly SOI data from 1900-2008

$$\Delta S^T(\mathrm{k}) = \frac{\sum_{t=1}^{T_o} |S^T(\mathrm{t+k}) - \mathrm{SOI}(\mathrm{t+k})|}{T_o} \tag{9.14}$$

where $T_o = 1308$, the final month in the SOI data series. The indicator described by (9.14) can be regarded as the average difference between the SOI data and the forecast. We can further analogously define proficiency indicators based on the three Bayesian-based forecasts (9.6)–(9.8), viz.

$$\Delta S^{BAY}(\mathrm{k}) = \frac{\sum_{t=1}^{T_o} |\hat{z}(t+k) - SOI(t+k)|}{T_o} \tag{9.15}$$

$$\Delta S^{BAY}_{\mathscr{U}}(\mathrm{k}) = \frac{\sum_{t=1}^{T_o} |\hat{z}_{\mathscr{U}}(\mathrm{t+k}) - \mathrm{SOI}(\mathrm{t+k})|}{T_o} \tag{9.16}$$

$$\Delta S^{BAY}_{\mathscr{D}}(\mathrm{k}) = \frac{\sum_{t=1}^{T_o} |\hat{z}_{\mathscr{D}}(\mathrm{t+k}) - \mathrm{SOI}(\mathrm{t+k})|}{T_o}. \tag{9.17}$$

9.6 Comparing TSF and BBT

In this section we provide a full comparison of the means of $\overline{\Delta S^T}(k)$ and the Bayesian indicators $\overline{\Delta S^{BAY}}(k), \overline{\Delta S^{BAY}_{\mathscr{U}}}$ and $\overline{\Delta S^{BAY}_{\mathscr{D}}}$. We compared the predictability power of the TSF and BBT in three distinct scenarios: (1) A strong $\mathscr{U}$ event; (2) a strong $\mathscr{D}$ event; and (3) a mixed episode of both $\mathscr{U}$ and $\mathscr{D}$ events. From Equations (9.14)-(9.17) we have the differences between the forecast and actual data, as given in Tables (9.1) and (9.2).

As expected, ΔS^T increases rapidly with n, as the exponent term in the series in Equation (9.12) will dominate the behaviour of the indicator. The difference increases in a power-law fashion with increasing M and n. The indicators $\overline{\Delta S^{BAY}_{\mathscr{U}}}$ and $\overline{\Delta S^{BAY}_{\mathscr{D}}}$ in Table (9.2) all increase substantially more gradually than ΔS^T. Also, as expected, TSF provides better forecasts when variations of recent SOI values are small.

We then compared the TSF and BBT-based indicators over three separate events recorded in the SOI data set, demarcated by rectangular regions as shown in Figure

	n						
M	1	2	3	4	5	6	7
1	11.27	18.11	24.65	31.50	37.98	44.61	51.73
2	15.30	37.23	70.17	115.15	171.14	237.77	316.18
3	17.54	59.97	152.61	316.42	570.05	933.69	1 428.00
4	18.41	80.33	265.67	686.56	1 492.50	2,868.90	5,039.20
5	18.68	94.77	390.98	1,241.70	3,223.60	7,230.50	14,545.00
6	18.74	103.08	505.57	1,930.40	5,950.70	15,530.00	35,700.00

Table 9.1 The differences in the forecast and the actual data when using $\overline{\Delta S^T}$.

	n						
polarity	1	2	3	4	5	6	7
Mixed	8.25	9.73	13.37	22.33	41.05	79.60	158.24
$\mathscr{U}$	3.56	3.50	3.71	5.25	10.02	21.27	45.17
$\mathscr{D}$	2.97	2.90	3.47	5.95	12.28	26.17	56.26

Table 9.2 The differences in the forecast and actual data using $\overline{\Delta S^{BAY}}$, $\overline{\Delta S^{BAY}_{\mathscr{U}}}$, and $\overline{\Delta S^{BAY}_{\mathscr{D}}}$.

(9.6 (a)). In particular, we selected three types of episodes: the first is an *Up* scenario where the SOI values are strongly positive; the second is a *Down* episode where the SOI values are more strongly negative; and the third is a mixture of the two.

The performance of these indicators is displayed in Figure (9.6 (b)). By the nature of the BBT — constructed using an algorithm of mean values of past points — we see, somewhat unsurprisingly, that the BBT-based indicators perform better than TSF (with $M = 2$) for essentially all $n > 1$ and for the *Down* and *Mixed* scenarios. The conditional $\mathscr{U}$ and $\mathscr{D}$ indicators also yield considerably better forecasts than TSF, which diverges rapidly for the *Down* and *Mixed* scenarios. In the *Up* scenario, TSF (with $M = 2$) performs better initially, since in this case, values of the SOI just prior to the initial point are close to constant. However, the BBT *Up* forecast adapts better to the trend and after 7 months it is performing as well as TSF.

A selection of TSF curves for $M \in (1,6)$ is given in Figure 9.7, showing the marked divergent behavior of S^T for increasing k. In particular, as $M \to 6$ we see the forecasts predict SOI values which are well beyond four times that of the maximum SOI value observed to date.

9.7 Discussion

ENSO represents the most important source of the Earth's climate variability with time scales on the order of a few months to years. Forecasting this climate variability has become important for economic, sociological and health reasons.

A six-month forecast is shown in Figure 9.8. The analysis shows fairly good forecast results for the first few months and a breakdown of predicability power after only a few months. The BBT-based forecasting methods relied on the indicator

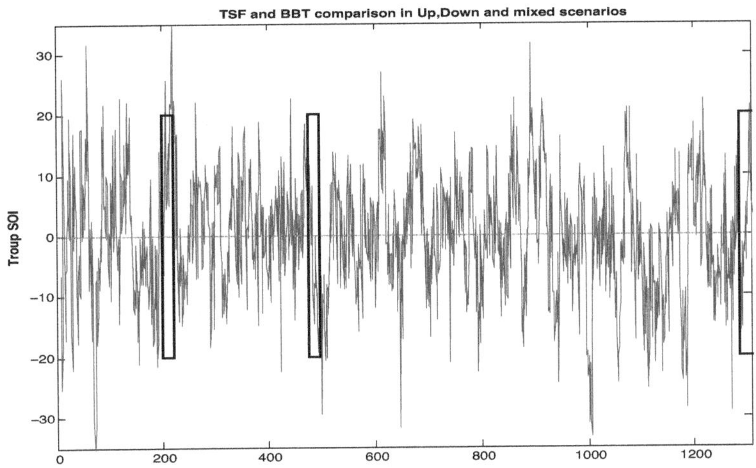

(a) SOI for the year 2005. The three rectangular areas indicate sequences forecast using TSF and BBT.

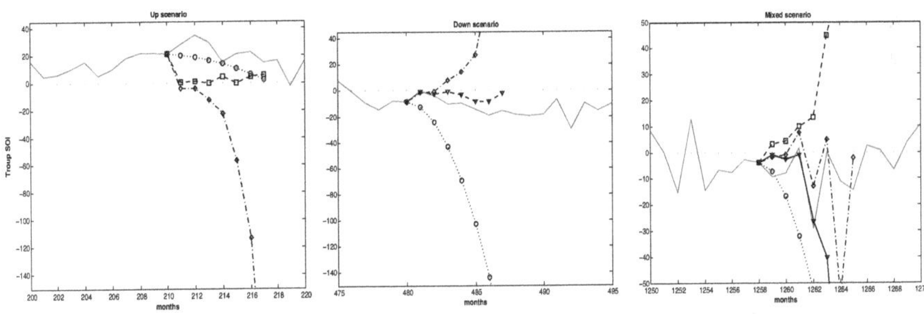

(b) The TSF (circles) and BBT (diamonds) for three episodes: $\mathscr{U}$, $\mathscr{D}$ and mixed.

Fig. 9.6 Comparing TSF (S^T, $M = 2$, circles) and BBT (S^{BAY}(diamonds), $S^{BAY}_{\mathscr{U}}$(squares), and $S^{BAY}_{\mathscr{D}}$(triangles)) for an Up scenario, Down scenario and a Mixed scenario (year 2005).

persistence and is based on a single indicator, the SOI. The control method, TSF, used the previous months data and the approximation to the SOI derivative in a Taylor series expansion. Both the TSF ($M = 2$) and BBT are compared in Figure (9.8). As expected, both methods eventually break down, however the BBT-Up scheme correctly adapts to the *Up* episode of this particular segment of SOI data.

Clearly the BBT outperforms the control method TSF for SOI predictions greater than 1 month. We further remark that when $M = 1$ (i.e. linear predictions) the TSF method does indeed perform better than the BBT approach. This improvement is not due to the predictive power of TSF when $M = 1$, rather it is only due to the nature of the definition of the SOI . Had SOI been defined with a scaling factor larger than 10 in Equation (9.1), then the TSF ($M = 1$) predictive power would decrease. On

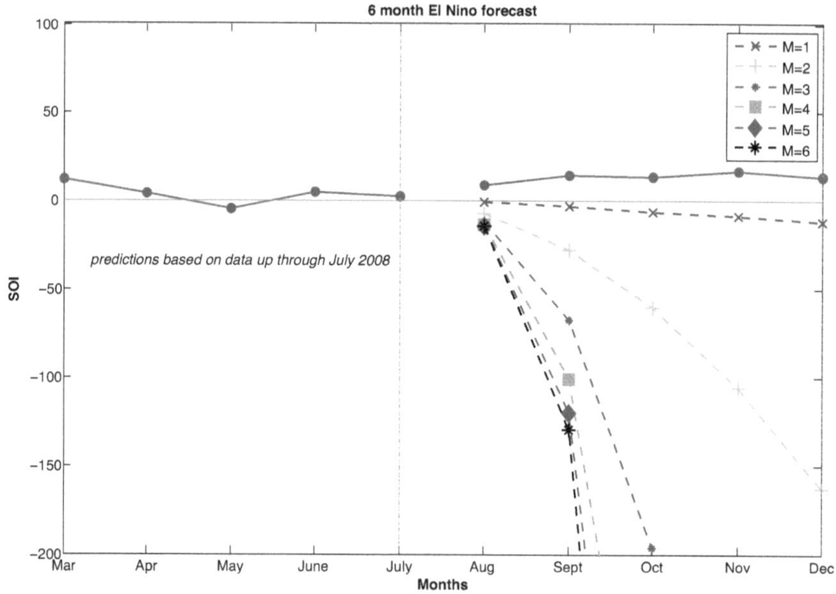

Fig. 9.7 A 6 month forecast from Jul-Dec 2008, using S^T with $M \in (1,6)$. The solid line (-•-) is the actual SOI data.

the other hand, the BBT predictions conditioned on $\mathscr{U}$ and $\mathscr{D}$ events are robust to the arbitrary constant 10 in the SOI definition.

Numerous literature including (11), (14) and (29), on the other hand, points to the same conclusion for schemes that are much more complicated . For example, sophisticated numerical methods including modeling of oscillations via equatorial wave dynamics in coupled ocean-atmosphere systems have also shown no better forecasting results than about 6 months (27). The simplified schemes are promising especially in the area of economic SOI-based forecasting where single indicator models are a natural tool for standard statistical-financial models.

Acknowledgements The authors would like to acknowledge the partial support from the Australian Research Council under grant DP0987148, as well as from the Phenomics and Bioinformatics Research Centre, at the University of South Australia.

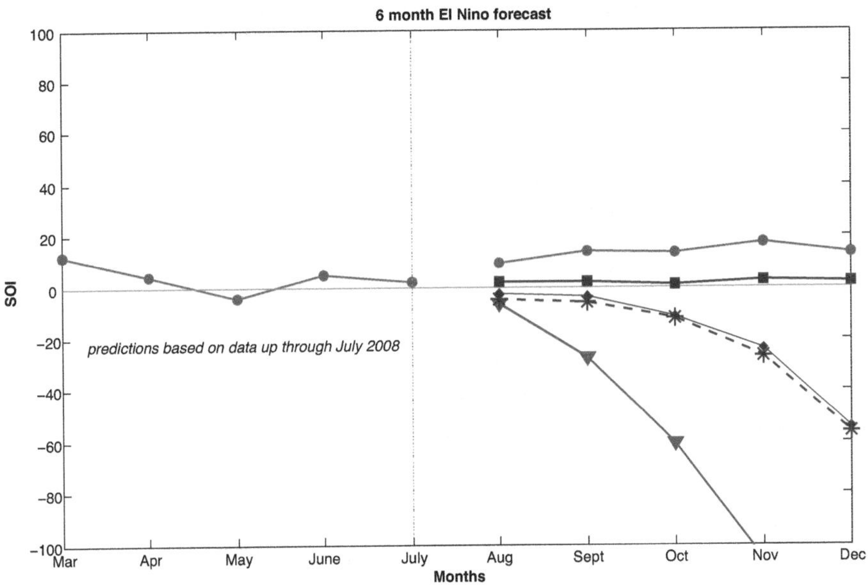

Fig. 9.8 SOI data (2008) compared to TSF ($M = 2$), a S^T ($\triangledown$) forecast and BBT forecasts S^{BAY} ($\diamond$), $S^{BAY}_{\mathscr{U}}$ ($\square$) and $S^{BAY}_{\mathscr{D}}$ ($*$). The line (-•-) is the actual SOI data used to construct the forecasts from June through November 2008.

References

[1] Barnett, T.P., Latif, M., Graham, N., Flugel, M., Pazan, S., White, W. (1993): ENSO and ENSO-related predictability. Part I: Prediction of equatorial Pacific sea surface temperature with a hybrid couple ocean-atmosphere model. *J. Climate*, **6**, pp. 1545-1566.

[2] Brier G. W.: Verification of forecasts expressed in terms of probability, *Monthly Weather Review*, **78**, pp.1-3 (1950)

[3] Bureau of Meteorology (The), Commonwealth of Australia: *Elnino Statistics*, `http://www.bom.gov.au/climate/glossary/elnino/elnino.html,cited2008.`

[4] Chagnon, S.A.: Economic Impacts of Climate Conditions in the United States: Past, Present and Future, *Climatic Change*, **68**, pp.1-2 (2005)

[5] Chiera, B.A.: Australian Sea Levels, ENSO and Green house Forcing: Modelling and Predictability Analysis, Ph.D. Thesis, University of South Australia, (2000)

[6] Coelho, C.A.S., Pezzulli, S., Balmaseda, M., Doblas-Reyes, F. J., D. B. Stephenson, D.B.: Forecast calibration and combination: A simple Bayesian

approach for ENSO, *Journal of Climate*. **17** (7), pp.1504-1516 (2004)
[7] Department of Natural Resources and Mines (State of Queensland). `http://www.nrm.qld.gov.au/longpdk/lpsoidat.htm`, cited 2008
[8] Duffie, D., Richardson, H.R.: Mean-variance hedging in continuous time, *Ann. Appl. Prob.* **1**, pp.1-15 (1991)
[9] Follmer, H., Sondermann, D.: Hedging of non-redundant contingent claims. *Contributions of Mathematical Economics in Honor of Gerard Debreu*, H. Hidenbrand, A. Mas-Colell, (eds.) pp. 205-223. North Holland, Amsterdam (1986)
[10] Gordon A.H.: Forced oscillations with damping proportional to velocity on a beta plane : the Southern Oscillation connection, *Environmental Modeling and Assessment*. **1**, pp.243-261 (1996)
[11] Gray, W. M., Sheaffer, J.D., Knaff, J.A.: Influence of the stratospheric QBO on ENSO variability, *J. Meteor. Soc. of Japan*, **70**, 5, pp.975-995 (1992)
[12] James, R.C.: El Niño. *Mathematics Magazine*. **25**, 21-31, 269–272 (1952); **26** pp.21-31 (1952)
[13] Katz R.W.: Sir Gilbert Walker and a connection between El Niño and Statistics *Statistical Science*, Vol 17, No. **1**, pp.97-112 (2002)
[14] Knaff, J. A.: Evidence of a stratospheric QBO modulation of tropical convection, Dept. of Atmos. Sci. Paper No. 520, Colo. State Univ., Ft. Collins, CO, p. 91 (1993)
[15] Landscheidt T.: Forecast of global temperature, El Nino, and cloud coverage by astronomical means, In: Bate, R. : Global Warming. The continuing debate (eds.) The European Science and Environment Forum (ESEF) pp.172-183 Cambridge (1998)
[16] Landscheidt, T.: Long-range forecasts of solar cycles and climate change, In: Rampino, M. R., Sanders, J. E., Newman, W. S. and Kunigsson, L. K., (eds.) : *Climate. History, Periodicity, and predictability* pp. 421-445 New York, van Nostrand Reinhold (1987)
[17] Landscheidt T.: Solar oscillations, sunspot cycles, and climatic change, In : McCormac, B. M., Hsg. (eds.): *Weather and climate responses to solar variations* pp. 301-304. Boulder, Associated University Press (1983)
[18] Maity, R., Nagesh Kumar, D.: Bayesian dynamic modeling for monthly Indian summer monsoon rainfall using El Niño Southern Oscillation (ENSO) and Equatorial Indian Ocean Oscillation (EQUINOO), J. Geophys. Res., **111** (2006)
[19] Mason, S. J., Mimmack, G.M.: Comparison of some statistical methods of probabilistic forecasting of ENSO. *J. Climate*. **15**, pp.8-29 (2002)
[20] NOAA, Climate Prediction Center, `http://www.cpc.noaa.gov/CitedFebruary2009`
[21] Patt A.: Communication probabilistic forecasts to decision makers: A case study of Zimbabwe. BCSIA, Environment and Natural Resources Program, Kennedy School of Government, Harvard University. Discussion paper 2000-19, `http://environment.harvard.edu/geaCited2000`

[22] Peixoto, J. P., Oort, A.H. 1992: Physics of climate. *American Institute of Physics*, New York (1992)
[23] Pudvokin, M. Veretenenko, S.: Cloudiness decreases associated with Forbush decreases of galactic cosmic rays. *J. Atm. Terr. Phys.* **57**, pp.1349-1355 (1955)
[24] Quinn W. H., Neal T. V., de Mayolo,S.E.A.: El Niño occurrences over the past four and a half centuries. *Journ. of Geophys. Res.* **92**, 14, pp.449-461 (1987)
[25] Ramage C. S.: El Niño. *Scient. American.* pp.59-61 (1986)
[26] Ruizdeelvira, A., Ortizbevia, M.J.: Application of statistical techniques to the analysis and prediction of ENSO: Bayesian oscillation patterns as a prediction scheme, *Dynamics of atmospheres and oceans.* **22** (1-2), pp.91-114 (1995)
[27] Schopf, P. S., Suarez, M.J.: Vacillations in a coupled ocean-atmosphere model. J. Atm. Sci. **45**, pp.549-566 (1988)
[28] Solow, A.R., Adams, R.F., Bryant, K.J., Legler, D.M., O'Brien, J.J., McCarl, B.A., Nayda, W., Weiher, R.: The Value of Improved ENSO Prediction to U.S. Agriculture. *Climatic Change.* **39**(1), pp.47-60 (1998)
[29] Stammer, D., Wunsch, C., Giering, R., Eckert, C., Heimbach, P., Marotzke, J., Adcroft, A., Hill, C.N., Marshall, J.: The global ocean circulation during 1992-1997, estimated from ocean observations and a general circulation model. *J. Geophys. Res.*, **107** (C9) (2002)
[30] Svensmark, H., Friis-Christensen, E.: Variation of cosmic ray flux and global cloud coverage, a missing link in solar-climate relationships, *J. Atm. Sol. Terr. Phys.* **59**, p. 1225 (1997)
[31] Thompson, C.J., Battisti, D.S.: A Linear Stochastic Dynamical Model of ENSO. Part II: Analysis. *Journal of Climate*, **14** pp.445-465 (2001)
[32] Torrence, C., Webster, P.: The annual cycle of persistence in the El Niño - Southern Oscillation. *Quart. J. Roy. Meteor. Soc.* **124**, pp. 1985-2004 (1988)
[33] Trenberth, K.E.: The Definition of El Niño, *Bull. Amer. Met. Soc.* **78**, pp.2771-2777 (1997)
[34] Troup A.J.: The Southern Oscillation, *Quarterly Journal of Royal Meteorological Society.* **91**, pp.490-506 (1965)
[35] Wang, F., Jackson, C. S., Stoffa, P. L., Flügel, M., Chang, P.: A Stochastic Bayesian Approach to Identify the Dynamical Regimes of ENSO. American Geophysical Union. Fall Meeting 2004 Abstract # A14B-06 (2004)
[36] Zhang, Y., Wallace, J.M.: Is Climate Variability over the North Pacific a Linear Response to ENSO? *Journal of Climate.* **9**(7), pp.1468-1478. (1996)

Chapter 10
Generation of Synthetic Sequences of Electricity Demand with Applications

J.W. Boland

Abstract We have developed a model to generate synthetic sequences of half hourly electricity demand. The generated sequences represent possible realisations of electricity load that can have occurred. Each of the components included in the model has a physical interpretation. These components are yearly and daily seasonality which were modelled using Fourier series, weekly seasonality modelled with dummy variables, and the relationship with current temperature described by polynomial functions of temperature. Finally the stochastic componentwas modelled with ARMA processes. The temperature series was modelled in a similar fashion. The stochastic modelling was performed to build probability distributions of the outputs to calculate probabilistic forecasts.As one application several summers of half hourly electricity demand were generated and from them the value of demand that is not expected to be exceeded more than once in ten years was calculated. Additionally, the bivariate temperature and demand model was used in software designed to optimise the orientation of photovoltaic cells to match demand.

Keywords: Half hourly electricity demand, Fourier series, Multiple regression, ARMA, Stochastic modelling.

10.1 Introduction

The analysis presented here relies heavily on Magnano and Boland (2007), Magnano, Boland and Hyndman (2008), and Boland and Zekanovic (2008). In the electricity demand market the analysis of future capacity and reliability of power systems is essential. The increasing dependence on technology in different sectors

School of Mathematics and Statistics, CIAM, ISST, University of South Australia, Mawson Lakes Boulevard, Mawson Lakes, Adelaide, South Australia 5095, AUSTRALIA.

J.A. Filar and A. Haurie (eds.), *Uncertainty and Environmental Decision Making*, International Series in Operations Research & Management Science 138, DOI 10.1007/978-1-4419-1129-2_10,

makes power system reliability of increasing concern. The cost of unserved energy can be valued at thousand of dollars per MWh, (NEMMCO 2005). Consumers assess the reliability and security in terms of continuity and quality of the energy delivered. This is dependent upon the performance of all parts of the supply chain including generation, high voltage transmission and local distribution. To ensure customers receive a reliable supply:

- there must be enough generating plant capacity available to meet the demand;
- the power system needs to maintain its integrity;
- there must be enough transmission capability available to convey the electricity distribution network;
- the distribution networks must be capable of delivering the energy to individuals.

An accurate forecast of the demand at any time is critical to achieve reliability in each of these aspects. Reliability analysis is performed using market simulation software. To perform this analysis it is necessary to use traces of electricity load as input data. Presently this study is likely to be performed using historical traces. Usually traces associated with extreme temperature events are used. However, since these extreme events occur only rarely, there is a scarcity of load traces associated with these extreme temperature values available to perform reliability analysis. In Australia the National Electricity Market (NEM) requires the amount of unserved energy should remain below 0.002% over a specified period. The National Electricity Market Management Company (NEMMCO), needs to determine the reserve margin necessary to meet reliability standards. To do this, the expected future total customer demand needs to be known. In South Australia the demand is very volatile and depends upon extreme summer weather conditions and the behaviour of customers in these situations. As a result forecasting needs to encompass a range of possible outcomes to address future uncertainty. In the design and operation of modern power systems, the projections are made not only for the most likely future outcome but also for bounds on the likely demand. NEMMCO has set its reserve margins against the 10% probability of exceedance (PoE) of demand (NEMMCO 2005). This value is interpreted as the estimated maximum level of demand that is not expected to be exceeded more than once in ten years. The prediction of this extreme peak demand therefore becomes very important in planning for the national electricity market. The analysis presented in this paper provides a method to generate sequences with the same statistical properties as the real data to use them as input data in market simulation software and to calculate probabilistic load forecasts which will enable an assessment of the adequacy of the system in meeting demand for generation over subsequent years to be performed.

As is shown below, the model was built using temperature and electric load data. Given that it does not include either economic or demographic variables it is considered the first step of a two step model to calculate long run forecasts. This model is important on its own because it describes the main factors that drive the demand behaviour. The sequences generated from the model can be used to generate input data in market simulation software. In this work we present a model that is very effective to recreate sequences of electricity demand with similar characteristics to the real. At the same time the distribution of half hourly electricity demand is obtained. This distribution will be used to calculate reliable probabilistic forecasts of long term electric load. Given that the model identifies the intra-day cyclic effect for different daily mean temperatures, it can also be used to cater for climate change effects.

The aim of this work is twofold: to develop a model which describes the half hourly electricity demand behaviour, this tool is presented in Section 10.2; and to describe a procedure to generate sequences of demand using this model applied to a summer period, explained in Section 10.3. Each synthetic sequence represents a possible realisation of half hourly electricity demand behaviour. A demand distribution can be determined by the generation of several years of data. This distribution enables the calculation of probabilities of occurrence of different demand levels. The formal distribution of yearly peak electricity demand is also presented in the corresponding section.

Over more than the last two decades a wide variety of approaches has been taken to forecast short-term electricity load. The main categories can be classified as classical approaches, artificial intelligence based and a combination of both (see Bunn *et al* (1985) for an early review of the former and Mextanitis *et al* (2003) for one of the latter). The most well known statistical methods used are seasonal integrated moving averages (SARIMA), mostly used as a benchmark model [Soares and Souza (1006), Taylor *et al* (2006)], multiple regression [Ramanhatan *et al* (1997), Pardo *et al* (2002), Mirasgedi *et al* (2006) and McSharry *et al*(2005)], time-varying splines (Harvey and Koopman 1993), exponential smoothing (Taylor 2003), and principal component analysis [Taylor *et al* (2006), Manera and Marzullo (2005)]. Some of the works mentioned above are a combination of more than one method.

Ramanhatan *et al* (1997) modelled each hour separately. This approach is the basis of several works that follow the same criterion but introducing changes to the former. Soares and Souza (2006) developed a univariate stochastic model which includes a stochastic trend, dummy variables to model the weekly pattern and the influence of holidays and explains the seasonality by generalised long memory processes using GARMA models. Other authors use non linear relationships with temperature for each hour of the day and SARIMA models to explain serial correlation in the noise. These relationships with temperature were modelled using non linear parametric (Sesworahardjo *et al* 2006) and nonparametric approaches (Liu *et al* 2006).

Our approach in the assessment of load demand uses classical techniques to develop the model. We aim to obtain a model whose components describe the main drivers of electricity demand. However, in contrast to most of the models, the purpose is not to obtain accurate short term forecasts but to obtain a model that enables the generation of possible traces of demand that could have occurred and eventually, to use the model to calculate long run probabilistic forecasts instead of traditional ones. Traditional forecasting can be described as estimating values for a given period. They assess the uncertainty surrounding the middle forecast through the use of confidence bounds. However, the probability associated with the upper and lower limits are unknown. A probabilistic forecast can be defined as the forecast that is given by the probability distribution. Probabilistic forecasting gives the probability associated to each possible value of demand. Our model was developed on the same basis as previous work done to forecast half hourly electricity demand one step ahead (Boland and Magnano 2007). However, some of the components have been estimated differently due to the difficulty in synthetically generating components with the same characteristics of the real data. We will discuss in Subsection 10.2.6 why we can not use the same model used in Boland and Magnano (2007). In this work simulation techniques are applied using a time series approach.

Through this process the electricity demand distribution is calculated and also the different half hourly traces are obtained. If we were interested only in obtaining the distribution of yearly peak load, it could be thought that a simple method to obtain this distribution would be to fit a distribution to the historical yearly peak demand. However, due to the small number of observations it would be difficult to fit the appropriate distribution. and the trace of load associated with this value of demand would not be available

Although a wide range of techniques have been successfully developed for short term forecasts, as far as we can ascertain, this kind of approach that uses time series modelling to produce possible traces of electricity load and eventually probabilistic forecasts has not been broadly taken. Analogous approaches to describe electricity load behaviour were performed by Mirasgedis *et al* (2006). and McSharry *et al* (2005). Mirasgedis *et al* (2006) modelled monthly and daily data and use real temperature to estimate the model and typical meteorological year data to analyse different scenarios of temperature. Multiple regression models were used to estimate the model using a time series approach and autocorrelated errors. McSharry *et al* (2006) model daily data and used traces of surrogated temperature and the model was specifically designed to obtain probabilistic forecasts. This work is the only one that was designed to calculate probabilistic forecasts based on a time series approach. In our approach, we use real temperature to estimate the model and synthetic sequences of half hourly temperature to generate the electricity demand sequences. These sequences of half hourly temperature have similar statistical characteristics to the real temperature. We model half hourly data instead of daily because this was the requirement specified by the Electricity Supply Industry Planning Council (ESIPC). The electricity market in Australia operates on a half-hourly bid system by generators. Therefore, there is a need for modelling demand on this finer time scale. As mentioned previously, these traces will be used in market simulation software and to do so we need to obtain the trace on a half hourly resolution. Working with a finer time scale adds complexity to the model due to the intra-day variation. We analysed in detail the load profiles for different types of days and their relationship with daily average temperature. Our model adds further to previous approaches in the sense that these daily load profiles need to be modelled. This identification of different cyclical intra-day patterns for different daily average temperature gives a flexibility that also allows for catering for climate change.

The half hourly demand model consists of deterministic and stochastic components. The deterministic behaviour is given by yearly cycles, weekly seasonality, intra-day cycles, relationships with current temperature and associations with each period of day. The stochastic component is given by the relationship of the current electricity demand with its immediate past values. This paper is structured as follows. Section 10.2 describes in detail how the demand model was constructed. Section 10.3 explains the synthetic generation procedure, demonstrates an example application and presents a discussion of the results. Section 10.4 describes the daily temperature model, needed for the application to optimisation of orientation of solar cells for load matching. Section 10.5 presents validation of the demand model. In Section 10.6, we show how the procedures developed in the analysis can be applied. Section gives concluding remarks.

10.2 Half hourly electricity demand model

The model was estimated using three years of half hourly data to analyse the deterministic and stochastic components that drive half hourly electricity demand. The data used spans the period from 01/07/2002 to 30/06/2005, as shown in Figure 10.1, with no apparent trend in the data.

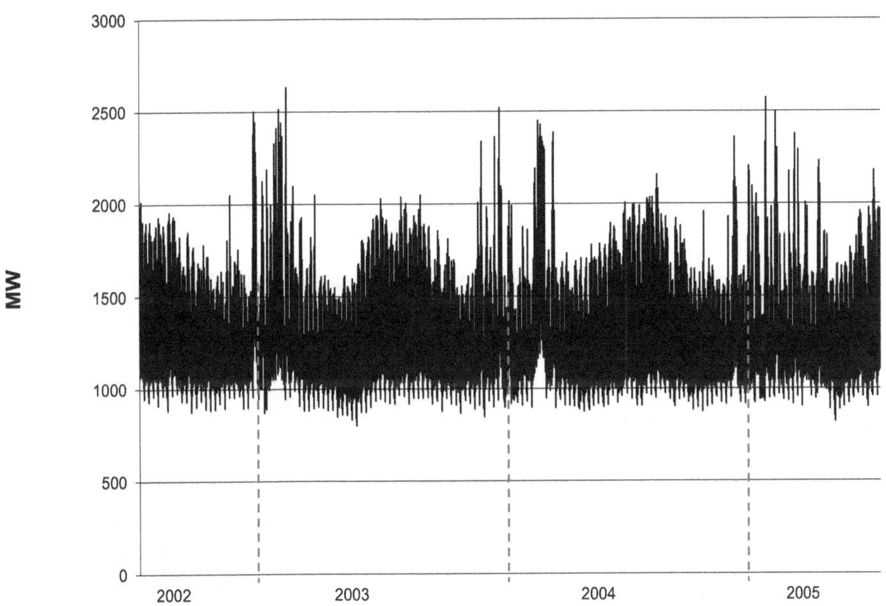

Fig. 10.1 Half hourly electricity demand (01/07/2002 - 30/06/2005) - Dashed lines represent the end/beginning of a year

The electricity demand shows peaks in both summer and winter. In the case of winter the magnitude of the variability is lower but the time series still demonstrates higher levels than in autumn and spring days. This variability is significantly larger when the temperature is high. This behaviour can be seen in Figure 11.2 where one year of half hourly data is presented.

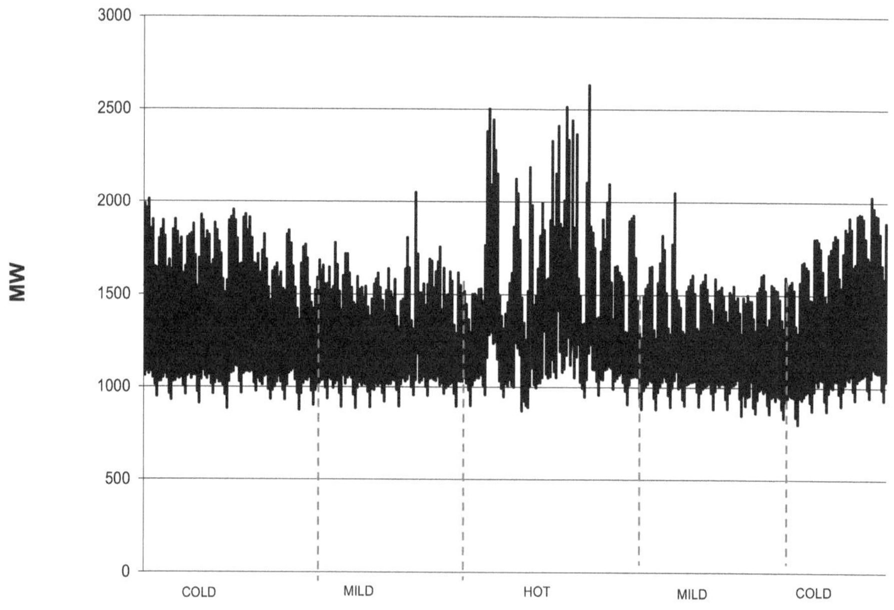

Fig. 10.2 Half hourly electricity demand (01/07/2002 - 30/06/2003) - Dashed lines represent changes in demand variability related to changes in temperature

To identify other characteristics of the data shorter periods such as weeks and days were also analysed. The electricity demand presents similarities within weeks of the same season and also within days of the same season. The model to generate synthetic sequences was developed only for the summer period which was considered to be from December to February. This is the critical period in South Australia when the supply could possibly not meet the demand. South Australia experiences high temperatures in summer and because of the high association between electricity and temperature, blackouts and brownouts may occur. In contrast, winter is not considered as a critical period in South Australia as the minimum temperature in South Australia is not low enough to cause high demand. The historical average minimum temperature in this season is 8 ° C.

The analysis of this time series was performed in the following way. Yearly, weekly and daily seasonality were estimated respectively and then, the relationship with the current period of temperature was analysed. These components were estimated with the techniques used in Boland and Magnano (2007) which are explained in detail below. Then, the errors obtained after eliminating these components from the original data were analysed by period of the day. Differences in means and standard deviations were identified depending on the period to which the data belongs. These errors were modelled using an autoregressive moving average *(ARMA)* model after these errors had been transformed to ensure normality.

10.2.1 Data structure

The analysis was performed using the South Australian data over three years as mentioned before. The generation of electricity in South Australia is from a mix of fuel types consisting of sub-bituminous coal, distillate, natural gas and dual-fuel firing capability and also wind generation. The objective of the half hourly demand model was to analyse the factors that drive electricity demand. Real temperature over the same period was used to perform the study. In order to model the typical weekday and weekend behaviour each public holiday was replaced by the same day of the previous week. Christmas week was replaced by an average of the same day of the two weeks before and after with the corresponding temperatures. This allows us to analyse a continuous record of data without these infrequent deterministic variations. Alternatively, dummy variables could have been used but this method was used instead because the final usage of the model is to analyse peak electricity demand. Public holidays and Christmas week do not represent a threat when peak load is being analysed.

10.2.2 Yearly cycles

The levels of demand vary according to the season. In summer and winter the demand is higher as a consequence of the air conditioning and heating appliances. This behaviour would suggest the existence of two cycles per year. To analyse the number of significant cycles within a year, i.e. the fundamental period, the power spectrum function was calculated. The power spectrum is the Fourier transform of the autocovariance function and shows how the variance of the stochastic process is distributed with frequency. The power spectrum in Figure 10.3 shows six significant cycles within the data range which mean that there are two significant cycles per year. There are other significant frequencies around the values 156 and 313 which refer to weekly seasonality and at 1095 and higher describing daily seasonality. However these cycles are not included at this stage since only yearly cycles are being identified. These other significant frequencies are described in Sections 2.3 and 2.4 respectively.

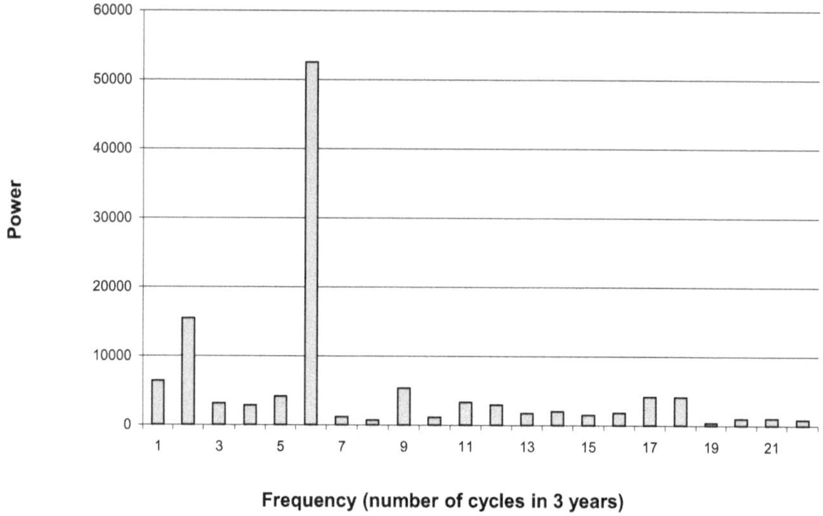

Fig. 10.3 Power spectrum showing six cycles in three years

This behaviour was modelled using Fourier series. The formula that describes the twice yearly cycles on a half hourly basis has the following form:

$$\zeta_t = \overline{Y} + \delta_{p1} \cos\left(\frac{2\pi t p}{17520}\right) + \delta_{p2} \sin\left(\frac{2\pi t p}{17520}\right) \quad t = \overline{1,17520} \tag{10.1}$$

where $\overline{Y}$ is the mean half hourly demand, δ_{p1} and δ_{p2} are the amplitude coefficients estimated from the data, p represents the number of yearly harmonics, in this case $p = 2$. The yearly cyclic estimation for one year's data can be seen in Figure 10.4. The estimated component was subtracted from the original series. Then only the summer months were analysed, since this is the period of interest. The next step was to estimate different levels of demand according to the day of the week.

10.2.3 Weekly seasonality

In general weekdays register higher levels of electricity demand than weekends. Also within weekdays there are different levels registered. Usually the demand on Fridays is lower than that on other weekdays. At the same time the level of demand on Saturdays is higher than on Sundays. This seasonality was modelled using dummy variables. When the seasonality is considered deterministic, the sum of the

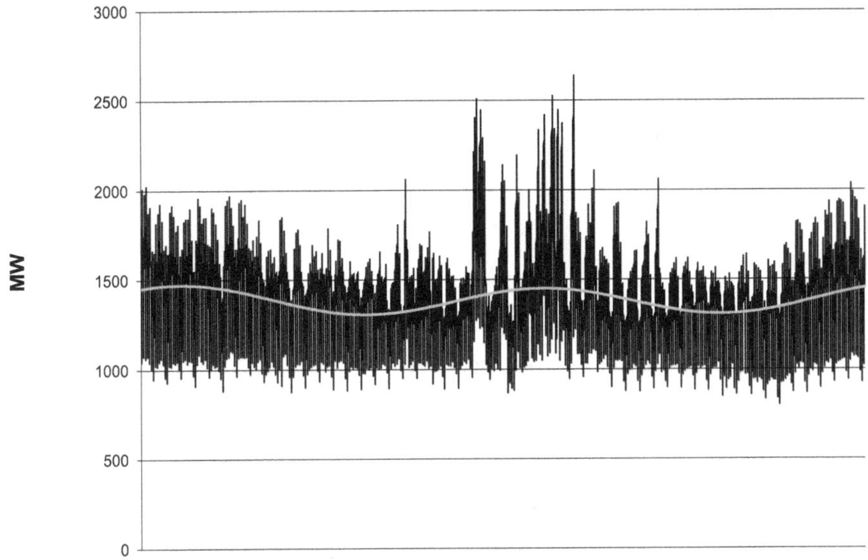

Fig. 10.4 Half hourly electricity demand and Fourier series fit

seasonal effects must be zero (Blacona *et al* 2000). This means:

$$\sum_{j=1}^{7} \alpha_j = 0 \qquad (10.2)$$

Or equivalently $\sum_{j=1}^{6} \alpha_j = 0$, thus:

$$\alpha_j = -\sum_{j=1}^{6} \alpha_{t-j}, \quad t = 7, 8, \ldots \qquad (10.3)$$

where α_1 means that, on average, on Monday the level of demand is α_1 MW above (it would be below if α_1 were negative) the weekly average once yearly seasonality has been removed. Analogous interpretation can be made for the other weekly effects. Table 10.1 shows the estimated weekly effect for each day of the week. The highest levels of demand occur on Mondays and Tuesdays. The lowest levels occur in weekends, with Sunday being lower than Saturday as it was expected. Figure 10.5 shows the model for one week after adding the weekly seasonality contribution to the yearly trend. There is not much agreement as yet, but we will be adding more predictors and we wish to show the contribution of each.

The estimated weekly seasonality was also eliminated from the data. The next step was the analysis of intra-day patterns.

Table 10.1 Weekly effects

Day	Monday	Tuesday	Wednesday	Thursday	Day	Friday	Saturday	Sunday
Coefficient	88.2	92.8	50.1	57.3	Coefficient	34.8	-150.6	-172.6

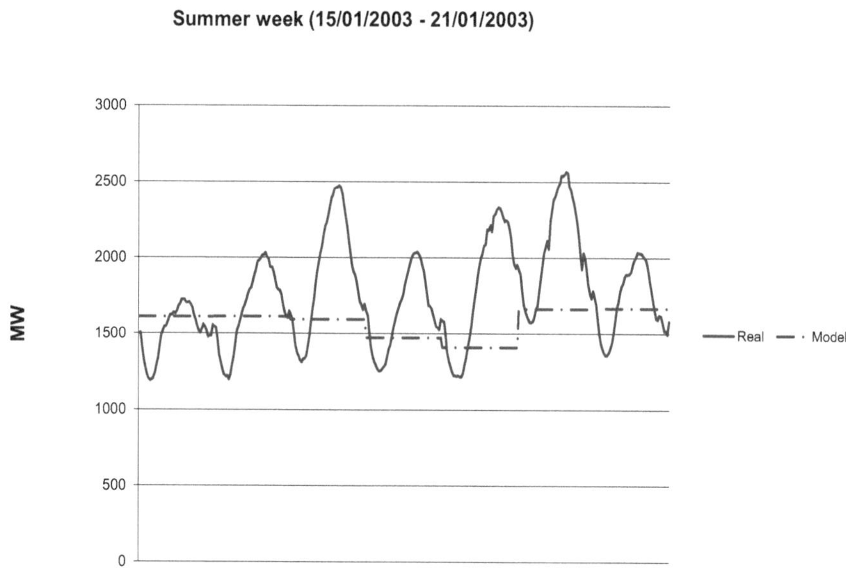

Fig. 10.5 Model with yearly and weekly seasonality

10.2.4 Intra-day cycles

The electricity demand behaviour within the day has a significant deterministic component. As was shown in Boland and Magnano (2007), on warm days, the level of demand reaches its lowest point early in the morning, around 4.00 am and its highest in the afternoon between 2.00 pm and 5.00 pm. Smaller but significant peaks also occur in other periods of the day. For example, in the morning when the shops open and people start their activities the level of demand rises. Similarly, in the evening when people get back home and turn on the home appliances the demand increases. On mild days these domestic patterns become more evident and consequently the afternoon peak is smoother. Fourier transformations were used to identify the intra-day cycles. Weekends and weekdays patterns are similar but the level of the former is lower. Firstly, following the criteria used in Boland and Magnano (2007), four Fourier series were estimated depending whether the daily mean temperature is above or below 22 °C and also whether the data to be estimated corresponds to weekend or weekday. The Fourier series are given by:

$$\varsigma_t = \overline{Z} + \sum_{p \in P} \delta_{p1} \cos\left(\frac{2\pi t p}{48}\right) + \delta_{p2} \sin\left(\frac{2\pi t p}{48}\right) \quad t = \overline{1,48} \tag{10.4}$$

where $\overline{Z}$ is the mean of the residual after eliminating from the original series the yearly and weekly seasonality already estimated, δ_{p1} and δ_{p2} are the amplitude coefficients estimated from the data, P represents the set of significant harmonics. The results of the estimation for days with daily mean temperature above 22 °C can be seen in Figure 10.6. Weekdays and weekends on warm days have three significant harmonics whose sum represents the average behaviour. The lower level usually occurs around 4.30 am while the higher level of demand occurs around 4.00 pm. Figure 10.7 shows the average demand behaviour for days with daily mean temperature below 22 °C. Weekdays and weekend in mild days have seven significant harmonics whose sum represents the domestic behaviour aforementioned. The lower level is around 5.00 am and during the day there are peaks early in the morning, in the middle of the day, early in the afternoon and at night.

The Fourier series approach used to describe intra-day profiles showed a suitable description of electricity load behaviour. However, further exploratory analysis showed that the average load profile of a day whose daily mean temperature was 22.5 °C differs from another one whose daily mean temperature is 26 °C, for example. Thus, we extended the approach shown in Boland and Magnano (2005) by analysing groups of days according to different daily mean temperature. The days were divided into days with daily mean temperature less than or to equal to 20 °C, between 20 °C and 22 °C, between 22 °C and 24 °C, between 24 °C and 26 °C and greater than 26 °C, assuming that the daily profiles within the groups show more homogeneity than the groups analysed in Boland and Magnano (2005). All the intervals mentioned include the upper limit. From now on, we refer to these groups as Groups I to V respectively. This approach, although it requires the estimation of many parameters, captures the variability in the intra-day load profile as a function of daily mean temperature. The improvements in the model after adding intra-day cycles to a summer week can be seen in Figure 10.8. As was done with the previous components identified, intra-day cycles were also eliminated from the original data. The resulting data was used to find the relationship with current temperature.

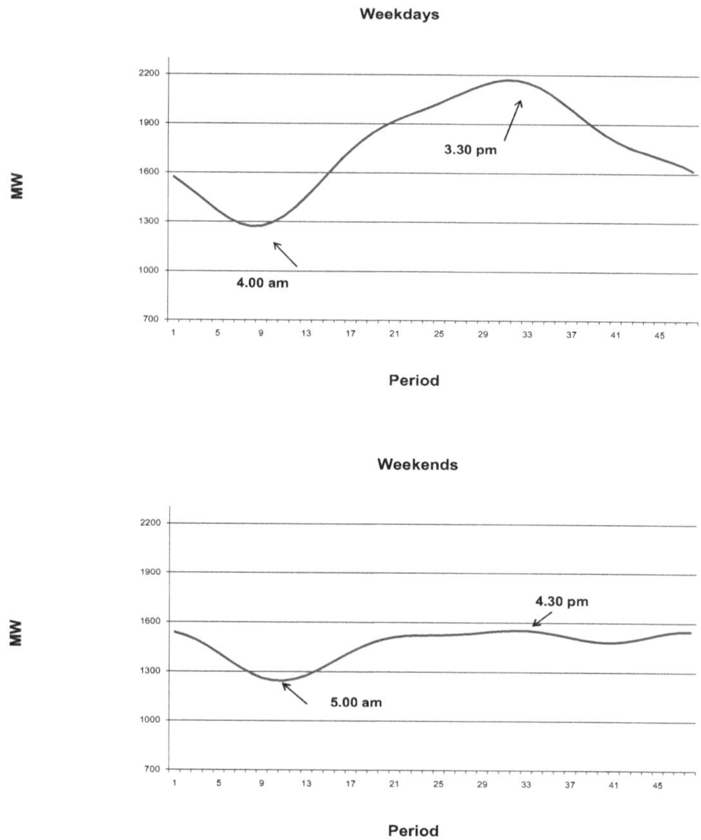

Fig. 10.6 Intra-day cycles - Daily mean temperature above 22 $^\circ$ C - weekdays and weekends

10.2.5 Relationship between demand and current temperature

The data obtained after eliminating the yearly cycles, weekly seasonality and intra-day cycles from the original series, $e_{3t} = y_t - (\zeta_t + \alpha_t + \varsigma_t)$, was analysed to find its relationship with current temperature, x_t. The analysis was performed by groups of days as defined in Section 10.2.4. There was a weak relationship with temperature for the days whose daily mean temperature was below 22 $^\circ$ C, i.e Groups I and II. The explanation for this result is that most of the variability within each group was mainly modelled by the intra-day cycles. The relationship between load and current temperature was model with polynomial regression for groups whose daily mean temperature was above 22 $^\circ$ C.

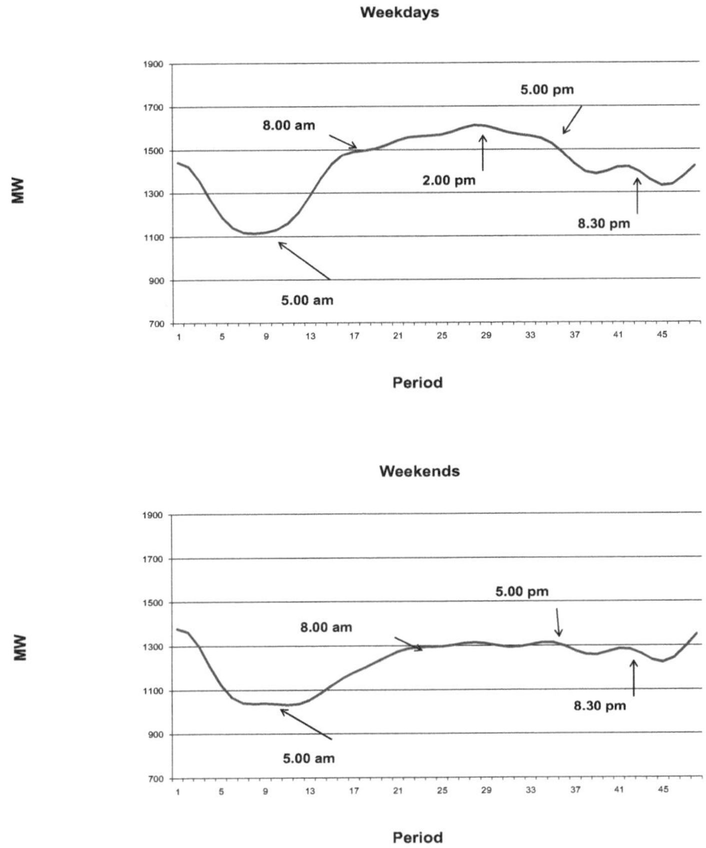

Fig. 10.7 Intra-day cycles - Daily mean temperature below than or equal to 22 ° C - Weekdays and weekends

Consequently three polynomial functions, two quadratic and one cubic, were estimated. In most of the cases the R^2 is relatively low. However, it should be noted that at this stage we are modelling the residuals of a model whose components already explain a significant amount of the variability of the electricity demand. In the previous step, when the daily profile was modelled the effect of temperature was already considered as the intra-day profile was modelled conditional on the daily mean temperature. The highest coefficient of determination is in Group V and its value is 24%.

The relationships described can be seen in Figure 10.9. Alternatively, the relationships between demand and temperature at different periods within the day - from 1 to 48 - were also analysed under the assumption that the relationship is stronger closer to the peak time. However, after assessing this alternative, which incorporates

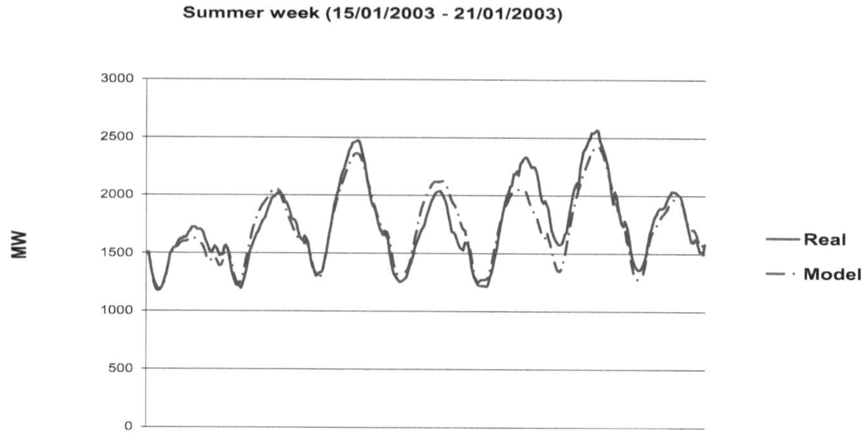

Fig. 10.8 Model with yearly and weekly seasonality and intra day cycles

different relationships depending on the time of the day, no extra benefits were obtained and the number of parameters was higher. However, it was evident that in the periods from 3.00 am to 7.30 am the effect of this relationship was negligible. For this reason, the model does not include the temperature effect for the periods from 3.00 am to 7.30 am. Although it was shown in the literature that the effect of temperature on demand is different according to the time of the day (Liu *et al* 2006) the reason why we did not obtain greater improvements is because we accounted for this variability using different Fourier series for each group of temperature. After modelling the intra-day profile in this way it is expected that the actual relationship between the residuals of the model and current temperature to be weaker.

Figure 11.8 shows the estimated model with the incorporation of temperature effects for a summer week. After analysing the relationship between demand and current temperature this component was eliminated and the remaining data, $\varepsilon_{4t}^* = y_t - (\zeta_t + \alpha_t + \varsigma_t + \tau_t)$, was divided by period of the day.

10.2.6 Analysis of e_{4t} per period during summer

After eliminating the previous estimated component the residual

$$\varepsilon_{4t}^* = y_t - (\zeta_t + \alpha_t + \varsigma_t + \tau_t)$$

was analysed. In Boland and Magnano (2005), ε_{4t}^* was modelled by a multiple regression on the moving averages of the previous two days of temperature and ε_{4t}^*. The model was estimated in this way to eliminate the autocorrelation of order 48

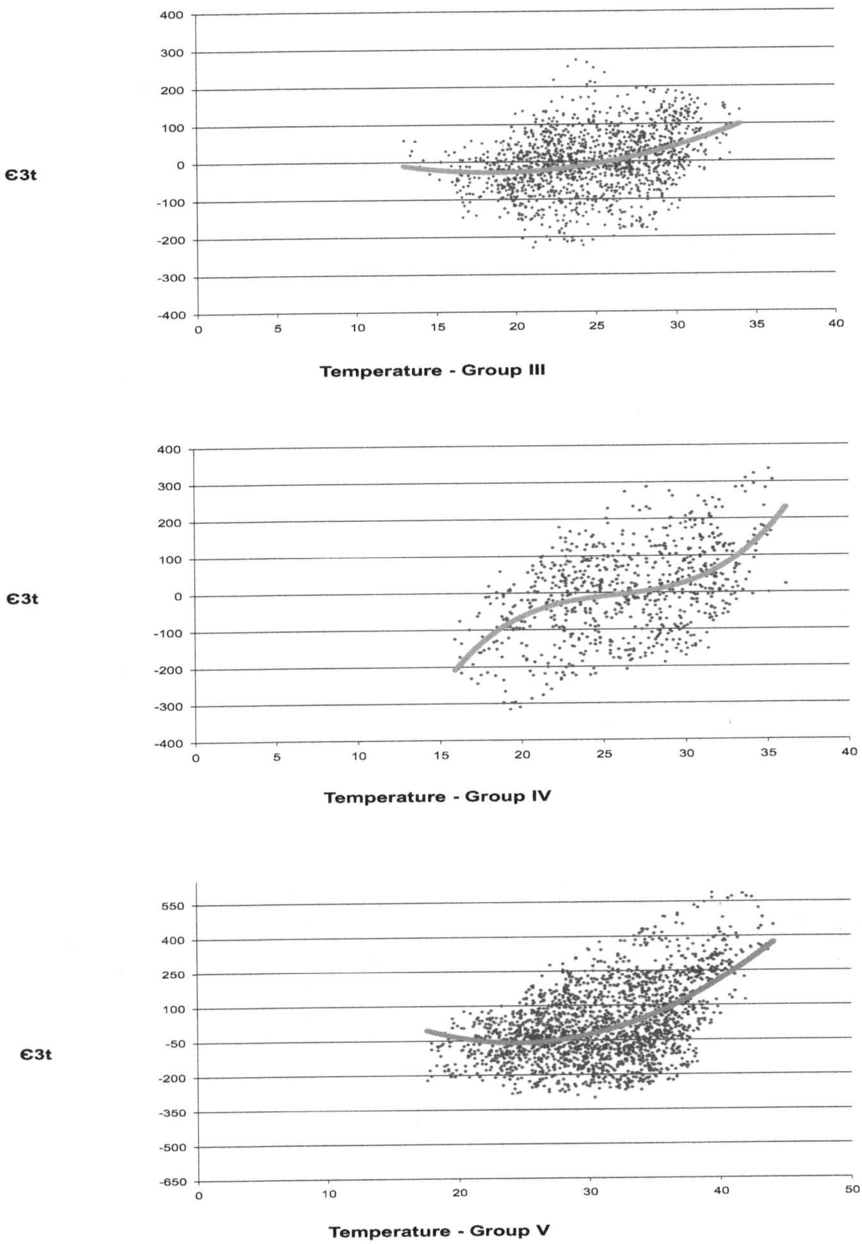

Fig. 10.9 Relationship between e_{3t} and current temperature x_t

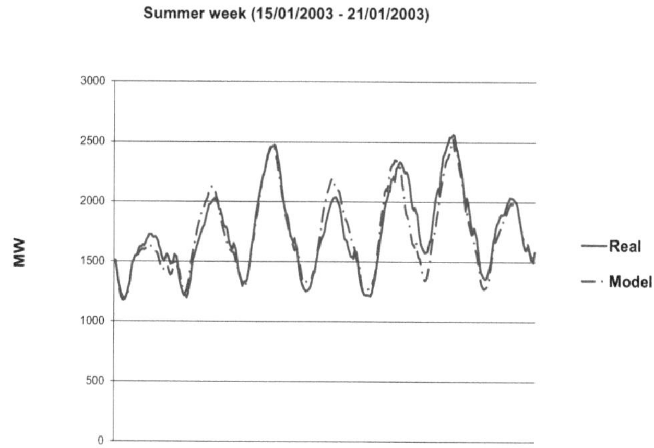

Fig. 10.10 Fit of the combination of yearly and weekly seasonality, intra-day cycles and current temperature to the real data.

(the effects of the previous day) which was highly significant. In the synthetic generation process the components of the model are created in a reverse process, generating first the stochastic components and then adding the deterministic ones (the synthetic generation process is explained in Section 10.3.2). Thus, the distribution of each generated and real component should be similar. When $\hat{\varepsilon}^*_{4t}$ was generated following the method used in Boland and Magnano (2005), the distributions of ε^*_{4t} and $\hat{\varepsilon}^*_{4t}$ were different. This difference was due to the fact that the past values of ε^*_{4t} - the difference between the real and the generated data - was not available. For this reason an alternative approach was chosen to model ε^*_{4t}. In this case the data was divided into 48 groups corresponding to the period of day to which the data belongs. This approach was inspired by the work done by Ramanathan *et al* (1997), who modelled the 24 hours of the day separately. Similar to the estimation of daily cycles and relationships with current temperature, the analysis was performed for five different groups of days according to daily mean temperature. Means and standard deviations were calculated for each period and group of days. The higher levels of standard deviations are observed for groups IV and V, e.g. the groups that correspond to higher daily mean temperatures. In general, within the 48 periods - except for Group I - the standard deviations increase between the periods 16 and 37. In some groups - Groups II, III and V- this behaviour is more evident than in others.

The standardization of the ε^*_{4t} subtracting the mean and dividing by the standard deviation corresponding to each period was performed to reduce the variability corresponding to each period of day. This is:

$$\varepsilon^*_{4t} = \frac{\varepsilon_{4t} - \varphi_{ij}}{\upsilon_{ij}} \tag{10.5}$$

where $i = \overline{1,48}$, φ_{ij} and υ_{ij} are the mean and the standard deviation at period i, day group j. At this stage, different standard deviations were used for each group. A possible improvement would be to analyse if grouping these parameters in order to reduce the number of parameters used in the model would produce the same results.

The next step in the construction of the model was the analysis of ε^*_{4t}. The distributions of ε^*_{4t} for each group are symmetric (Figure fig12 shows the distribution of ε^*_{4t} for Group IV). However, the null hypothesis of normality is rejected in the five groups ($p < 0.01$).

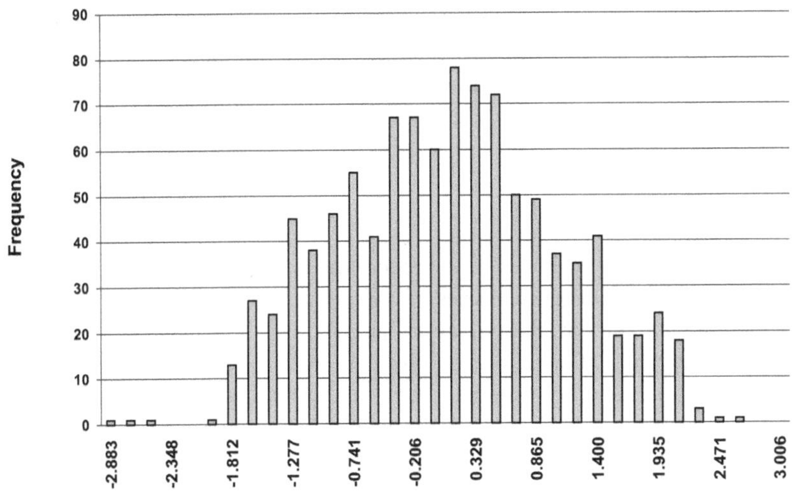

Fig. 10.11 ε^*_{4t} histogram Group IV

Since this model was designed to be used as a tool to generate synthetic sequences of half hourly demand, a transformation to a normally distributed variable was performed to the data. This transformation enables the creation of synthetic sequences with the same distribution as the original series (Sfeir 1980). Again the analysis was done by group of days separately. The transformation was performed in the following way. The empirical cumulative distribution function (*CDF*) for each group of days was obtained and for each value ε^*_{4t} its cumulative probability, $F(\varepsilon^*_{4t})$ was calculated. Then, $F(\varepsilon^*_{4t})$ was assumed to be the cumulative probability of a standard normal variable z_{4t}, i.e. $F(\varepsilon^*_{4t}) = F(z^*_{4t})$. Afterwards, the inverse function of the standard normal cumulative distribution function was calculated to obtain the corresponding value of z_{4t}. In other words the cumulative probability of the empirical data was assumed to correspond to the cumulative probability of a standard normal variate. This transformation can be shown in the following steps:

$$\varepsilon_{4t}^{*} \rightarrow F\left(\varepsilon_{4t}^{*}\right) \rightarrow F\left(z_{4t}\right) \rightarrow z_{4t} \tag{10.6}$$

At this stage the series was analysed as a single series given that, otherwise, one discontinuity is likely to occur every 48 time steps because the model depends on the daily mean temperature. Thus, every time the daily mean temperature belongs to a different group a discontinuity in the series is present. The autocorrelation function (*ACF*) and the partial autocorrelation function (*PACF*) of z_{4t} were analysed to fit ARMA models (Box and Jenkins 1976) to the transformed data. The small but significant partial autocorrelations are observed in the partial autocorrelation function (PACF) in Figure 10.12. They show that there could be some seasonal pattern remaining that has not been explained by the model. However, we proceed assuming that the seasonal component has been correctly specified. Different ARMA models were assessed under the assumption that the series is stationary and the model will be assessed for synthetic generation purposes. The models were chosen based on Akaike's (*AIC*) and Schwarz's (*SBC*) information criteria. The model that best fit the data is an ARMA(2,1).

$$\left(1-\phi_1 B-\phi_2 B^2\right) z_{4t}=\left(1-\theta_1 B\right) a_t \tag{10.7}$$

After analysing the residuals of the model the Box-Ljung test was performed to test for serial correlation of the series of residuals. The Q statistics show that there are no significant autocorrelations up to lag 17, then the autocorrelations are significant but they are still small - see Figure 10.13. The significant autocorrelations indicate that there are other patterns present in the residuals that have not been captured by the model. Although we tried to identify the reason of this behaviour, we only could identify one of them. There is a significant partial autocorrelation of order 336 which indicates that the weekly seasonality was not completely described by the dummy variables. Given that this model is used for synthetic generation rather than short term forecasting purposes, a_t is close enough to white noise to attempt to generate synthetic sequences and see if we get reasonable results.

The final model can be written as:

$$y_t=\zeta_t+\alpha_t+\varsigma_t+\tau_t+\hat{e}_{4t}^{*} \times \upsilon_t+\varphi_t+a_t \tag{10.8}$$

where ζ_t, α_t, ς_t and τ_t describe yearly seasonality, weekly seasonality, intra-day cycles and relationship with current temperature respectively, $\hat{e}_{4t}^{*}$ is an ARMA process, the values φ_t and υ_t are the mean and standard deviation of period of the day corresponding to the estimated observation and it was assumed $a_t \overset{iid}{\sim} \left(0, \sigma_a^2\right)$.

The mean absolute percentage error (*MAPE*) was calculated within the sample using Eq. 9 and the value was 0.75 which is considered to reflect a very accurate model. Also the out of sample (*MAPE*) was calculated and its value was 2.4%. This difference between the in-sample and out-of-sample assessment could be due to the problems in modelling z_t mentioned in the previous section. It is worth mentioning that even though we are not performing short-time forecasting, 2.4% reflects a very good result.

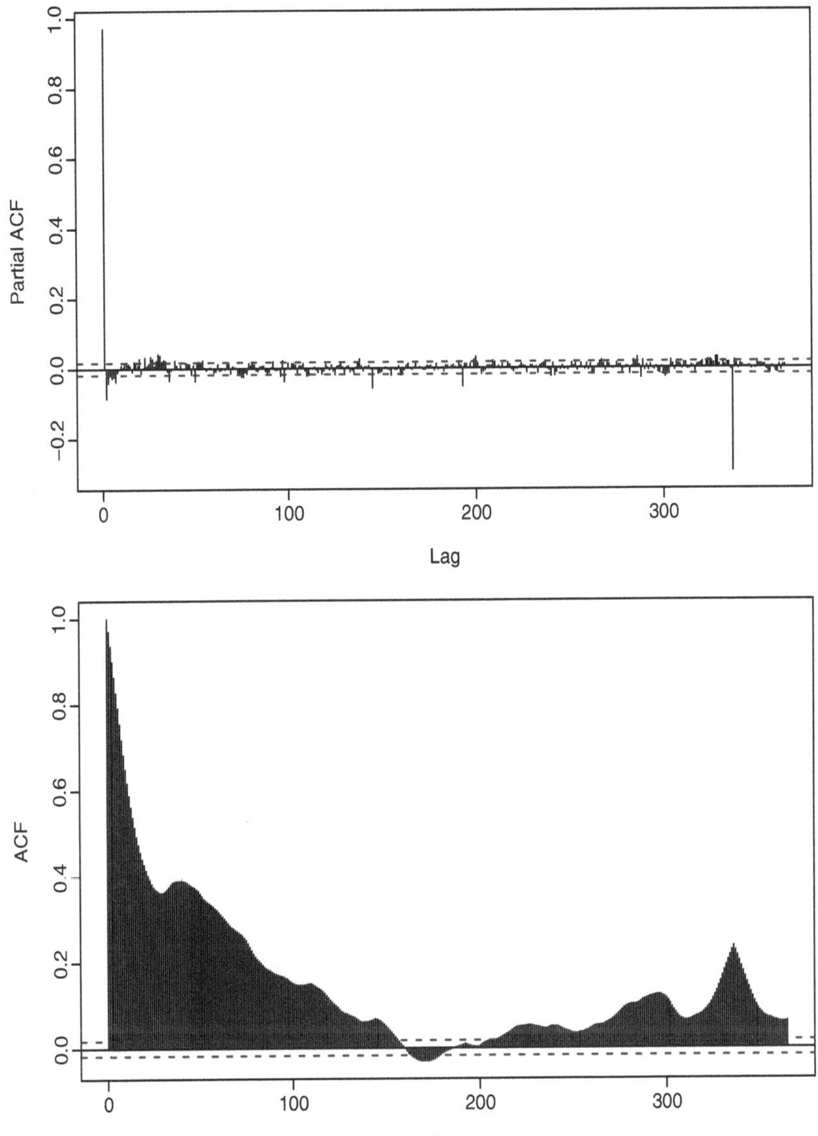

Fig. 10.12 ACF and PACF of z_4

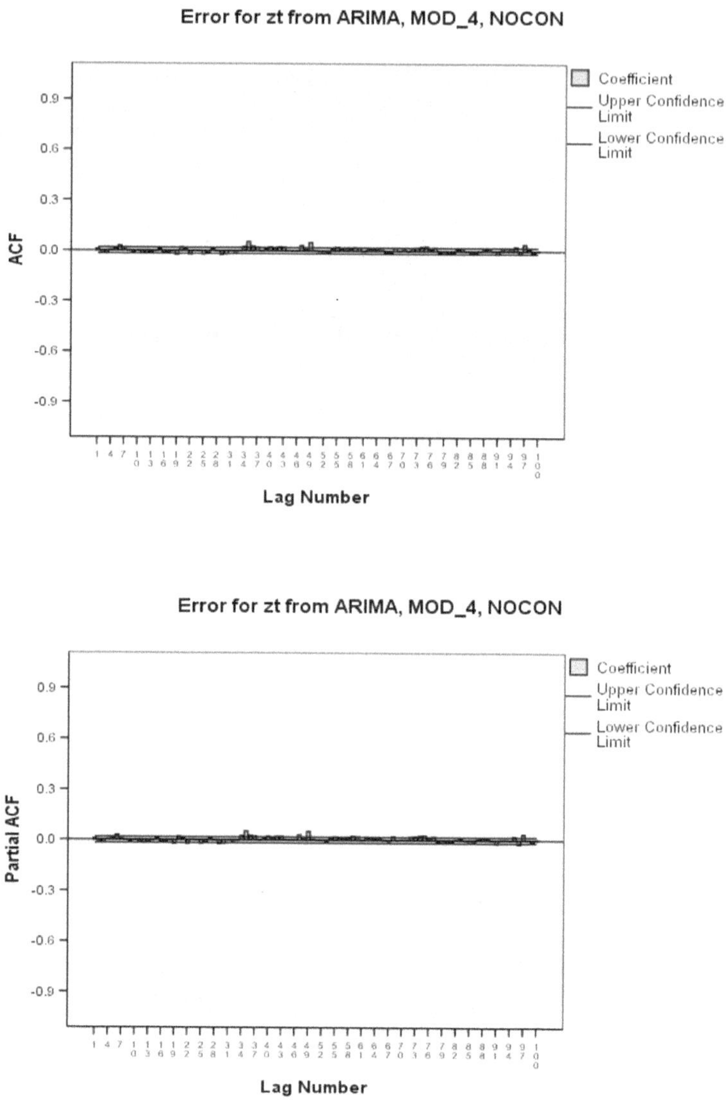

Fig. 10.13 ACF and PACF of a_t

$$MAPE = \sum_{i=1}^{n} \frac{|y_i - \hat{y}_i|}{y_i} \times 100 \tag{10.9}$$

10.3 The generation process

The real sequence of half hourly electricity demand used in the estimation of Eq. (10.8) is one of the possible realisations that could have occurred. Different values of half hourly electricity demand would have been produced by different scenarios of temperature and the influence of different random effects. This analysis aims to simulate several realisations or traces of half hourly electricity demand to use them as input data in market simulation software and to generate the frequency distribution of.the demand. The frequency distribution enables the calculation of probability of occurrence of different levels of demand. In this section Eq. (10.8) is used as a tool to generate synthetic sequences of half hourly electricity demand. The synthetic generation was done reversing the process described in previous sections. First the white noise, a_t, were generated. This generation involves the identification of the distribution of the real a_t. Then, the ARMA process was generated and $\hat{\varepsilon}^*_{4t}$ values were obtained. Afterwards the $\hat{\varepsilon}^*_{4t}$ terms were calculated simply by multiplying by the standard deviation and adding the mean of the period. Finally, the rest of the deterministic components were added to the generated data given as result the final generated value, $\hat{y}_t$.

Equation 8 depends on deterministic components, current temperature and random effects. The deterministic components and the relationship with current temperature only depend on the daily mean temperature used in the generation and they are deterministic functions. The generation of random effects obviously is not deterministic and differs in each sequence generated. The validation was done through descriptive analysis and statistical tests. The generation process is explained in detail in Section 10.3.2.

10.3.1 Final error distribution

In order to start the synthetic generation, we need to obtain the distribution of the final errors, which in this work are denoted as a_t. The histogram of a_t shows that the data distribution is symmetric and leptokurtic. Figure 10.14 displays the histogram of a_t with the corresponding normal curve. Thus, even though z_{4t} is normally distributed, a_t is not, which is a possible result.

The tails are fatter than the ones in the normal distribution. These characteristics suggest that the data follow a double exponential distribution whose probability density function (*pdf*) is given by:

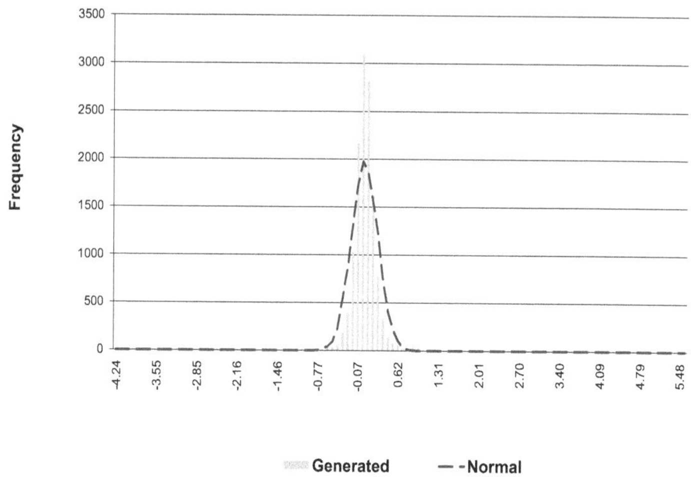

Fig. 10.14 Generated a_t and the normal curve

$$f(x) = \frac{e^{-\left|\frac{x-\mu}{\beta}\right|}}{2\beta} \tag{10.10}$$

The parameters of the double exponential distribution were estimated using maximum likelihood. Once the distribution of the final error was identified the synthetic generation was performed.

10.3.2 Steps of the synthetic generation process

The generation is presented in 3 steps. In the first step the final errors $a_t$are generated. In the second step the $a_t$are used to generate $\hat{z}^*_{4t}$ which follows an ARMA process. Then these values are transformed into $\hat{\varepsilon}^*_{4t}$. In the third step the values of the half hourly demand data are obtained. Each step is described in detail below.

- Step 1: Generation of a_t

The generation of a summer trace involves the generation of 90 days of half hourly data, i.e. the generation of 4320 observations. Firstly 4320 random numbers of a Uniform distribution on (0, 1) were generated. These values are assumed to be the cumulative probabilities of the double exponential distribution, $F(a)=P(X < a)$, which have been generated randomly. The values of a corresponding to the cumulative probabilities $F(a)$ were calculated using the following formulae:

$$\hat{a} = \begin{cases} \mu + \beta \ln(2F(a)) \text{ if } F(a) \leq 0.5 \\ \mu - \beta \ln(2(1 - F(a))) \text{ otherwise} \end{cases} \tag{10.11}$$

Figure 10.15 shows the real and the generated a_t represented by a grey bars and a dashed black line respectively.

- Generation of $\hat{\varepsilon}^*_{4t}$

Using the $a_t$obtained in the previous step the ARMA processes, $\hat{z}^*_{4t}$, which normally distributed are created. The process is generated in the following way:

$$\hat{z}_{41} = \hat{a}_1$$

$$\hat{z}_{42} = \phi_1 \hat{z}_{41} - \theta_1 \hat{a}_1 + \hat{a}_2$$

$$\hat{z}_{43} = \phi_1 \hat{z}_{42} + \phi_2 \hat{z}_{41} - \theta_1 \hat{a}_2 + \hat{a}_3$$

$$\vdots$$

$$\hat{z}^*_{4t} = \sum_{j=1}^{2} \phi_j \hat{z}_{4t-j} - \theta_1 \hat{a}_{t-1} + \hat{a}_t \tag{10.12}$$

The standard deviation of the generated process is smaller than 1 and this is related to the fact that a_t follows a double exponential distribution instead of a normal distribution. Following the inverse path used in the model estimation, the $\hat{\varepsilon}^*_{4t}$ estimates are obtained in the following way. The cumulative probability of $\hat{z}^*_{4t}$was calculated using the mean and standard deviation of the generated process. Then, each $F(\hat{z}^*_{4t})$ is assumed to be the *CDF* of $\hat{\varepsilon}^*_{4t}$, i.e. $F(\hat{z}^*_{4t}) = F(\hat{\varepsilon}^*_{4t})$. This assumption allows the identification of $\hat{\varepsilon}^*_{4t}$. We assumed that the $P(x < \hat{\varepsilon}^*_{4t}) = F(z_{4t})$ then from the empirical *CDF* of ε^*_{4t} the value of $\hat{\varepsilon}^*_{4t}$ is obtained. The process can be summarised in the following way:

$$\hat{z}_{4t} \rightarrow F(\hat{z}^*_{4t}) \rightarrow F(\hat{\varepsilon}^*_{4t}) \rightarrow \hat{\varepsilon}^*_{4t} \tag{10.13}$$

Figure 10.16 shows the histogram of the real ε^*_{4t} and one generated realisation of $\hat{\varepsilon}^*_{4t}$ represented again by grey bars and dashed black lines respectively.

- Step 3: Generation of $\hat{y}_t$

In this step the generated half hourly electricity demand value is created in the following way. Firstly, $\hat{\varepsilon}^*_{4t}$ is obtained by multiplying each $\hat{\varepsilon}^*_{4t}$ by the standard deviation, υ_t, and adding the mean, φ_t, corresponding to each period. As we explained before, the mean and the standard deviation only depend on the period within the day and on the daily mean temperature. The next step was the addition of the remaining components which are the relationship with current temperature, τ_t, intra-day cycles, ς_t, weekly seasonality, α_t, and yearly seasonality, ζ_t.

Each synthetic generated value is given by:

$$\hat{y}_t = \hat{e}^*_{4t} \times \upsilon_t + \varphi_t + \tau_t + \varsigma_t + \alpha_t + \zeta_t \tag{10.14}$$

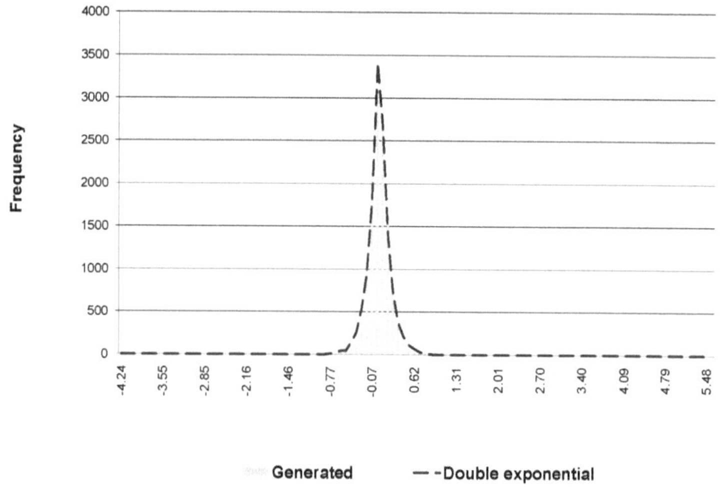

Fig. 10.15 Real and generated a_t

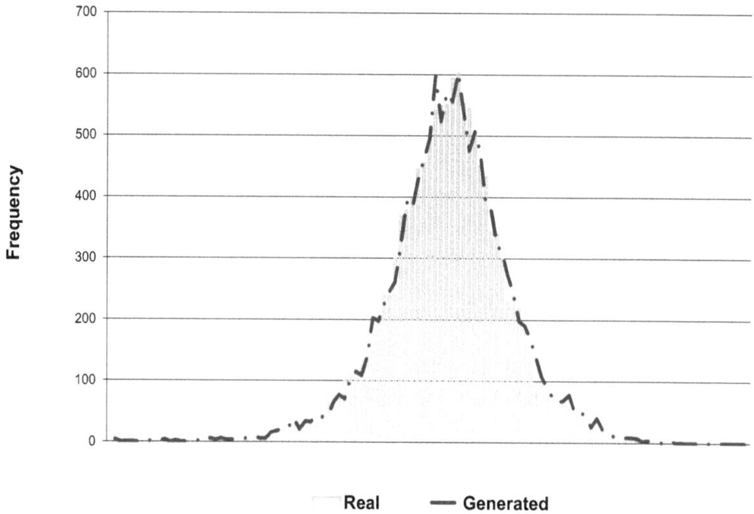

Fig. 10.16 Real and generated ε_{4t}

Figure 10.17 shows the histogram of the half hourly demand that occurred in the summer 2002-03 in South Australia and one realisation of a generated summer. In the generated series it was noted that some variability was introduced in the periods between 1.30 am and 8.00 am that usually resulted in a misspecification of the intra-day shape. It is clear from Figures 10.6 and 10.7 that the demand has a strong and deterministic pattern in this period - the demand strictly decreases until around 5.00 am and then strictly increases. The addition of the generated noise misrepresented the V-shape characteristic of this period. Thus, to follow strictly this pattern no generated errors were included in this period. A procedure that could be further investigated is the resampling of the errors from each of the periods from 1.30 am to 8.00 am. However, it should be also investigated how the autocorrelation function is affected since the ARMA process generates the relationship of current demand with immediate past demand.

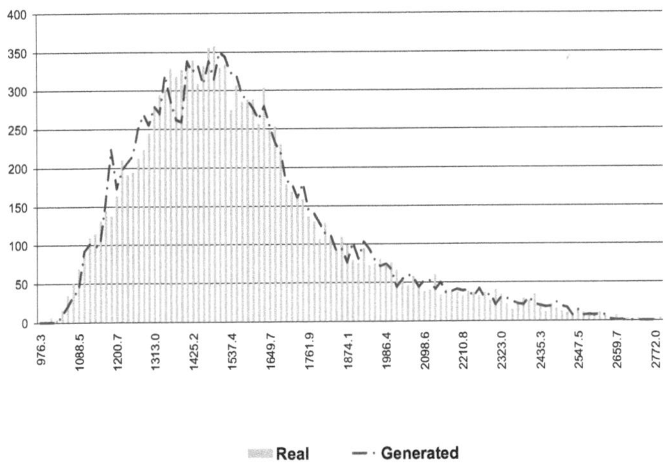

Fig. 10.17 Real and one realisation of generated half hourly electricity demand

10.4 Daily mean temperature model

Since the half hourly demand synthetic generation algorithm is dependent on the daily mean temperature, we describe how we model that variable. This includes a description of how we can produce synthetic sequences of daily mean temperature that are used as input to the demand synthetic generation.

First we compute the daily mean temperatures by simply taking the mean of the 48 half-hourly observations for each day giving 3650 daily means for the ten years of data. We also compute the standard deviation of the half-hourly temperatures for each day of the year giving 365 standard deviations.

Our daily mean model is given by

$$\bar{x}_j = \mu(j) + \sigma(j)\bar{x}_j^*,$$

where $\mu(j)$ describes the annual cycle for the mean temperature and $\sigma(j)$ describes the annual cycle for the standard deviation of temperature. These cycles are given by first-order Fourier series approximations:

$$\mu(j) = a + \delta_1 \cos\left(\frac{2\pi j}{365}\right) + \delta_2 \sin\left(\frac{2\pi j}{365}\right) \tag{10.15}$$

$$\sigma(j) = b + \gamma_1 \sin\left(\frac{2\pi j}{365}\right) + \gamma_2 \cos\left(\frac{2\pi j}{365}\right). \tag{10.16}$$

We estimate Equation 10.15 and Equation 10.16 using ordinary least squares applied using the daily means and daily standard deviations respectively. One year of daily temperatures and the estimated yearly cycle $\hat{\mu}(t)$ are shown in Figure 10.18. The lower panel of Figure 10.18 shows the fitted curve $\hat{\sigma}(t)$ along with the daily standard deviations.

Then we construct the standardized residual series, $\hat{x}_t^* = (x_t - \hat{\mu}(t))/\hat{\sigma}(t)$. Exploratory analysis showed that the distributions of $\hat{x}_t^*$ were different for each month. Thus the analysis of the standardised residuals was performed for each month separately.

For each month separately, we transform the standardized residuals to normality using the transformation $z_t = \Phi^{-1}(\hat{F}(x_t^*))$ where $\hat{F}$ is the empirical distribution function and Φ^{-1} is the inverse of the standard normal cumulative distribution function. Then z_t will have a standard normal distribution.

These residuals (again taking each month separately) were then modelled using ARMA(p,q) models:

$$\phi_p(B)z_t = \theta_q(B)a_t \tag{10.17}$$

where $\phi_p(B) = (1 - \phi_1 B - \phi_2 B^2 - \cdots - \phi_p B^p)$ and $\theta_q(B) = (1 - \theta_1 B - \theta_2 B^2 - \cdots - \theta_q B^q)$, B is the *backward shift operator* which is defined by $Bx_t = x_{t-1}$ and a_t is Gaussian white noise with mean zero and variance σ_a^2.

After analysing the autocorrelation function (ACF) and partial autocorrelation function (PACF) of the transformed data, different ARMA models were fitted to z_t. For each month, the selection of the best model was done based on Akaike's information criterion defined as

$$AIC(m) = n\ln(\hat{\sigma}_a^2) + 2M \tag{10.18}$$

where n is the effective number of observations (which is equivalent to the number of residuals that can be calculated from the series), $\hat{\sigma}_a^2$ is the maximum likelihood estimate of σ_a^2, and $m = p + q$ is the total number of parameters in the model. The optimal model is that for which AIC(m) is minimum.

The synthetic generation of daily mean temperatures is formed by reversing the above modelling procedure:

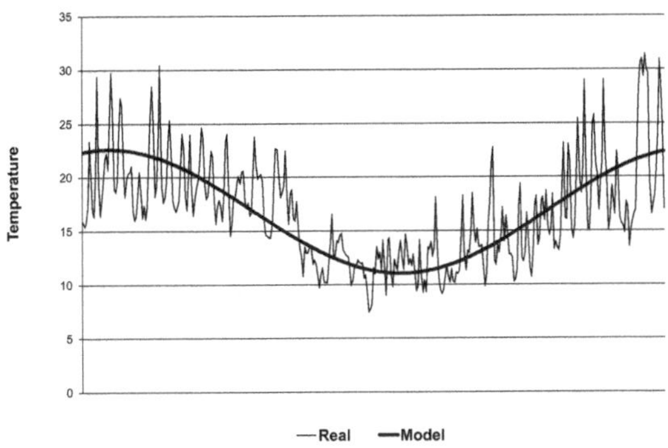

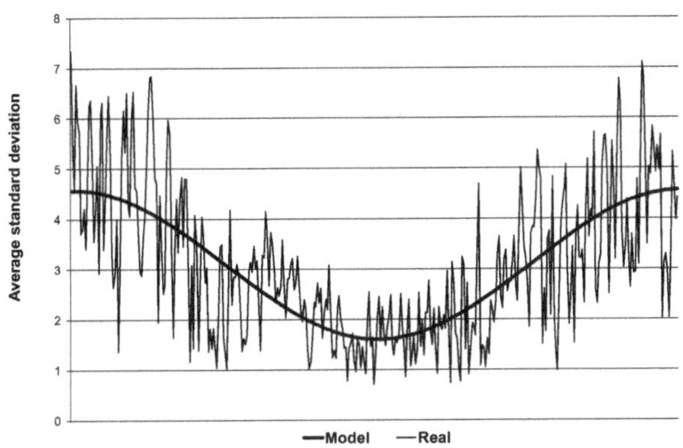

Fig. 10.18 Top: Daily mean temperatures for January–December 2005, and the estimated annual cycle $\hat{\mu}(t)$. Bottom: Daily standard deviations, and the estimated annual cycle $\hat{\sigma}(t)$.

$$\bar{x}_j = \hat{\mu}(j) + \hat{\sigma}(j)\hat{F}^{-1}(\Phi(z_t)), \tag{10.19}$$

where z_t are simulated from the ARMA process given by Equation 10.17. The errors for the ARMA model were randomly generated from a normal distribution.

10.4.1 Validation of the model

To validate the model, descriptive statistics and graphs were used to compare real and generated values. The descriptive statistics in Table 10.2 show that both the real and generated data have similar characteristics. This similarity is also shown in the empirical distribution functions of real and generated daily mean temperature presented in Figure 10.19. Finally, we compared the ACF of the real and generated daily mean temperature in Figure 10.20. Both autocorrelations follow a similar pattern.

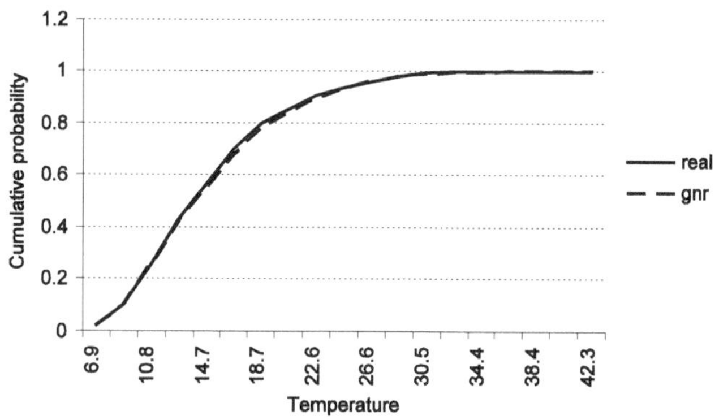

Fig. 10.19 Cumulative distribution functions of real and generated daily mean temperatures.

Table 10.2 Descriptive statistics for real and generated daily mean temperatures by month.

	Dec		Jan		Feb	
	Real	Gnr	Real	Gnr	Real	Gnr
Mean	20.9	20.5	23.2	23.4	23.2	23.5
Median	19.7	19.1	22.1	22.3	22.4	22.8
St.dev	4.5	4.3	4.8	5.0	4.7	4.8
Skewness	0.9	0.9	0.6	0.4	0.5	0.4
Kurtosis	0.2	0.1	−0.7	−0.9	−0.5	−0.7

Based on the descriptive statistics and on the graphs presented, we conclude that the daily model presented generates traces of daily mean temperature with statistical properties similar to the ones of the real data.

10.5 Verification of the generation algorithm for half hourly demand

Histograms of the half hourly demand that occurred in the summer 2002-03 in South Australia and one realisation of a generated summer are shown in Figure 10.17. The graph shows that both histograms are close. The discrepancies depicted are expected to occur because we are generating data with the same physical characteristics of the real data but with random inputs in the model. Figure 10.21 shows the output

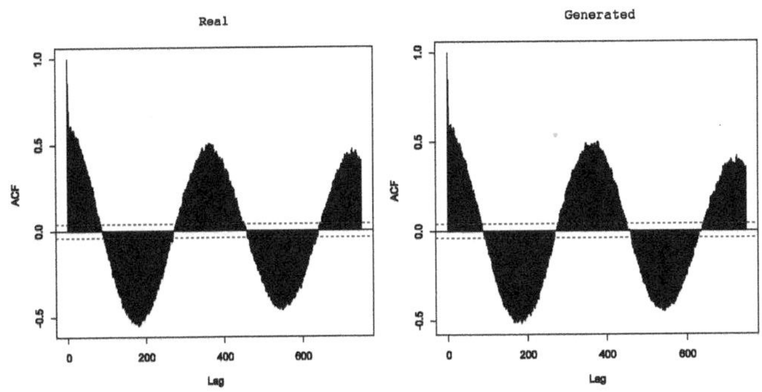

Fig. 10.20 ACF of real and generated daily mean temperatures.

in a time series form. Usually the comparison of real versus generated via a trace is not appropriate because the same pattern is not expected to result for the same period given that different temperature traces are used in each generation. However, for comparison purposes Figure 10.21 shows two traces generated with the same temperature trace.

To perform the validation of the model, the real traces used to estimate the model and several years of generated summers are compared using descriptive statistics and statistical tests. The model developed was used to generate 500 traces of half hourly electricity demand using 50 summers of synthetic half hourly temperature.

The half hourly temperature model was developed identifying the intra-day pattern of the half hourly temperature and accounting for the thermal inertia present in the daily average temperature described by some days in a row of increasing temperature before a cool change occurs. Magnano, Boland and Hyndman (2008) describe how the half hourly model encompasses not only persistence on a sub-diurnal time scale through an autoregressive process, but also dependence on the general trend from past days. Each trace of synthetic temperature was used to generate ten years of synthetic demand. First we compared the *CDFs*. Figure 10.22 shows that the *CDFs* of the real and generated data are very similar. The higher discrepancies are around the value 1850 MW.

Then, the real and generated data were analysed by intervals. Table 10.3 shows the percentage of occurrence of different levels of demand for both the real and the generated set of values. In general, the real and the estimated percentage of occurrence are similar. Descriptive statistics were calculated and the results for both set of data are also very close, see 10.3 . The higher differences occur in the percentile 95 where the generated underestimates the real data by about 30 MW. The similarity between the statistics is another indication that the generated data is a good representation of the real.

As part of the validation process two-sample Kolmogorov-Smirnov (KS) tests have been performed to test the hypothesis that two samples belong to the same population. The tests were performed for each half hour. The KS test is based on the

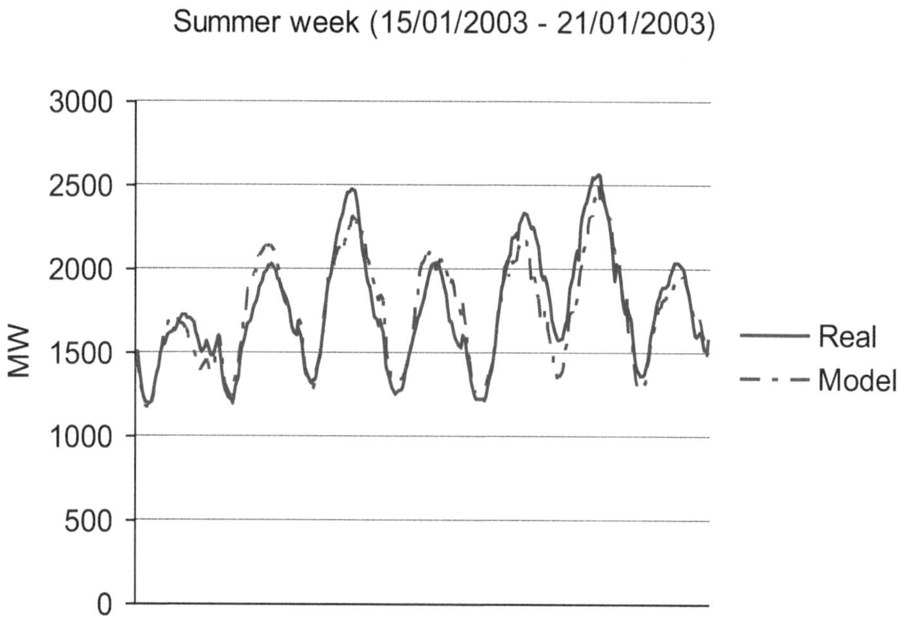

Fig. 10.21 Real and generated half hourly electricity demand - 15/01/2004 - 20/01/2004

maximum vertical distance between the cumulative distribution functions of both samples, in this case the statistic used is $\sup_i |F_r(y) - F_g(y)|$ where $F_r(y)$ and $F_g(y)$ are the *CDFs* of the real and generated sample respectively. The null hypothesis was not rejected in 44 of 48 tests - the null hypotheses were rejected between 1.30 am and 3.30 am.

10.6 Applications

We will describe two applications of the model, the probability of exceedance of extreme demands, and optimisation of orientation of photovoltaic cells to match electricity demand.

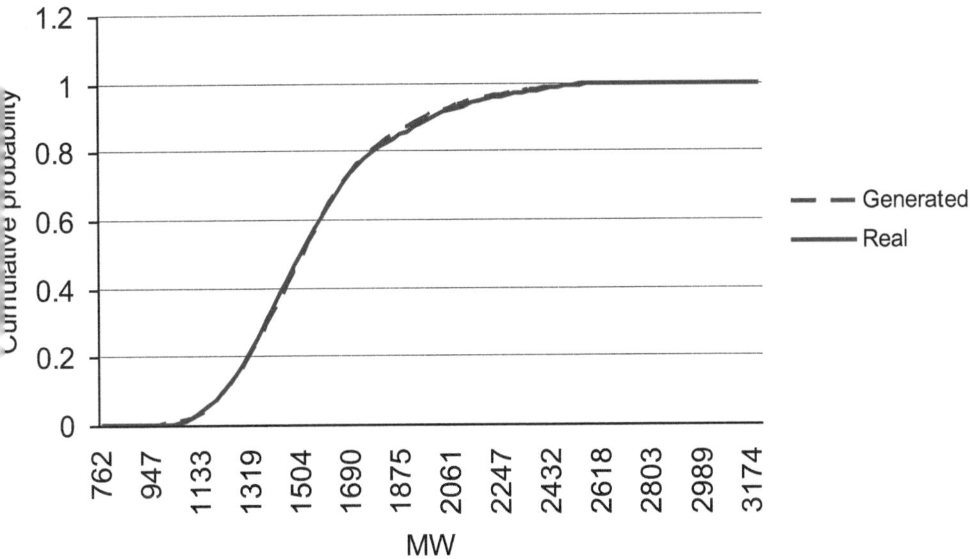

Fig. 10.22 Distribution function of real and generated half hourly electricity demand

Table 10.3 Comparison of the percentage of occurrence of different levels of half hourly demand in various intervals

MW	Real	Gnr
(0,1100]	2.091	2.096
(1100,1250]	10.887	10.730
(1250,1400]	21.304	20.119
(1400,1550]	22.600	23.243
(1550,1700]	17.677	18.502
(1700,1850]	9.252	10.380
(1850,2000]	6.304	6.099
(2000,2150]	4.259	3.873
(2150,2300]	2.600	2.495
(2300,2450]	1.782	1.530
(2450,2600]	0.980	0.708
(2600,2750]	0.231	0.183
(2750,2900]	0.031	0.035
(2900,3050]	0.000	0.003
(3050,1400]	0.000	0.000
(3200,3350]	0.000	0.000

10.6.1 Yearly peak demand

As it was mentioned at the beginning of this work, from the synthetic generations we are able to estimate probability distribution functions. Different distributions could be estimated. We could calculate the half hourly electricity demand distribution, the daily peak demand distribution or the yearly peak demand distribution. As an example we have calculated the yearly peak distribution which is shown in Figure 10.23. To do this from each of the 500 generated summers the highest half hourly value corresponding to the yearly peak was selected. Then, the empirical cumulative distribution function (CDF) of this variable was obtained.

As a particular example, we have calculated the value of demand that is not expected to be exceeded more than once in ten years, 10% PoE. This value was calculated as the value of the yearly peak CDF whose cumulative probability is 90%. The value obtained is 2870 MW. This is a sensible value since the historical maximum demand is 2833 MW which occurred in the summer of 2000-01. The actual yearly peak loads from the years used to develop the model-summers 2002-03, 2003-04 and 2004-05-are 2787 MW, 2596 MW and 2671 MW respectively. Based on the past evidence it can be concluded that the 10% PoE value is a reliable value.

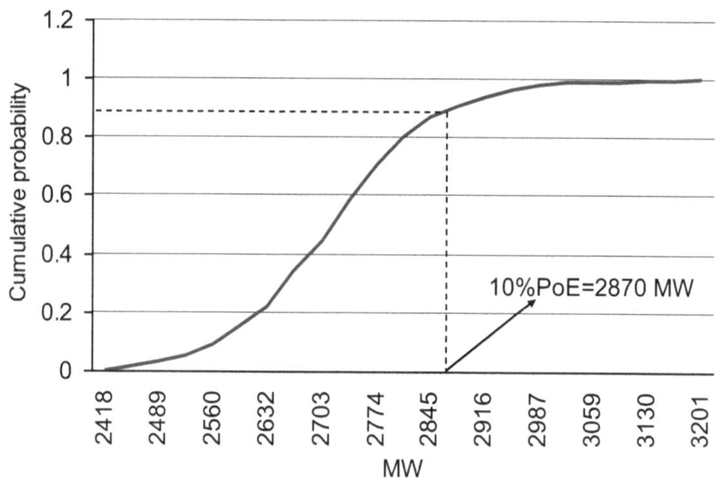

Fig. 10.23 CDF of generated yearly peak demand

10.6.2 Optimisation of orientation of photovoltaic cells to match electricity demand

The performance of any solar system depends on the solar radiation available to it. It is known that as the sun angle changes throughout the year, the amount of light falling directly on the array changes, and therefore the energy output. The sun angle is higher in the sky during summer and lower in the sky during winter. Also, the movement of the sun across the sky from the East in the morning to the West in the afternoon and the orientation of the solar panels affects the total energy output. Other factors can include the type of solar cells, the amount of cloud cover and the latitude of the location. This section will investigate how the performance of the collectors, photovoltaic cells, depends on its slope and azimuth.

The slope (tilt) is the angle at which the collectors are mounted relative to the horizontal. A slope of 0 degrees corresponds to a horizontal plane, while slope of 90 degrees corresponds to a vertical plane. Slope is always positive.

The azimuth, also called orientation, is the direction towards which the PV panel face. There are different starting positions for measuring azimuth. For example, if azimuth is measured from South it can take values: 90° for an east-facing array, 180° for a north-facing array, and 270° for a west-facing array.

Solar radiation that reaches the earth's surface can be split into two categories:

1. direct solar radiation and
2. diffuse solar radiation.

Direct solar radiation $Ih_{direct}(t)$ is radiation that passes through the atmosphere directly to the earth's surface. Diffuse solar radiation $Ih_{diffuse}(t)$ is scattered or reflected to the surface. The sum of these two components incident on a horizontal plane at the Earth's surface is referred to as the global solar radiation $Eh(t)$.

The following mathematical expression represents the above connection between solar radiation components:

$$Eh(t) = Ih_{direct}(t) + Ih_{diffuse}(t) \quad (10.20)$$

However direct and diffuse solar radiation on horizontal surface are different when they are measured on tilted surface. According to Lunde (1980) direct solar radiation on a tilted surface (I_{direct}) was calculated using trigonometric operations while Perez (1990) estimates diffuse solar radiation ($I_{diffuse}$) by correlation relationships between the diffuse on the horizontal and diffuse on the tilted surface. This implies a new formula for calculating global solar radiation on a tilted surface $E(t$.

$$E(t) = I_{direct}(t) + I_{diffuse}(t) \quad (10.21)$$

On the other hand, not all solar radiation is converted to electricity. Therefore, the useful energy collected by a solar cells $E_u(t)$ can be defined as:

$$E_u(t) = \varepsilon(Ih_{direct}(t) + Ih_{diffuse}(t)) \tag{10.22}$$

where ε is the efficiency parameter equal to 0.15, as an initial estimate. The efficiency parameter varies depending on the type of solar cells. It is known that if the temperature of the body of PV cells increases the performance of cells decreases. For example, if the temperature rises over 25°C the output efficiency of crystalline PV cells decreases by 0.5 percent per degree Celsius (Stapleton *et al*, 2004). For this preliminary study, it is assumed that efficiency is independent of type of cell and temperature. Direct and diffuse radiation are some functions of azimuth and slope depending on time period t:

$$I_{direct}(t) = f(\beta, \gamma) \tag{10.23}$$

$$I_{diffuse} = g(\beta, \gamma) \tag{10.24}$$

where γ is the azimuth and β is the slope of the collector.

Furthermore, the aim is to minimise the unmet load $u(t)$. Unmet load is the electrical load that the power system is unable to serve (Homer, 2005). It occurs when the electrical demand $L(t)$ exceeds the supply of photovoltaic cells $E(t)$.

$$u(t) = L(t) - E_u(t) \tag{10.25}$$

Representative load profiles over the day for Adelaide were used, corresponding to the season of the year (Magnano and Boland, 2007).

For mathematical formulation of this model it was assumed that there is limited storage for the solar energy, so if more electricity is produced than the load, it gets discarded. Note that this assumption is based on the whole grid and not on a single household basis.

Finally, the optimum azimuth and slope are found by minimizing the sum of the set of the values representing the hourly unmet load for a particular time period. As an example, we will perform the optimisation over the summer season, the season of highest electricity demand in South Australia.

$$min_{\gamma,\beta} U(t) = \sum_{i=1}^{n} u(t), \; u(t) > 0 \tag{10.26}$$

Here, n is the number of hours in summer. The aim of this section is to introduce some basic ideas of stochastic programming. Stochastic programming is an approach for modelling optimization problems where some of data may be subject to significant uncertainty (Shapiro and Philpott, 2002). In real life most parameters are likely to be random at the time when decision should be made. For example temperature, electricity demand and the performance of PV cells are random variables as they change through time and we are uncertain of their values in the future. Thus, in optimisation problems, the solution and the optimal objective value is random when dealing with random variables so the question is to find the optimal value under uncertainty.

Synthetic daily time series for temperature and solar radiation are constructed by adding uncertainty or white noise to the model generated by Fourier transforms. In a similar manner to the methodology used for half hourly demand and temperature, the daily temperature and solar radiation can be modelled in the form:

$$R_{Yt} = \frac{Y_t - \mu_{Yt}}{\sigma_{Yt}} \tag{10.27}$$

In essence, we have once again subtracted the mean function and standardised by dividing by the standard deviation function.

We then model the residuals as an $AR(1)$ process for solar radiation and an $AR(2)$ for temperature.

$$R_{St} = \alpha RS,t-1+Z_{St} \tag{10.28}$$

$$R_{Tt} = \theta_1 R_{T,t-1} + \theta_2 R_{T,t-2} + Z_{Tt} \tag{10.29}$$

To begin with we would like to determine distributional properties for the white noise $Z_{\Gamma t}$. From histogram analysis of temperature and solar radiation with a superimposed normal curve it was obvious that the $Z_{\Gamma t}$ are not normally distributed. Consequently since the original series are physically bounded from both sides it was concluded that a Beta distribution is the distribution that fits the data the best.

The construction of the stochastic inputs has been performed using an algorithm developed by Boland (2008):

1. Firstly we use the inverse transform method to generate the noise as random variates with the Beta distribution. Three hundred and sixty-five uniformly distributed random numbers $U \sim U[0,1]$ were generated, then the sequence of white noise terms $Z_{\Gamma t} = B^{-1}(U), t = 1, 2, \ldots, 365$ was calculated.
2. Next, using the $Z_{\Gamma t}$ from the previous step the temperature and solar ARMA processes were created. Given the values of the coefficients for each process the values of $R_i, i = 1, ..., 365$, are calculated using the recursive formulae (10.28), (10.29).
3. Lastly by rearranging Equations (10.27)or (10.28) we get the following general formula for generating a synthetic sequence for solar radiation and temperature: $Y_t = \mu_{Yt} + \sigma_{Yt} R_{Yt}$. As a result 365 observations were created for the temperature and solar radiation daily synthetic time series.

The spreadsheet model we created for optimising the orientation and slope for a specific set of inputs is programmed in such a way as to accept as input hourly values for each day of a year, ie 8760 observations. Thus we have to form hourly synthetic data. This is obtained by matching a day in the daily synthetic data sets with a corresponding day from 38 years of data from the Australian Climatic Database (ACDB) (Energy Partners et al, 2005) for solar radiation and from 25 years of data from the Magnano database (Magnano, Boland and Hyndman, 2008) for temperature and taking its hourly data values. We generated 250 realisations of half hourly electricity demand using 25 summers of synthetic half hourly temperature, ie 10 realisations for each temperature set.

A more precise algorithm for this generation follows:

1. Using the previous algorithm, generate synthetic solar radiation S_{si} and temperature T_{si} sequences for each day of a year.
2. Calculate daily averages of solar radiation and temperature from the corresponding databases.

Solar Radiation:

- From the ACDB, calculate the daily average global radiation for Adelaide for each of the days over the forty years $\overline{S}_{ij}, i = 1,\ldots,365; j = 1,\ldots,40$
- For each day in a year find the average real data value (out of the 40 possible) that most closely matches the synthetically generated value $\overline{S}_{ij} \approx S_{si}$.
- Take hourly values from day i and year j from Australian Climatic Database for diffuse solar irradiation on a horizontal plane (Wh/m2) and Direct solar irradiation on a plane normal to the beam (Wh/m2) to build a synthetic solar year. As a result we will have 8760 observations for diffuse and direct radiation.

Demand:

- For each day in a summer season match synthetic T_{si} and averaged $\overline{T}_{ij}$ temperature values where for days $i = 1,\ldots,59$ in January and February and $i = 334,\ldots,365$ for December.
- Take hourly demand values from day i and randomly chose a demand realisation for temperature realisation T_{si} from the Magnano database to build synthetic summer demand. As a result we will have 2160 demand observations.

We have produced the model that generates synthetic radiation and demand data that shows similar behaviour to the real. These synthetic data were used as inputs into the spreadsheet model. As stated before, the spreadsheet model is based on minimising the unmet load over the summer period in Adelaide. It has been run to generate 500 optimal slope and azimuth sets. Figures 10.24 and 10.25 present distributional properties of azimuth and slope respectively.

From the graphs above it appears that the optimal azimuth is 18 degrees while the slope appears to be bi-modal with peaks around 14.8 and 15.4 degrees from the horizontal. Inspection of the output indicates that the lower angle corresponds to higher demand years and vice versa.

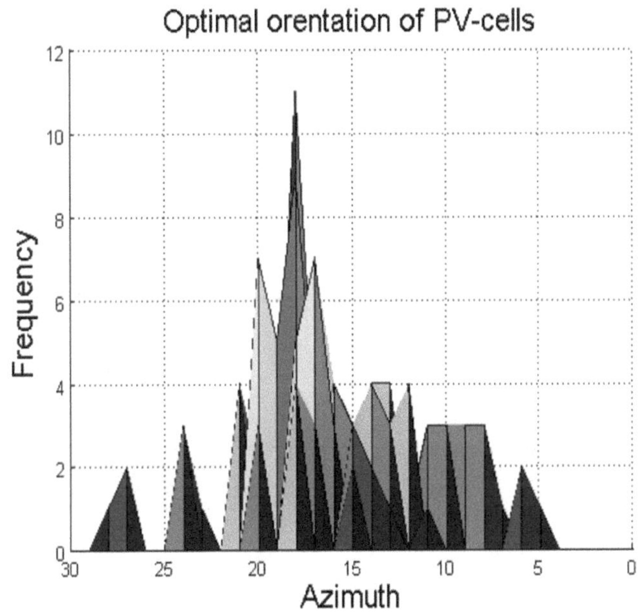

Fig. 10.24 Azimuth distribution

10.7 Conclusion

The objective of this work was to build a model that generates possible traces of half hourly electricity demand that could have occurred to use them as input data in market simulation software and to calculate probabilistic forecasts. Both objectives have been achieved. Though the probabilistic forecasts are static and do not include the effects of future economic movements, this could be added.

To assess the suitability of the model in generating synthetic sequences many comparisons and tests between real and generated data were performed. The results showed that the model is successful in generating data with similar characteristics to the real. We would like to emphasise that the model is very attractive since its terms explain the physical behaviour of the series. For example, the profile of a specific weekday under different scenarios of temperature can be identified.

We have demonstrated two separate types of applications of use of the model. One is in the identification of extreme values of demand. The other is in an area that will become increasingly attractive because of the introduction of the mandated

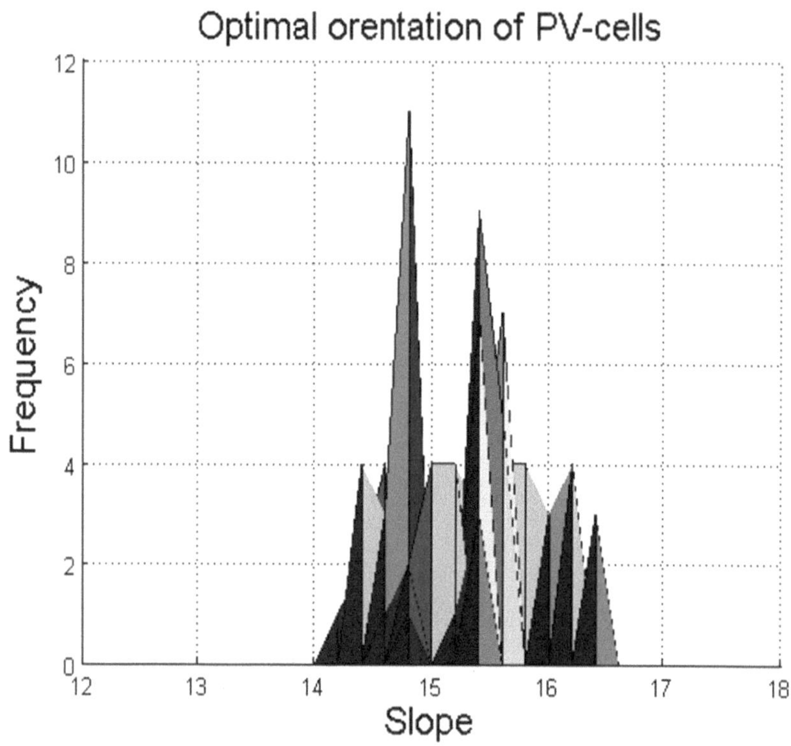

Fig. 10.25 Slope distribution

renewable energy target (20% renewable electricity by 2020) and an emissions trading scheme, set to begin in Australia in 2010. The optimisation of the orientation of solar cells to match electricity demand will aid in setting up solar farms to generate electricity.

Extensions of the model are being undertaken in different aspects. The first one is the inclusion of demographic and economic variables aforementioned. The second is the analysis of sensitivity of demand on temperature. Given a year of half hourly temperature, several realisations of half hourly demand can be generated. Thus, generating these sequences with two different scenarios of temperature - mild and hot average summer temperature - the increase in the 10% PoE load for a given increase in average summer temperature can be determined by analysing the generated empirical distributions. Finally, based on the model presented, we are analysing how climate change affects demand. We will do that by altering the temperature sequences in a way that captures climate change effects.

Acknowledgements We are grateful for the financial support, help in interpreting results and supply of data of the Electricity Supply Industry Planning Council of South Australia.

References

[1] Blacona M, and Abril C (2000) Modelo Estructural de Espacio de Estado para la demanda diaria promedio de energia electrica en la Republica Argentina. http://www.aaep.org.ar/espa/anales/resumen$_$00/blacona$_$abril.htm

[2] John Boland (2008) Time Series Modelling of Solar Radiation, in *Modeling Solar Radiation at the Earth's Surface*, Viorel Badescu (Ed.), Springer Berlin Heidelberg.

[3] Boland J. and Magnano L. (2005) Modelling Weekday Half Hourly Electricity Demand. *Proceedings of the International Conference of Applied Simulation and Modelling*. Paper 469-092.

[4] Boland J. and Zekanovic, Z. (2008) Optimising the orientation of photovoltaic cells for load matching, *Proceedings of the 3rd International Solar Cities Congress*, 17-21,February 2008, Adelaide, Australia.

[5] Box G.E.P. and Jenkins G.M. (1976) *Time Series Analysis: Forecasting and Control*. San Francisco: Holden-Day.

[6] Bunn D. W and Farmer E. D. (1985) *Comparative models for electrical load forecasting*. New York: John Wiley & Sons.

[7] Energy Partners Australia (2005) *Development of Climate Data for Building Related Energy Rating Software*, Energy Partners Australia, Canberra.

[8] Harvey A. and Koopman S. J. (1993) Forecasting Hourly Electricity Demand Using Time-Varying Splines. *Journal of the American Statistical Association* 88: 1228-1236.

[9] Liu J.M., Chen R., Liu L-M, Harris J.L. (2006) A semi-parametric time series approach in modeling hourly electricity load. *Journal of Forecasting* **25**, pp. 337-359.

[10] Magnano L. and Boland J. W. (2007) Generation of synthetic sequences of electricity demand: Application in South Australia, *Energy*, **32**, pp. 2230-2243.

[11] Magnano L., Boland J. and Hyndman R. (2008) Generation of synthetic sequences of half-hourly temperature, *Environmetrics*, **19**, pp. 818-835.

[12] Manera M. and Marzullo A. (2005) Modelling the load curve of aggregate electricity consumption using principal components. *Environmental Modelling & Software***20**,pp. 1389-1400.

[13] Metaxiotis K., Kagiannas A., Askounis D. and Psarras J. (2003) Artificial intelligence in short term electric load forecasting: a state-of-the-art survey for the researcher. *Energy Conversion and Management*, **44**, pp. 1525-1534.

[14] McSharry P. E., Bouwman S. , and Gabriel Bloemhof. (2005) Probabilistic Forecasts of the Magnitude and Timing of Peak Electricity Demand. *IEEE Trans. Power Syst*, **20(2**, pp. 1166-1172.

[15] Mirasgedis, S. and Sarafidis, Y. and Georgopoulou, E. and Lalas, D.P. and Moschovits, M. and Karagiannis, F. and Papakonstantinou, D. (2006) Models for mid-term electricity demand forecasting incorporating weather influences. *Energy*, **31**,pp. 208-227.

[16] NEMMCO (2005) Statements of Opportunities. National Electricity Market Management Company (NEMMCO). Sydney, Australia. `http://www.nemmco.com.au/nemgeneral/soo$\_$2005.htm`

[17] Pardo A., Meneu V., and Valor E. (2002) Temperature and seasonality influences on Spanish electricity load. *Energy Economics*, **24**, pp. 55-70.

[18] Perez R, et al. (1990) Modelling daylight availability and irradiance components from direct and global irradiance, *Solar Energy*, **44, No.5**.pp. 271-289.

[19] Ramanathan R., Engle R., Granger C.W.J., Vahid-Araghi F., Brace C. (1997) Short-run forecasts of electricity loads and peaks, *International Journal of Forecasting*, **13** pp. 161-174.

[20] Shapiro A and Philpott A (2002) A Tutorial on Stochastic Programming, *The Committee on Stochastic Programming*, `http://stoprog.org/index.html?introductions.html`, Viewed on 28 Aug 2007.

[21] Stapleton, G. , Milne, G. (2004), Photovoltaic systems,*The Institute for Sustainable Futures, University of Technology*, Sydney, `http://www.greenhouse.gov.au/yourhome/technical/fs47.htm`

[22] Sisworahardjo N.S., El-Keib A.A., Choi J., Valenzuela J., Brooks R. and El-Agtal I (2006) A stochastic load model for an electricity market, *Electric Power Systems Research*, **76(6-7)**, pp. 500-508

[23] Soares L.J. and Souza L.R. (2006) Forecasting electricity demand using generalized long memory. *International Journal of Forecasting*, **22**, pp. 17-28.

[24] Taylor J. (2003) Short-term electricity demand forecasting using double seasonal exponential smoothing. *Journal of the Operational Research Society*, **54**, pp. 799-805.

[25] Taylor J. W., Menezes L. M. and McSharry P. E. (2006) A comparison of univariate methods for forecasting electricity demand up to a day ahead, *International Journal of Forecasting*, **22(1)**, pp. 1-16.

Chapter 11
Models for Improving Management of Biosolids Odors

Steven A. Gabriel, Sirapong Vilalai, Prawat Sahakij, Mark Ramirez , Chris Peot

Abstract We describe recent modeling efforts to identify the factors that lead to high biosolids odor levels associated with advanced wastewater treatment plants (AWTP). These factors can be broken down into two groups: (i) those that are beyond the control of the AWTP such as ambient temperature, (ii) those that are controllable such as the number of centrifuges in operation, the amount of lime used, etc. We summarize our findings relative to different statistical models we developed that predict biosolids odor levels based on either subjective or analytic measurements from the District of Columbia Water and Sewer Author (DCWASA). These models take into account a host of factors to predict biosolids odor levels and are then used to generate a relevant probability distribution for odor levels using Monte Carlo simulation. Such probability distributions will guide AWTP managers relative to where to send the biosolids products taking into account the likelihood of high levels and thus indirectly, possible complaints from those living or working near the reuse sites. We also describe our recent efforts in also optimizing the operations of the AWTP and distribution network to balance both biosolids odors and costs. The resulting multiobjective optimization models are computationally challenging due to their size and non-convexities and we discuss some of the salient features as well as representative results. We extend this discussion by noting how to handle stochasticity directly into such optimization models.

11.1 Introduction, Overview, and Literature Review

The U.S. Environmental Protection Agency (EPA) through recent rules (Fed. Register, 1993; EPA, 1994ab) regulates the reuse of biosolids products produced at ad-

Steven A. Gabriel, Sirapong Vilalai, Prawat Sahakij
University of Maryland, College Park, Maryland, 20742 USA

Mark Ramirez , Chris Peot
District of Columbia Water and Sewer Authority, Washington, DC 20032 USA

J.A. Filar and A. Haurie (eds.), *Uncertainty and Environmental Decision Making*,
International Series in Operations Research & Management Science 138,
DOI 10.1007/978-1-4419-1129-2_11,

vanced wastewater treatment plants (AWTPs). Recycling these biosolids is a desirable option as compared to incineration or landfill but can generate complaints from the surrounding communities where the biosolids are applied. Despite these malodorous aspects, land application of biosolids is substantial and is slated to reach 70% in 2010 (Oleszkiewicz and Mavinc, 2002).

Biosolids managers have a goal of producing the least smelly product at a reasonable cost. While it is known that the bad odors mostly originate from reduced sulfur and nitrogenous compounds (Mosier et al., 1977), what is not known with certainty is how best to reduce these odors for a given day as well as how to plan sufficiently ahead of time to balance both odors and costs. Typical management decisions include how much of certain chemicals to add (e.g., lime, polymer), how much equipment to have operating (e.g., centrifuges), as well as a host of other tools.

The challenging task is to come up with a plan that is the least costly but reduces the odors sufficiently. This task thus requires a quantitative knowledge of how much odor can be reduced by the application of each of these tools in standard amounts as well as the associated costs. Statistical models to predict odor levels based on key independent variables are crucial in this step. These models can be based on analytical laboratory experiments odors (D'Amato and DeHollander, 1999; Rosenfeld 1999; Sostrand et al., 2000; Kim et al., 2001, 2002; Murthy et al., 2001, 2002; Novak et al., 2002; Gabriel et al., 2005) or subjective measures based on odor panels or field odor inspectors (Vilalai, 2003; Vilalai et al., 2007; Gabriel et al., 2006b).

The next step is to employ a multiobjective optimization model to find the Pareto optimal (Cohon, 1978) levels of these independent variables that lead to the best compromise of odors and costs. Development of such optimization models was the focus of the work in (Gabriel et al., 2007; Sahakij et al. 2007; Sahakij, 2008ab).

However, many of the inputs to the statistical and optimization models are in reality uncertain since they dependent on complex systems such as the weather (e.g., ambient temperature), subject to human error (e.g., excessive chemical dosing), or hard to predict (e.g., when certain equipment fails). Thus, one most consider that odor levels being a function of these influential variables is itself a random variable with some probability distribution.

In this paper, we present recent work on both statistical models to predict odor and a sample odor probability distribution (Section 11.2). Then, in Section 11.3 we describe recent multiobjective optimization models to balance odors and costs and describe how the probabilistic and optimization aspects can be brought together. Section 11.4 is a summary of the paper and a description of future research directions.

11.2 Summary of Statistical Modeling Work

As described above, statistical models were developed to predict biosolids odor levels using both subjective and objective factors. In what follows, we briefly describe three of these models and then use the results to generate a probability distribution for biosolids odor levels.

11.2.1 Statistical modeling to forecast odor levels of biosolids applied to reuse sites

The first model connected biosolids odor data at the reuse site with potentially odor-causing factors and ambient conditions at the AWTP. Due to limited information available on biosolids odor levels, subjective field odor data from inspectors were taken as an approximation of biosolids odor level from AWTP on any particular day regardless of the weather and field conditions at the application site.

Field inspectors identified the biosolids field odor level as either 0 (negligible), 3 (slight), 6 (moderate), or 9 (strong). Since each field inspector generally performed multiple odor assessments on various field sites each day, biosolids field odor data of an inspector on any particular day were averaged to represent the inspector's score for that day.

Data related to operations at the AWTP such as the level of sludge at the bottom of the secondary sedimentation basin, the percentage of gravity-thickened sludge in the blend tank, the number of centrifuges in operation, the amount of chemical additions, etc. were analyzed to see their connection with the field odor data. Additionally, ambient conditions such as wind speed, temperature, snowfall, and rainfall were also considered in the analysis. Linear regression was used to develop the associated statistical models (Kutner et al., 2005).

To provide a more comprehensive model, dummy variables were used relating to when particular phenomena were present. Dummy variables of selected factors at the 80th percentile were created to investigate the contributions of these extreme conditions to biosolids odor production. Examples include when the sludge blanket level was greater than 10 feet, when the $FeCl_3$ additions were more than 1290 gallons, or when the average temperature was greater than 77 degrees Fahrenheit. Consequently, a dummy variable was set equal to one if the phenomenon was present and equal to zero otherwise. In addition, interaction variables were used to capture the effects of two variables, for example a low level of lime and a high sludge blanket level.

The selected model used field odor data from one particular inspector during the winter period of 2002 was able to explain 82 percent of the variation in the odor data (adjusted $R^2 = 0.82$) and included the following variables:

- Two variables related to the dewatering process: number of centrifuges in service and number of belt filter presses in service
- Three variables related to chemical additions: amount of lime dose on day d-1[1], a dummy variable for when the polymer amount was greater than 200.5 lbs/dry

[1] The index d-1 represents the day before the odor data values were measured.

ton on day d-1, and a dummy variable for when the lime amount was less than 200.5 lbs/dt on day d-1,
- One variable associated with operations at the secondary process: sludge blanket level d-1 and;
- One variable related to ambient conditions: minimum temperature on day d-1.

The resulting statistical model was the following:

$$Y_d = 3.89 + 0.03X_1 + 0.98X_2 - 0.47X_3 - 1.91X_4 - 0.01X_5 + 0.56X_6 + 1.13X_7, \quad (11.1)$$

where

$Y_d =$ Inspector C's biosolids odor score on day d
$X_1 =$ the minimum temperature on day d-1, degrees Fahrenheit
$X_2 =$ the sludge blanket level at secondary east side on day d-1, feet
$X_3 =$ the number of belt filter presses in service on day d-1
$X_4 =$ the number of centrifuges in service on day d-1
$X_5 =$ the amount of lime addition on day d-1, lbs/dry ton
$X_6 =$ a dummy variable for when polymer addition is greater than 200 lbs/dry ton
$X_7 =$ a dummy variable for when lime addition was less than 308 lbs/dry ton

Note that all variables were collected on day d-1. Therefore, this model predicted biosolids odor at field site on the following day after data on independent variables were collected. All variables had the correct coefficient signs and were statistically significant at the 10% level except the amount of lime addition (p-value= 0.25). Figure 11.1 depicts prediction results from the model as compared with the actual field odor data for 2001-2002.

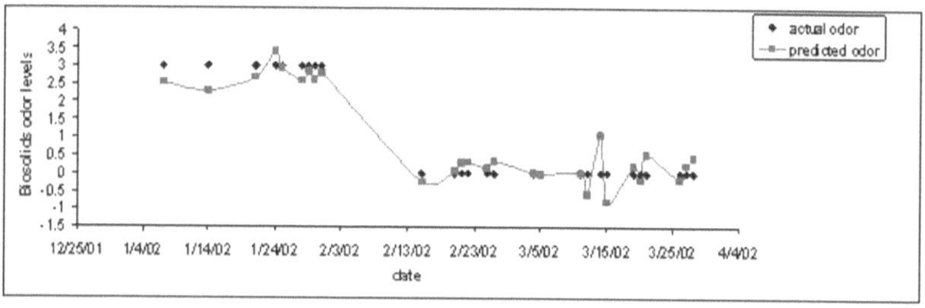

Fig. 11.1 Predicted and actual biosolids odor levels from model using inspector "C" data (Gabriel et al., 2006b)

Findings from this model provide useful information for biosolids odor management and were as follows:

- Controllable variables (i.e., the amount of lime and polymer additions and the sludge blanket level) had significant impacts on biosolids odor production. Controlling the sludge blanket level and applying appropriate amounts of lime and polymer would result in acceptable biosolids odor levels.

- Equipment used in the dewatering process such as belt filter presses and centrifuges contributed to odor reduction. A sufficient number of centrifuges operating relative to the amount of solids needed to be dewatered was important. This can be accomplished by providing regular maintenance to maintain the full dewatering capacity in the dewatering process.

More information on this model can be found in (Gabriel et al., 2006b; Vilalai , 2003).

11.2.2 Prediction of Dimethyl Disulfide levels from biosolids using statistical modeling

The second statistical model unlike the first one used a dependent variable that was analytically determined, namely concentrations of odorous chemical compounds from biosolids samples at the AWTP. Concentrations of various chemical compounds from biosolids were analytically assessed by multidimensional gas chromatography-mass spectrometry (GC-MS) (Aripse, 2005). Of the several odorous compounds extracted, dimethyl disulfide (DMDS), a major volatile sulfur compound contributing to biosolids odor, was eventually selected as the dependent variable (Gabriel et al., 2005).

Besides the operational variables considered in the first model, data on properties of sludge in the sludge blending system were also collected. These variables included the oxidation reduction potential (ORP)[2], temperature, pH, and H_2S concentrations from sludge in the gravity thickener (GT) process the dissolved air floatation (DAF) process, and the blend tank (BS).

The final model that was statistically signficant was the the following:

$$Y = -41.311 - 0.075X_1 - 0.104X_2 + 19.919X_3 + 19.456X_4, \tag{11.2}$$

where

$Y=$ DMDS ppmv/DMDS odor threshold ppm
$X_1=$ ORP from gravity thickener sample (ORP GT), (mv)
$X_2=$ ORP from DAF sample (ORP DAF), (mv)
$X_3=$ amount of sludge dewatered per centrifuge, (gallons/day/centrifuge)
$X_4=$ Blend ratio[2]

The model explained 79.44 percent of the variation in DMDS concentrations (adjusted $R^2 = 0.7944$). Figure 11.2 illustrates the predicted outcomes from the model compared with the actual values of DMDS concentrations for 2003-2004.

[2] ORP was used as an indicator of oxidation and reduction state of sludge as well as used as a means to measure anaerobic conditions of the sludge sample. A lower value of ORP means a greater anaerobic state of the sample.

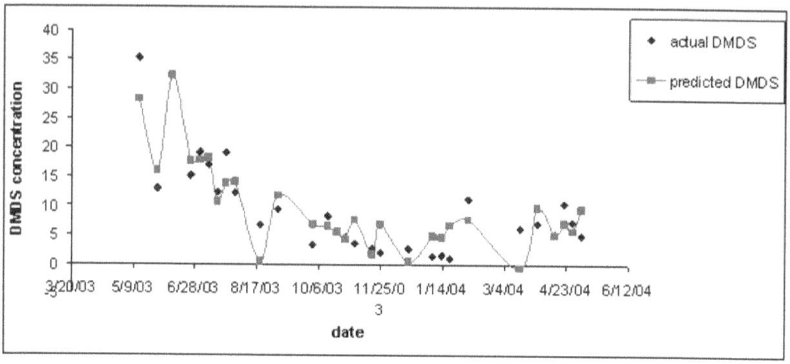

Fig. 11.2 Predicted and actual DMDS concentrations from DMDS model (Gabriel et al., 2005)

This model helped to show that ORP for sludge from GT and DAF needs to be monitored. Also, the capacity of the dewatering process as identified by the variable "amount of sludge dewatered per centrifuge" indicates that a proper amount of sludge to be dewatered needs to be maintained relative to the number of centrifuges available. Lastly, this analysis showed that it is key to manage the ratio of GT solids in the blend tank to a level less than 60 percent [3] for low biosolids odor levels. Further details on the DMDS model can be found in (Gabriel et al., 2005).

11.2.3 A combined subjective and objective odor model

The third model considered biosolids odors from both an analytical as well as a subjective perspective. This approach was different from the first two models in that odors from biosolids samples were assessed both by an odor panel of people as well as an odor-analyzing machine. How humans interact with biosolids odor was crucial to understand to develop a biosolids odor prediction model that reduces impacts of biosolids odor to communities. People can respond differently to distinct odor types and intensity. Thus, an odor panel of trained individuals was employed to assess biosolids odor levels taken from samples at DCWASA. In our work, a detection threshold (DT) was used as the dependent variable and it represented biosolids odor levels to be perceived by people. DT is the diluted concentration of odor where half of the odor panel can detect it correctly thus relating somewhat to the odor level where the majority of people in a community might first detect biosolids odors (McGinley and McGinley, 2002).

H_2S concentrations, another analytical odor measurement, were collected in addition to DT values. H_2S concentrations were more convenient to assess by a hand-

[3] A blend ratio of 60 percent or higher was identified in one of DMDS models (Gabriel et al., 2005) as a level influencing high biosolids odors.

held odor analyzer and were taken as an alternative to the odor panel data. Odor measurements in this model were taken at 24 hours after the samples were collected. The purpose was to take into account a 24-hour lag effect of biosolids odors relating to retention in bunkers at the plant before being sent out to reuse sites.

Independent variables considered in this model were similar to the first two models: operational variables at the plant and properties of sludge in the sludge blending system. However, additional variables related to properties of biosolids (i.e., percent solids and temperature) were also considered. These variables indicated properties of the final product.

The final model identified two variables associated with biosolids properties (i.e., percent solids and temperature of biosolids sample), two variables associated with the sludge blending system (i.e., the pH of sludge from the gravity thickener and percent of the gravity thickener sludge in the blend tank), and two variables associated with the operations at the plant (i.e., return-activated sludge concentration (RAS) and number of centrifuges running). The following is the resulting equation to predict DT levels:

$$Y = -12936.392 - 141.544X_1 + 256.656X_2 + 1828.693X_3 - 200.721X_4 + 31.349X_5 + 0.159X_6, \quad (11.3)$$

where

$Y =$ DT concentration, odor unit (ou)
$X_1 =$ percent solids of biosolids, %
$X_2 =$ temperature of biosolids, degrees Fahrenheit
$X_3 =$ GT pH
$X_4 =$ number of centrifuges running
$X_5 =$ percentage of gravity thickener sludge in the blend tank, %
$X_6 =$ concentration of return activated sludge in secondary process, mg/l

In terms of how biosolids managers could use equation (11.3), they could either apply appropriate actions to variables that were controllable (e.g., number of centrifuges) and/or employ monitoring methods for the remaining variables. Therefore, we suggest that the percentage of gravity thickener sludge in the blend tank should be maintained at 60 percent or less to guarantee acceptable biosolids odor levels. In addition, providing a sufficient number of centrifuges operating and guaranteeing regular maintenance are important to controlling biosolids odors in the dewatering process. The remaining variables including GT pH, temperature of biosolids, percent solids of biosolids, and RAS concentration at the secondary process should be monitored and reported to biosolids management on a daily basis.

11.2.4 Odor probability distribution and simulation analysis

Equations (11.1), (11.2) and (11.3) were based on predicting biosolids odor levels from a set of influential independent variables whose values were taken to be deterministic. In reality, each of these independent variables' values could be uncertain due to a variety of factors. Thus, if one had the probability distributions for these independent variables, then, using any of these three equations, one could generate a probability distribution for biosolids odor levels using any of the dependent variable measures discussed above. Using Monte Carlo simulation (Clemen and Reilly, 2001) this would lead to a probability distribution for biosolids odor levels.

Equation (11.3) was selected to show how to incorporate uncertainty into a biosolids odor production model. The first step then is to fit appropriate distributions for these independent variables. The software @Risk (Palisade Corporation, 2001) was employed to derive theoretical probability distributions for these independent variables based on the chi-square goodness-of-fit test. Figure 11.3 illustrates an example after fitting a probability distribution to data for the variable GT pH. The closest fit was the logistic distribution.

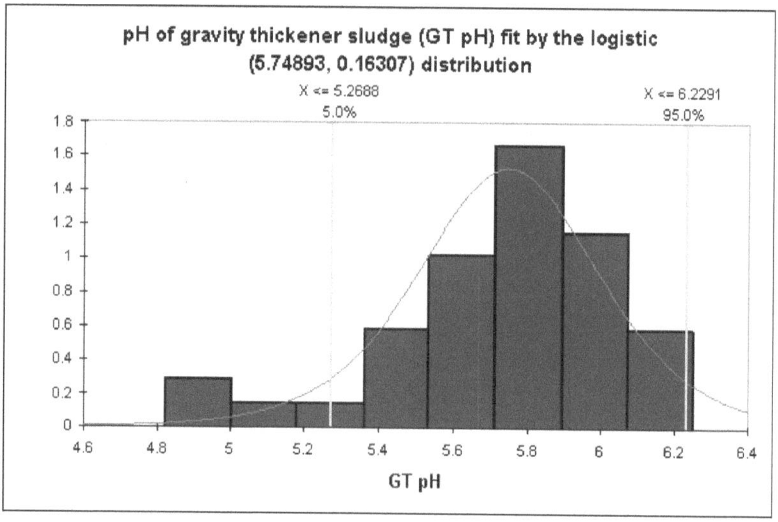

Fig. 11.3 An example from deriving probability distribution to fit input variables

Probability distributions were eventually determined for all the independent variables in the model (11.3) with the number of centrifuges running modeled as a discrete probability distribution based on historical values. For example this meant a 12% probability for four centrifuges was used, a 25% probability for five centrifuges, etc. Theoretical probability distributions for all the independent variables and their parameters derived are summarized in Table 11.1.

Using theoretical probability distributions in Table 11.1 and Monte Carlo simulation to sample biosolids odor values from equation (11.3) until the convergence was met, the probability distribution for biosolids odors shown in Figure 11.4 was produced.

Summary results from the Monte Carlo simulation were compared to actual values collected and are shown in Table 11.2.

Table 11.1 Theoretical probability distributions assigned to independent variables in simulation process

Independent variables	Distributions	Parameters	Units
Percent solids of biosolids	Lognormal	mean = 28.455, SD = 2.765	%
Temperature of biosoids	Lognormal	mean = 107.750, SD = 2.89	degrees Fahrenheit
GT pH	Logistic	mean = 5.751,SD = 0.296	-
Percentage of gravity thickener sludge in the blend tank on day $d-1$	Logistic	mean = 56.97,SD = 13.573	%
Concentration of returned activated sludge on day $d-1$	Triangular	min = 557.39, most likely = 5100, max = 16014	mg/liter

Note: SD = Standard deviation

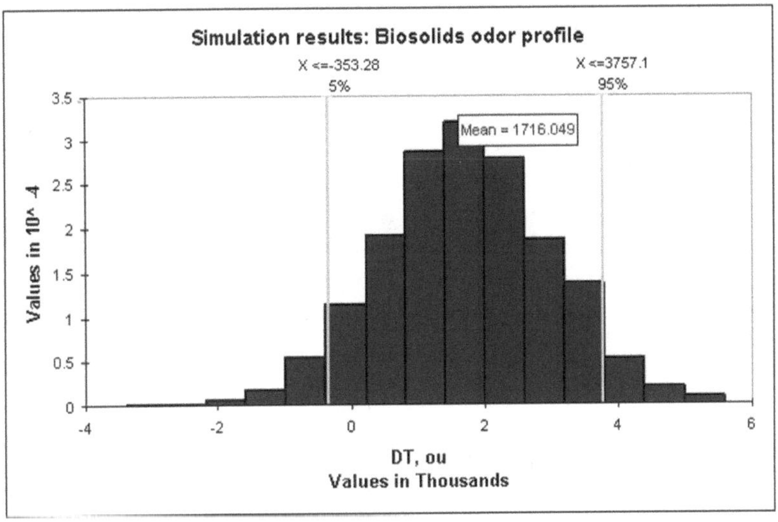

Fig. 11.4 Biosolids odor profile from simulation process

Table 11.2 Summary statistics from simulation process and from data collection

	Summary Statistics	
	Simulation result	Data collection
Maximum	5565	8694
Minimum	-3371	240
Mean	1716	1582
Median	1696	825
Standard deviation	1272	1697

It is interesting to note that the average biosolids odor level from the simulation was higher than its historical counterpart (1716 vs. 1582 ou) but the situation was reversed for the maximum values (5565 vs. 8694). This may be a result of some outliers in the actual data. Indeed, the maximum value of 8694 ou from the actual data collected can be considered as a rare event since there was only one observation where DT was greater than 7000 odor ou. For the average values, these results indicate perhaps that after running the simulations for a large number of iterations, the average DT level was greater than the average DT from the data collected but only by a small amount (e.g., 100 ou). As shown in Figure 11.4, the simulation produced DT values that were evenly spread taking into account that the median and mean values were close to each other. This is in contrast to the historical data for which the median value of 825 ou was far less than the mean value of 1582 ou. This was presumably the result of a fewer number of data points collected compared to those simulated and a higher standard deviation.

In addition, the probability distribution of biosolids odors showed that there was 39 percent chance that the plant would produce biosolids with DT higher than 2000 ou.[4] This was based on simulated data at this particular AWTP plant. Better management of the percentage of GT solids in blend tank (specifically less than 60 percent) as well as operating a sufficient number of centrifuges and consequently reducing retention time of solids in the secondary sedimentation tank and the blend tank could reduce a chance of this odor incident. In summary, simulation technique provides a clearer picture of uncertainty for biosolids management as compared to current daily practices.

11.3 Summary of Multiobjective Optimization Work

The multiobjective optimization models we developed sought to simultaneously minimize biosolids odors as well as wastewater treatment and biosolids distribution costs. The sum of daily odor scores from the odor prediction equation (11.1) determined the total odor. The wastewater treatment and biosolids distribution cost constituted the total costs. The constraints contained conservation of flow, upper and lower bounds on processes, flow, or variables. In addition, logical conditions such as distribution must be at least 23 tons per shipment if positive or biosolids could not be shipped to fields that would bring on odors beyond their thresholds were also included. Moreover, we also utilized equation (11.1) among several odor prediction equations developed by Gabriel et al. (2006b), as linear constraints in our optimization models.

The constructed models are divided into two modes. In the first mode the percentage flows (to be dewatered) from the blend tank to DCWASA (F_d^{dc}) and the onsite dewatering contractor (F_d^k) were exogenously determined (Base Case). In particular, for our case study, we set $F_d^{dc} = 0.9$ ($F_d^k = 1 - F_d^{dc}$). In other words, the

[4] DT scores greater than 2000 ou were considered extremely high (75 percentile).

DCWASA plant would handle 90% of the flows to be dewatered. Other values of F_d^{dc} were tried when we performed a sensitivity analysis for F_d^{dc}. In mode two, : F_d^{dc} and F_d^k were endogenously determined by the optimization model. With a 31-day time horizon, 782 reuse fields, and three hauling contractors, the first mode resulted in a large-scale, mixed integer linear program with 219,394 constraints and 145,576 variables. In particular, there were 72,757 continuous variables, 72,757 binary ones, and 62 non-binary discrete variables relating to number of centrifuges and belt filter presses. The second mode yielded a mixed integer bilinear program where the bilinear terms were later approximated. The resulting approximation problem was a mixed integer linear program with 219,705 constraints and 146,413 variables. In particular, there were 73,408 continuous variables, 73,005 binary ones, and 62 non-binary discrete variables. Next, we discuss the optimization model results when F_d^{dc} was exogenously determined.

11.3.1 Mode 1: F_d^{dc} exogenously determined

In finding the Pareto optimal solutions, we employed both the weighting and constraint methods (Cohon, 1978). For the weighting method, we minimized

$$w^1[\text{ total odor}] + (1 - w^1)[\text{ total cost}]$$

as the objective function and each variation of the weight w^1 was counted as one subproblem. As for the constraint method, we constrained the total odor to a certain right-hand side value and minimized the total cost. Each subproblem was constructed by varying the right-hand side for the total odor constraint.

Using the specified set-up, we tried 99 subproblems for the weighting method and another 92 subproblems for the constraint method. For the weighting subproblems, we were able to solve only one subproblem to optimality which was also Pareto optimal. The solution time was 942 seconds. For the other 98 sub-problems, the solver could not finish within the preset maximum runtime of 1,800 seconds. This was due to the non-convex nature of the model and, possibly to the existence of "duality gap" points (Cohon, 1978).

The constraint method was better at finding Pareto optimal solutions presumably due to its superior ability to find duality gap points, however with a large computational burden (ReVelle and McGarity, 1997). Out of the 92 constraint subproblems, one problem was infeasible; another four problems could not be solved to optimality within a preset maximum runtime of 1,800 seconds, and the remaining 87 problems could be solved to optimality. However, only 27 optimal solutions were also Pareto optimal. The solution times varied from 40 to 1,545 seconds. The wide range of solution times is presumably due to the non-convex nature of the model. Including one Pareto optimal point obtained from the weighting method and the 27 Pareto optimal points from the constraint method, Figure 11.5 displays the 28 Pareto optimal points for the Base Case.

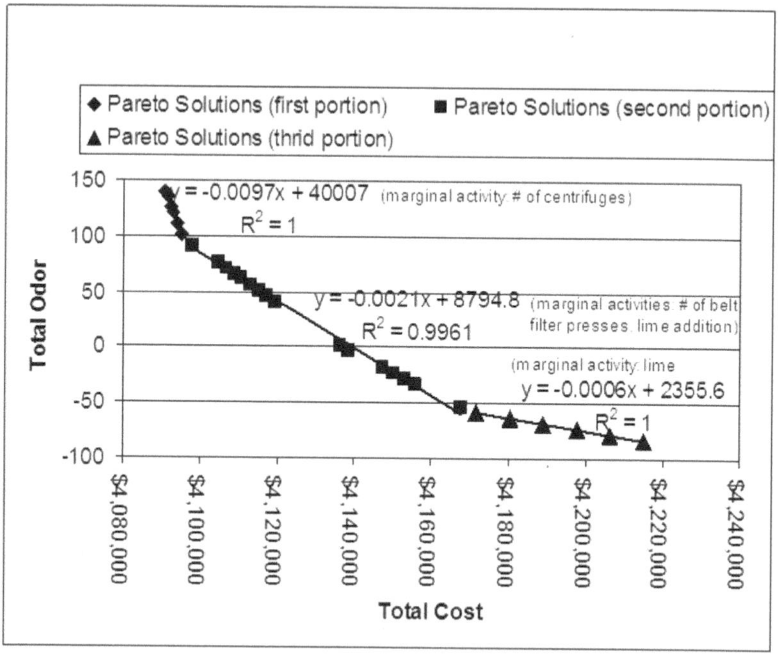

Fig. 11.5 Base Case's Pareto optimal points (Sahakij, 2008a)

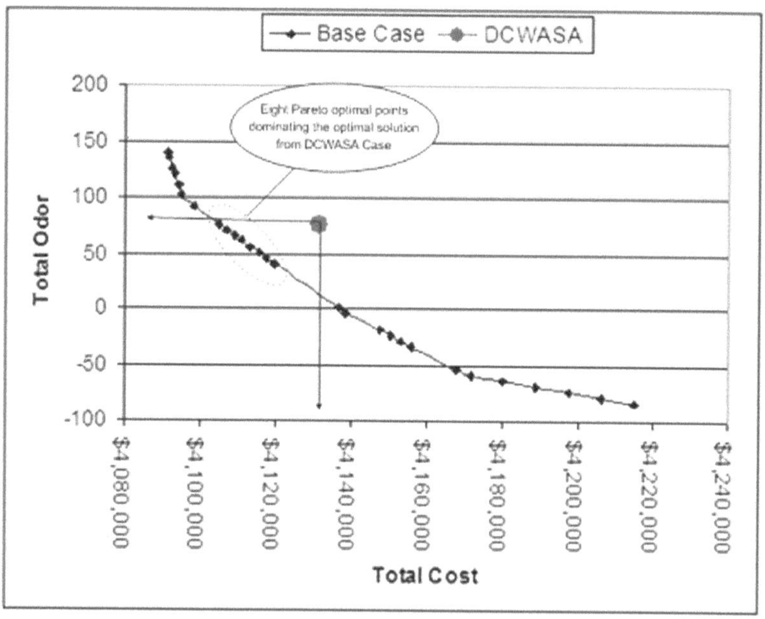

Fig. 11.6 Comparison of Base Case and DCWASA Case (Sahakij, 2008a)

First, note that certain odor scores were negative. This was due to the data in the time horizon selected going beyond the range of data used to generate the odor prediction equation where the odor scores were in the range $[0,9]$. It is also noted here that the negative odor scores don't represent the hedonic tone (McGinley et al., 2000) of the odor. In addition, these odor scores were relative and hence are still useful for management.

As indicated, the 28 Pareto optimal points may be divided into three portions. Going left to right, statistical regressions on the three portions were as follows.

- First portion: total odor $= -0.0097\,(\text{total cost}) + 40007$
- Second portion: total odor $= -0.0021\,(\text{total cost}) + 8794.8$
- Third portion: total odor $= -0.0006\,(\text{total cost}) + 2355.6$

All regressions had R^2 and adjusted R^2 values greater than 0.99 with statistically significant coefficients (i.e., t-statistics greater than 85 in absolute value). According to these equations, to reduce odor by one index point, one needs to pay on average $1/0.0097 = \$103$, $1/0.0021 = \$476$, and $1/0.0006 = \$1{,}667$, respectively, for the first, second, and third portions of this Pareto curve. Our analysis also indicates that the number of centrifuges in service was the marginal activity in reducing the biosolids odor level for the first portion. The marginal activities for the second portion were the number of belt filter presses in service and lime addition. Lastly, lime addition was the marginal activity for the third portion. Furthermore, combinations of activities that are optimal are also crucial and not immediately obvious. For example, decreasing lime additions and increasing the number of belt filter presses may be the optimal combination for one point on the Pareto curve.

Additionally, we compared the Pareto optimal total costs and total odors with historical values obtained by fixing values of all decision variables to values used by DCWASA and the on-site dewatering contractor during the time period in question (01/01/02 – 01/31/02) [5]. Figurc 11.6 displays the historical total cost and total odor values obtained from DCWASA Case versus the 28 Pareto optimal points.

As indicated, in Figure 11.6, eight Pareto optimal points dominate the historical total cost and total odor values. That is, moving to one of these eight Pareto points would have resulted in total costs and/or total odor reductions. In particular, the total costs would have decreased between \$11,869 and \$26,669 over a 31-day time frame by exercising combinations of adjusting lime additions, number of centrifuges, etc. according to the Pareto optimal solutions. Similarly, total odors would have decreased by 0.56 to 35.34 odor index points. This graph clearly shows that the historical tradeoff between costs and odor was not Pareto optimal and was in fact dominated by points obtained from running our models. This is important for biosolids managers when decisions must be made in the face of uncertainty and limited resources.

---

[5] However, due to an incompleteness of data, we were not able to fix the amount of wet tons applied to each reuse field. Otherwise, all other variables such as lime addition and number of centrifuges in service were fixed to historical levels.

11.3.1.1 Sensitivity Analysis for F_d^{dc}

As stated earlier, F_d^{dc}, the percentage of flow to be dewatered by DCWASA, was initially fixed to 0.9. A sensitivity analysis for $F_d^{dc}$was done by trying $F_d^{dc} \in \{0.1,0.2,0.3,0.4,0.5,0.6,0.7,0.8,0.9\}$and comparing the results. The analysis showed that all subproblems where $F_d^{dc} \in \{0.1\}$ were infeasible.Figure 11.7 displays the Pareto optimal points obtained when $F_d^{dc} \in \{0.2,0.3,0.4,0.5,0.6,0.7,0.8,0.9\}$. As indicated, the Pareto optimal points obtained when $F_d^{dc} \in \{0.2\}$ dominated other optimal points when $F_d^{dc} \in \{0.3,0.4,0.5,0.6,0.7,0.8,0.9\}$. This means that having DCWASA handle 20% of the flows was best in terms of the cost-odor tradeoff.

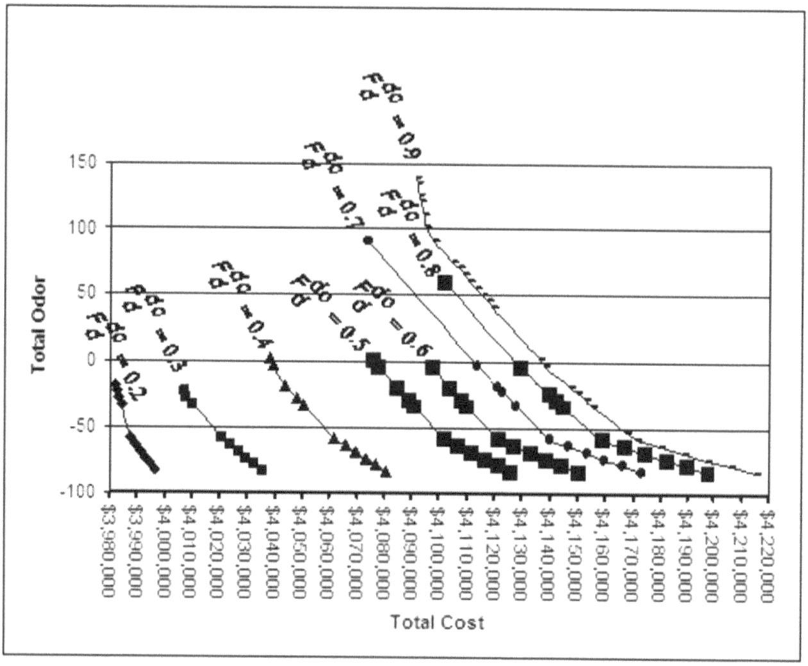

Fig. 11.7 Pareto optimal points when $F_d^{dc} \in \{0.1,0.2,0.3,0.4,0.5,0.6,0.7,0.8,0.9\}$

It is important to note here that these optimization results do not imply that the on-site contractor was more cost efficient than DCWASA in the dewatering process. In fact, these optimization results were driven by the nature of the odor prediction equation (11.1) where most odor reducing variables corresponded to the on-site contractor. Thus, when more flows were assigned to the on-site contractor, less total odors could be achieved.

11.3.2 Mode 2: F_d^{dc} endogenously determined

While the previous mode fixed the percentage of F_d^{dc} and then tried different values to find the overall best one, mode two endogenously determined F_d^{dc} as part of the model. This resulted in a bilinear term which led to approximation schemes and a more complex and computationally challenging model. In particular, the bilinear terms arose from the multiplication of two decision variables, i.e., lime additions (L_d^{dc}) and F_d^{dc}. In order to be able to solve the resulting optimization problem with existing standard solvers, the approximation of the term $L_d^{dc}F_d^{dc}$was needed. The approximation procedure consisted of two major steps: converting the bilinear term into the difference of two quadratic terms via Schur's decomposition (Horn and Johnson, 1985; Gabriel et al., 2006a) and the piecewise linearization of the quadratic terms resulting from this decomposition using special ordered set of type 2 (SOS2) variables (Beale and Tomlin, 1970; Beale and Forrest, 1976; Escudero, 1978). The resulting approximation problem was a large integer linear program where most subproblems could not be solved within a preset maximum run time. However, with the use of the Dantzig-Wolfe decomposition technique (Dantzig 1998; Dantzig and Wolfe 1960), we were able to improve lower bounds for our optimization problem and obtained several Pareto optimal points. Figure 11.8 indicates that the Pareto optimal points obtained when $F_d^{dc}$was determined exogenously were dominated by the ones obtained when $F_d^{dc}$was determined endogenously. This was undoubtedly due to the larger feasible region when $F_d^{dc}$was determined endogenously. In addition, this also indicates the benefit of solving a larger and more complicated problem instead of solving smaller and less complicated ones.

For more details on the model and other sensitivity results see (Gabriel et al., 2007; Sahakij et al., 2007ab; Sahakij, 2008).

11.3.3 Adding stochasticity to the optimization models

The optimization models described above are large-scale and complex but assume that all the data are deterministic. In reality many of the inputs related to the odor equation (11.1), costs, and other factors may be stochastic in nature. Consequently, a successor model to those presented should directly take into account those uncertain elements as opposed to just performing sensitivity analyses to gain insight into uncertainty. In what follows, we describe two possible future directions that could be taken to stochasticize the previously mentioned optimization problems.

The first direction would be to build a two-stage stochastic optimization with recourse (Birge and Louveaux, 1997). In these sorts of problems, typically the first stage relates to some planning level decisions and the second level to some operational level choices indexed by scenario. In the present context, the first-stage decisions might relate for example to how many centrifuges or belt filter presses should be used or how much lime or polymer should be applied as the default values. The

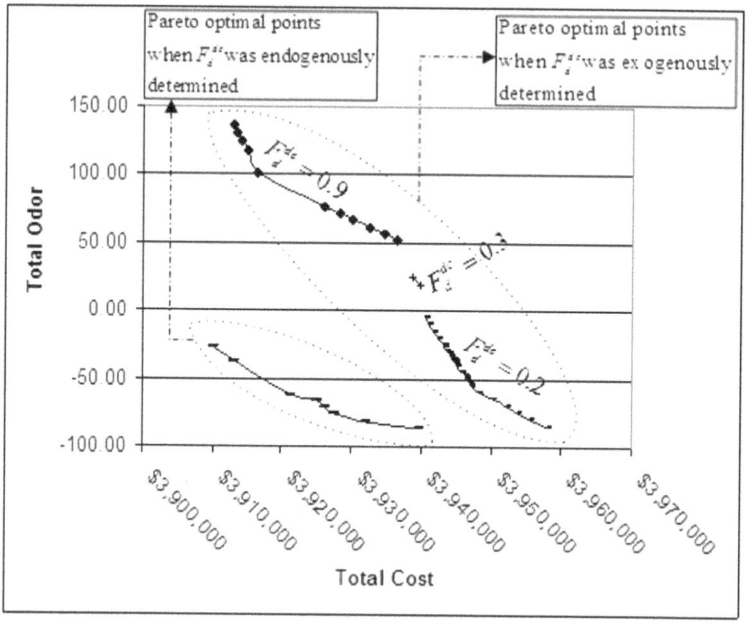

Fig. 11.8 Comparison of Pareto optimal points when F_d^{dc} was determined endogenously and exogenously (Sahakij, 2008a)

recourse or corrective actions that would be considered in the second stage could include renting more centrifuges/belt filter presses or relying more on an outsider constractor, applying additional chemical doses, adjusting blend ratios, or the like. The resulting problem would be a very large-scale mixed integer program but with a block diagonal structure relating to each of the scenarios for part of the coefficient matrix. Typically this problem would need to be solved multiple times to generate (an approximation of) the Pareto curve. The Benders method called the L-shaped method in the context of stochastic programming (Birge and Louveaux, 1997) could then be applied to the resulting stochastic mixed integer program for one iteration of the weighting or constraint method. Alternatively, Benders method could be run just once with given positive weights or odor constraint right-hand sides if just one Pareto point was to be determined. In the Benders setting, the master problem would contain decision variables independent of scenarios and the subproblem would include those scenario-specific variables. The benefits to biosolids managers of using this model would be both Pareto optimal tradeoffs between costs and odor as well as more robust guidance on operations since the model would endogenously take into account risk.

A second approach to combining uncertainty and optimization would be to employ a chance-constraints optimization model (Birge and Louveaux, 1997). In this setting there would be no recourse actions as outlined above. Instead, one would use a probabilistic constraint to ensure that the biosolids system being modeled would function appropriately with a certain likelihood. An example of such a probabilistic constraint would be

$$P[\text{odor level} \leq \text{threshold}] \geq \alpha \tag{11.4}$$

where the odor level would be random variable perhaps with a probability distribution as specified in Section 11.2. The threshold value would be a given maximum level supplied by the modeler and the parameter α would be the reliability level. For example, if this reliability parameter equaled 95%, this would mean that only 5% of the time would the predicted odor be allowed to exceed this limit. The biosolids management would then be able to run sensitivity analyses to measure the cost of certain desired odor reliability levels. The key would be to determine the deterministic equivalent (Birge and Louveaux, 1997) of equation (11.4) so that the resulting problem could be solved. Of course other probabilistic optimization approaches such as stochastic dynamic programming (Puterman, 1994) or Monte Carlo simulation (Clemen and Reilly, 2001) combined with optimization at each iteration could also be considered to directly combine uncertainty and optimization together in the context of biosolids management.

11.4 Summary and Future Directions

In this paper we have presented both statistical and optimization models to help biosolids managers reduce both odors and costs. The statistical models were based on both subjective and objectives measures of odor levels and took into account data from odor panels, field inspectors, and odor-analyzing machines. One of these statistical models was then applied in a large-scale, mixed integer optimization problem to determine Pareto optimal points that balanced odor and costs. These models proved to be computationally challenging to solve especially when certain bilinear terms were incorporated albeit in approximated form. Dantzig-Wolfe decomposition proved to be important in solving these problems involving bilinear approximations more efficiently. An analysis of the output revealed key marginal activities, (e.g., number of centrifuges in operation) as well as sensitivity to such things as the percentage of flows to be dewatered handled by the AWTP.

We also generated an odor level probability distribution from one of the statistical models taking into account estimated distributions for the independent variables. The probability distribution for odor levels is an important tool for biosolids managers in their operations and future work can include covarying distributions for the independent variables and further analyses useful for biosolids management. Lastly, we described two methods to incorporate probabilistic considerations (e.g., probability distribution for odor levels) with the mixed integer optimization models described. This would constitute an important direction of future work to aid biosolids managers in efficiently trading off costs and for in the presence of uncertainty.

References

[1] Aripse, S. C., (2005). "Measuring and developing a control strategy for odorous gases from solids handling processes of a large wastewater treatment plant." MS thesis, University of Maryland, College Park, MD.

[2] Beale, E.M.L., and Forrest, J.J.H. (1976). "Global optimization using special ordered sets." *Mathematical Programming*, 10, 52-69.

[3] Beale, E.M.L., and Tomlin, J.A. (1970). "Special facilities in a general mathematical programming system for non-convex problems using ordered sets of variables." *Proc., Fifth IFORS Conference*, Tavistock, London, 447-454.

[4] Birge, J.R., and Louveaux, L. (1997). *Introduction to Stochastic Programming*, Springer, NY.

[5] Clemen, R. T., and Reilly, T. (2001). *Making hard decisions with Decision-Tools*, Duxbury, CA.

[6] Cohon, J.L. (1978). *Multiobjective Programming and Planning*, Dover Publications, Inc., NY.

[7] Dantzig, G.B. (1998). *Linear programming and extensions*, Princeton University Press: Princeton, NJ.

[8] Dantzig, G.B., and Wolfe, P. (1960). "Decomposition principle for linear programs." *Operations Research*, 8(1), 101-111.

[9] D'Amato, R. M., II, and DeHollander, G. R. (1999). "Gaseous emissions from wastewater facilities." *Water Envir Research,* **71**(5), 715-720.

[10] Environmental Protection Agency (EPA) (1994). "Biosolids recycling: beneficial technology for a better environment." *Rep. No. EPA 832/R-94-009*, U.S. Environmental Protection Agency, Office of Water, Washington, DC.

[11] Environmental Protection Agency (EPA) (1994). "Plain English guide to the EPA part 503 biosolids rule." *Rep. No. EPA/832/R-93/003*, U.S. Environmental Protection Agency, Office of Wastewater Mgmt, Washington, DC.

[12] Escudero, L. (1978). "A comparative analysis of linear fitting for non-linear functions on optimization, a case study: air pollution problems." *European Journal of Operational Research,* 2, 398-408.

[13] Federal Register (1993). "Standards for the use or disposal of sewage sludge: final rules."*Federal Register*, 58(32), 9248-9415.

[14] Gabriel, S.A., Bertrand, R.G., and Sahakij, P. (2006). "A practical approach to approximate bilinear functions in mathematical programming problems by using Schurs decomposition and SOS type 2 variables." *Journal of the Operational Research Society*, 57, 995-1004.

[15] Gabriel, S.A., Sahakij, P., and Ramirez, M. (2007). "A multiobjective optimization model for processing and distributing biosolids to reuse fields." *Journal of the Operational Research Society*, 58, 850-864.

[16] Gabriel, S.A., Vilalai, S., Aprise, S., Kim, H., McConnell, L.L., Torrents, Peot, C., and Ramirez, M. (2005). "Prediction of Dimethyl Disulfide Levels from Biosolids Using Statistical Modeling" *Journal of Environmental Science and Health*, 40, 2009-2025.

[17] Gabriel, S.A., Vilalai, S., Peot, C., and Ramirez, M. (2005). "Statistical modeling to forecast odor levels of biosolids applied to reuse sites." *Journal of Environmental Engineering*, 132 (5), 479-488.
[18] Horn, R.A., and Johnson, C.R. (1985). *Matrix Analysis*, Cambridge University Press: Cambridge.
[19] Kim, H., Murthy, S. N., McConnell, L., Peot, C., Ramirez, M., and Strawn, M. (2001). "Examination of mechanisms for odor compound generation during lime stabilization." *Proc., WEFTEC 200,* Atlanta, GA.
[20] Kim, H., Murthy, S., McConnell, L., Peot, C., Ramirez, M., and Strawn, M. (2002). "Characterization of wastewater and solids odors using solid phase microextraction at a large wastewater treatment plant." *Water Sci and Tech,* **46**(10), 9-16.
[21] Kutner, MH., Nachtsheim, **C.**, Neter, J., and Li, W. (2005). Applied Linear Statistical Models, McGraw-Hill/Irwin 5th edition.
[22] McGinley, C. M., and McGinley, M. A. (2002). "Odor Testing Biosolids for Decision Making", *Proc., WEF Specialty Conference*,Austin, TX,
[23] McGinley, M. A., McGinley, C. M., Mann J., (2000). "Olfactomatics: Applied Mathematics For Odor Testing", WEF Odor / VOC 2000 Specialty Conference, Cincinnati, OH, 16-19 April 2000.
[24] Mosier, A. R., Morrison, S. M., and Elmond G. K. (1977). "Odors and emissions from organic wastes." *Soil for Management of Organic Waste and Waste Waters*, Soil Science Society of America, Madison, WI, 532-569.
[25] Murthy, S. N., Sadick, T., Baily, W., Peot, C., Tolbert, D., and Strawn, M. (2001). "Mitigation of Odors from Lime Stabilized Biosolids." *Proc., WEF Residuals and Biosolids Management Conference*, San Diego, CA.
[26] Murthy, S., Kim, H., McConnell, L., Peot, C., Bailey, W., Novak, J., and Glindemann, D. (2002). "Mechanisms for odour generation during lime stabilization." *Proc., IWA Biennial Conference*, Melbourne, Australia.
[27] Novak, J., Glindemann, D., Murthy, S. N., Gerwin, S. C., and Peot, C. (2002). "Mechanisms for generation and control of trimethyl amine and dimethyl disulfide from lime stabilized biosolids." *Proc., WEF Odor Conference*, Albuquerque, NM.
[28] Oleszkiewicz, J. A., and Mavinic, D. S. (2002). "Wastewater biosolids: an overview of processing, treatment, and management." *J of Envir Engrg Sci,* **1**, 75-88.
[29] Palisade Corporation, (2001). *Guide to Using @Risk: Risk Analysis and Simulation Add-In for Microsoft Excel*, Palisade Corporation, NY.
[30] Puterman, M.L. (1994). *Markov Decision Processes Discrete Stochastic Dynamic Programming*, John Wiley & Sons, Inc., NY.
[31] ReVelle, C., and McGarity, A. E. (eds) (1997). *Design and Operation of Civil and Environmental Engineering Systems*, John Wiley & Sons, NY.
[32] Rosenfeld, P. E. (1999). "Characterization, quantification, and control of odor emissions from biosolids applications to forest soil." PhD thesis, University of Washington, WA.

[33] Sahakij, P. (2008). "Multiobjective optimization models for distributing biosolids to reuse fields." Ph.D. Thesis, University of Maryland, College Park, MD.

[34] Sahakij P., Gabriel, S.A., Ramirez, M., and Peot C. (2007). "A multi-objective optimisation model for processing and distributing biosolids to reuse fields." *Proc., 12th European biosolids and organic resources conference*, Aqua Enviro, Wakefield, UK.

[35] Sahakij, P., Gabriel, S.A. , Ramirez, M. , and Peot, C. (2008). "Multi-objective Optimization Models for Distributing Biosolids to Reuse Fields: A Case Study for the Blue Plains Wastewater Treatment Plant" December 2007, *Networks and Spatial Economics*, published online.

[36] Sostrand, P., Tvedt, B., Eduard, W., Bye, E. and Heldal, K. (2000). "Hazardous peak concentrations of hydrogen sulfide gas related to the sewage purification process." *Amer Ind Hyg Assoc J*, **61**(1), 107-110.

[37] Vilalai, S., (2003). "Forecasting odor levels for biosolids product based on ambient conditions" MS thesis, University of Maryland, College Park, MD.

[38] Vilalai, S., Gabriel, S.A., Peot, C., and Ramirez, M. (2007). "Biosolids odor prediction models: sensory and analytical measurement approaches" *Proc., 12th European biosolids and organic resources conference*, Aqua Enviro, Wakefield, UK.

Index

FSC
www.fsc.org
MIX
Papier | Fördert
gute Waldnutzung
FSC® C083411

Kolibri 360 GmbH
Ferdinand-Jühlke-Straße 7
99095 Erfurt, Deutschland
produktsicherheit@kolibri360.de